CAMBRIDGE LIBRARY COLLECTION

Books of enduring scholarly value

Life Sciences

Until the nineteenth century, the various subjects now known as the life sciences were regarded either as arcane studies which had little impact on ordinary daily life, or as a genteel hobby for the leisured classes. The increasing academic rigour and systematisation brought to the study of botany, zoology and other disciplines, and their adoption in university curricula, are reflected in the books reissued in this series.

A Selection from the Transactions of the Royal and Horticultural Societies

Thomas Andrew Knight (1759–1838) was a distinguished British naturalist and botanist who is often regarded as the father of nineteenth-century horticultural science. From 1811 to 1838 Knight was the president of the Royal Horticultural Society and his interest in structural biology, plant physiology and plant breeding is evident in this collection of papers, published in 1841. On his country estate in Herefordshire, Knight devoted his time to research and writing, and carried out experiments on plants and trees. He published papers on his theories about such physiological problems as the ascent and descent of sap and how buds are produced. The main focus, however, is on Knight's own practical work: building greenhouses and experimenting with plant nutrition, fertilisation and the improvement of fruit trees by selective breeding (work later appreciated by Darwin). In an interesting chapter on animals, Knight relates his observations on the behaviour of bees and dogs.

Cambridge University Press has long been a pioneer in the reissuing of out-of-print titles from its own backlist, producing digital reprints of books that are still sought after by scholars and students but could not be reprinted economically using traditional technology. The Cambridge Library Collection extends this activity to a wider range of books which are still of importance to researchers and professionals, either for the source material they contain, or as landmarks in the history of their academic discipline.

Drawing from the world-renowned collections in the Cambridge University Library, and guided by the advice of experts in each subject area, Cambridge University Press is using state-of-the-art scanning machines in its own Printing House to capture the content of each book selected for inclusion. The files are processed to give a consistently clear, crisp image, and the books finished to the high quality standard for which the Press is recognised around the world. The latest print-on-demand technology ensures that the books will remain available indefinitely, and that orders for single or multiple copies can quickly be supplied.

The Cambridge Library Collection will bring back to life books of enduring scholarly value (including out-of-copyright works originally issued by other publishers) across a wide range of disciplines in the humanities and social sciences and in science and technology.

A Selection from
the Transactions
of the Royal and
Horticultural Societies

THOMAS ANDREW KNIGHT

CAMBRIDGE
UNIVERSITY PRESS

CAMBRIDGE UNIVERSITY PRESS

Cambridge, New York, Melbourne, Madrid, Cape Town,
Singapore, São Paolo, Delhi, Tokyo, Mexico City

Published in the United States of America by Cambridge University Press, New York

www.cambridge.org
Information on this title: www.cambridge.org/9781108037297

© in this compilation Cambridge University Press 2011

This edition first published 1841
This digitally printed version 2011

ISBN 978-1-108-03729-7 Paperback

THOMAS ANDREW KNIGHT Esqᵣᵉ. F.R.S. L.S. &c.

President of The London Horticultural Society.

S Cole. Printed by M & N Hanhart. R.J.

A SELECTION

FROM THE

PHYSIOLOGICAL AND HORTICULTURAL

PAPERS,

PUBLISHED IN THE

Transactions of the Royal and Horticultural Societies,

BY THE LATE

THOMAS ANDREW KNIGHT, ESQ.,

PRESIDENT OF THE HORTICULTURAL SOCIETY OF LONDON, ETC. ETC.

———◆———

TO WHICH IS PREFIXED,

A SKETCH OF HIS LIFE.

LONDON:

LONGMAN, ORME, BROWN, GREEN, AND LONGMANS.

MDCCCXLI.

INTRODUCTION.

DURING the life of the late MR. ANDREW KNIGHT, he was repeatedly urged by many scientific Horticulturists to collect together and republish all that he had written on Vegetable Physiology and Horticulture; and since his lamented death this wish has again been repeated to his family, and having been seconded by the advice of some of his friends whose pursuits have well qualified them to judge of the probable utility of such a work, it has been determined to offer to the public a selection from the Papers which at various periods he communicated to the Royal and Horticultural Societies, in a form that will render their contents available to many persons by whom the Transactions of these Societies are not attainable.

Vegetable Physiology is a branch of science which till very lately has not been very extensively cultivated; hence the number of persons competent to judge of the value of Mr. Knight's researches is necessarily limited : but the continual reference that is made to his papers in the works of M. De Candolle, Dutrochet, Du Petit-Thouars, Féburier, Keiser, and other foreign writers on similar subjects, by several of whom his experiments have been repeated and the results confirmed, shows that his labours are extensively known and appreciated on the continent of Europe.

Sir Humphrey Davy in his Lectures on the Chemistry of Agriculture, and Dr. Lindley in his Theory of Horticulture, together with many other writers among his countrymen and countrywomen, have by the adoption of his opinions afforded a gratifying proof of the estimation in which they hold both his theoretic views, and the practical results he deduced from them.

A taste for Horticulture has for some years been so universally cultivated, that all classes are familiar with Mr. Knight's name as a writer, and the extracts from his papers which are found in many of the periodical publications on Horticulture and Arboriculture of the present day, have caused the readers of these works to be in some degree conversant with the particular subjects on which he has treated; and though the value of the present work may be diminished by the task of editing it having unavoidably fallen to those who are ill qualified to do justice to the undertaking, they are still cheered by the hope that their imperfect attempt may, nevertheless, by making both Mr. Knight's character and his writings better known, be the means of demonstrating more fully to the world the constant and never-tiring exertions of his mind in the pursuit of knowledge, and its application to purposes of practical utility for the benefit of his fellow-creatures.

Mere discoveries in abstract science, and even their application to an increased production of animal and vegetable
food, may seem to the casual observer not entitled, from the
want of dazzling brilliancy, to more than secondary importance
and fame; but when he reflects on the growth of crime and
the insecurity of property resulting from the goading and
baneful influence of the *male suada fames* on a population
rapidly outgrowing its means of existence, he will then allow
that such labours as those of Mr. Knight are likely to exercise
a most beneficial influence on the moral as well as physical
welfare of society.

It is necessary to say a few words to explain why a work
requiring so little time or preparation as a selection of papers,
and the simple sketch of Mr. Knight's life prefixed to it, should
not have appeared at a much earlier period after his death.
And this it was the anxious wish of his family should have
been the case. They were, however, induced to concede their
own wishes on this point to the suggestions of a gentleman
who had kindly undertaken to furnish the memoir, and
who considered that the materials put into his hands were
sufficient to form a more pretending volume.

That gentleman having very lately declined to proceed with
the work on the ground of ill health, the original design has
again been adopted, and the very few letters or memorandums
of Mr. Knight that remain have been arranged by those
unused to write for the public eye, and whose judgment may
probably be biassed by the devoted respect and affection they

feel for the memory of the beloved parent whose character they have attempted to portray.

Under such circumstances, it is hoped this unpretending memoir will be received with indulgence.

Mr. Knight's family cannot omit this opportunity of acknowledging with thankfulness how much they owe to GEORGE BENTHAM, Esq., for the kindness which has led him to render them many and important services in the publication of this work: and to Dr. LINDLEY they are also indebted for assistance in the selection of the papers that have been deemed most desirable for the present volume. Their thanks are also due to the Councils of the Royal and Horticultural Societies for the loan of the copper-plates illustrating the papers here reprinted, and to several friends of Mr. Knight for communications on various subjects.

March, 1840.

CONTENTS.

PART II.

APPENDIX.

LIFE

OF

THOMAS ANDREW KNIGHT, ESQ.

Of the early history of the family from which Mr. Andrew Knight is descended but little is known with certainty. The records of Shrewsbury show that, from the reign of Henry VI. to that of Charles I, a family of the name of Knight resided in that town, and repeatedly filled its civic offices; and one of them, Thomas Knight, in 1509, was elected one of its representatives in parliament. A pedigree of the Shrewsbury Knights is preserved in the British Museum, with arms identical with those used by the present family : and the christian names borne by both bear a striking similarity. The name disappears from the Shrewsbury annals just at the time when Mr. Richard Knight, the great grandfather of the subject of this memoir, is known to have been residing on an estate of his own at Castle Green in the parish of Madeley, in the same county; and when it is recollected with what irregularity parish registers were kept during the civil wars, it is not surprising that the connecting link should not have been more exactly traced.

Mr. Richard Knight's eldest son, Francis, was born at Castle Green in 1640, where he succeeded his father and resided till his death. The second son, Richard, was born in 1658, and attained to considerable eminence in his day, from the success which attended his mercantile speculations, and the high character he established for independence and probity; and he deserves more especial notice here as the founder of the fortune of his family.

He early embarked in the iron trade, and worked a forge, the remains of which are still to be seen at the lower end of Coalbrooke Dale. This district was not at that period, as it is now, the great field of the iron trade of Shropshire, and he soon quitted it for a forge at Moreton, in the parish of Shawbury.

The smelting of iron at this time was carried on almost universally by means of wood charcoal in small furnaces, the bellows of which were worked by water-wheels, and were gene rally situated on the banks of streams, in the vicinity of large tracts of coppice wood. The scale on which these works were carried on, as compared with those of the present day, may best be understood by the fact that, in 1740, a few years before Mr. Richard Knight's death, there were only fifty-nine iron furnaces in the whole of England and Wales*, and the average quantity of metal produced by each was 5 tons 13 cwt. per week; while in Shropshire alone there were lately between fifty and sixty furnaces at work, each producing above seventy tons per week†! In 1740 there were only six furnaces in Shropshire, which together made two thousand tons per annum : of these Mr. Richard Knight had two, besides several forges; he had also one forge in Staffordshire, and shares in nearly the whole of the iron works of Worcestershire, and a furnace and forge at Bringewood near Ludlow, in Herefordshire.

Long before this time the manufacture of iron had begun to decline‡, owing to the increasing difficulty of procuring an adequate supply of fuel; which is not surprising, when it is known that a large furnace will consume in a year the produce of one hundred and twenty acres of coppice wood! The trade

---

* See Art. on Iron making in Supp. to Encyclop. Brit.

† Paper read at meeting of the Shropshire Nat. Hist. Soc. by Mr. T. Blunt.

‡ Dudley, who wrote in the reign of James I. states that there were at that time in England three hundred furnaces for the manufacture of pig-iron, making the astonishing quantity annually of one hundred and eighty thousand tons, though he says " the trade is falling into decaye."—See Supp. to Encyclop. Brit. —A curious old pamphlet, without date, but written since 1714, " On the Interest of Great Britain in supplying Herself with Iron," gives the whole quantity then made as 12,190 tons, and states that it had been 19,485 tons.

did not begin to revive till about 1750, when the use of pit coal* in blast furnaces became general. It seems, therefore, probable that Mr. Knight owed his extraordinary success solely to the efforts of his own powerful mind and the enlarged views by which his proceedings were directed, for it was evidently not during the prosperous days of the iron trade that he made his fortune, nor is it supposed that his original capital was at all considerable.

After he became a rich man, he never departed from the simplicity of his early habits. One of his few indulgences was that of riding a fine horse: and this, perhaps, may have been as much dictated by prudence as pleasure, for before the establishment of country banks large sums of money were necessarily transferred from place to place on horseback. One undoubted deviation from the unostentatious mode of living attributed to him has been handed down, in a magnificent silver punch-bowl, capable of containing nine quarts, with the contents of which it was his custom to regale himself and his friends. Many anecdotes are told of this old gentleman, which, after all due allowance has been made for the change of manners that the lapse of two centuries has made, still show that he must have been a person of very singular habits.

On one occasion a large quantity of Russian iron was advertised for sale at a certain inn in the city, and on the day appointed, Mr. Knight arrived there meanly dressed; and while waiting for the sale to commence, he volunteered his assistance to relieve a man who was employed in turning a spit on which a piece of beef was roasting. While so employed, he entered into conversation with the landlord, who told him that a great

* Fuller, in his " Worthies of England," printed in 1662, indulges in the following amusing anticipations on this subject :—" What we may call river or fresh-water coals, digged out in this county (Shropshire), at such a distance from Severn that they are easily ported by boat into other countries. Oh! if this coal could be so charked as to make iron melt out of the stone, as it maketh it in smiths' forges, to be wrought in the bars ! But Rome was not built in a day ; and a new world of experiments is lefte to the discoverie of posteritie."

iron-master from Shropshire, of the name of Knight, was, with many others, expected to be present at the sale. He remained incognito, and in the back ground, till the sale was nearly over, yet he managed to become the successful competitor; but from his shabby appearance the auctioneer hesitated to accept him, from a doubt of his responsibility to pay the amount. The sum which he cleared by this transaction is said to have been extremely large.

At another time, he lost his way in the dark on a common near Stourbridge, when he was conveying a very large sum of money in his saddle-bags, and he was at length himself admitted into the cottage of a collier, and his horse and bags, which he said were filled with nails, were placed in an adjoining shed. A wedding feast, which is always an occasion of much more gaiety among colliers than it is with agricultural labourers, was being celebrated in the cottage, when Mr. Knight joined the party, and danced in his boots, till the return of daylight enabled him to proceed on his journey. At parting he prepared to present a gratuity to his host for his entertainment, but it was declined on the ground that nothing was expected from so poor a man. He then made himself known, and presented the collier with five guineas.

Mr. Richard Knight removed to Bringewood Forge about the year 1698, of which he had taken a lease for twenty-one years from the second Lord Craven, and on the improvement of which he immediately expended between £20,000 and £30,000. Lord Craven's predecessor had, about thirty years before, purchased an extensive tract of land, including the forest of Mocktree and the chase of Bringewood, from the Earl of Lindsey, to whose father, the first earl, it had been granted by Charles I.* in reward for the services Lord Lindsey had rendered to the Royal cause during the struggle between the king and the parliament.

---

* In a paper, No. 354 of the Harl. MSS. in the British Museum, is a survey of the forests and chases of Mocktree and Bringewood, made in the reign of James I. from which the following is an extract : " These forests are stately grounds, and do breed a great and large deer, and will keep of red and fallow

Before the expiration of the lease, Mr. Knight had himself become the possessor of this portion of Lord Craven's estate, together with much other land adjoining, on part of which his grandson, Mr. Payne Knight, between the years 1773 and 1776, built the mansion of Downton Castle.

Previously to Mr. Knight's removal to Bringewood, he had married the daughter of Mr. Andrew Payne of Shawbury, and two sons had been born. The eldest, Richard, afterwards married Miss Powell of Stanage Park, county Radnor, and had a daughter, who became the wife of Thomas Johnes, Esq., of Havod, county Cardigan, and was mother to Colonel Johnes and the Rev. Samuel Johnes Knight.

From her the Johneses inherited Croft Castle, in Herefordshire, Stanage Park, and the Priory, near Cardigan.

Of the second son, Thomas, more will be said hereafter. There were also two younger sons, Edward and Ralph, who were ancestors of those branches of the family who settled at Wolverley, in Worcestershire, and Henley Hall, in Shropshire. Mr. Knight had several daughters, from one of whom is descended Mr. Samuel Rogers, the distinguished author of " The Pleasures of Memory ;" and another married Mr. — Spooner, of Warwickshire, and was the mother of the last race of that name.

Mr. Richard Knight died at his house at Downton, February 3rd, 1745, and was buried in the chancel of Burrington church, under an appropriate monument—a large slab of cast iron !

Mr. Thomas Knight was born in the year 1700, and entered the Church. In 1730 he was presented by the Lord Herbert to the livings of Ribbesford and Bewdley, in the county of Wor-

deer two or three thousand at least. Mem. That the forest and chase of Mocktree and Bringewood are near adjoining to the castle of Ludlow, the chief house of the Prince of Wales, out of which the President and Council had their timber for building, and wood and coals for their provisions, besides the pleasures of the game, till they were granted to Robert, Earl of Essex (by Queen Elizabeth), since when the Lord President and Council have been enforced to buy their timber, wood, and coals, which was a great charge to her Majesty, and is likely to be so to his Majesty."

cester, which he held till his death ; though, after his marriage, he resided at Wormesley Grange, near Hereford. He was a man of great simplicity and kindness of character, combined with superior ability, and his views on many subjects appear to have been in advance of the period in which he lived ; he was greatly beloved and respected by his neighbours, by whom his remarks and axioms were long remembered, and quoted to his children.

He died November 3rd, 1764, and was interred at Wormesley. He left two sons and two daughters : the eldest, Mr. Payne Knight, was born February 3rd, 1750 ; Thomas Andrew, the youngest son, was born at Wormesley Grange on the 12th of August, 1759, and was therefore only five years old at the period of his father's death.

The early education of both these brothers was much neglected, particularly that of the eldest, who never was at a public school, or at either of the Universities ; and the eminence to which he attained as a scholar, adds another to the many instances on record, of the manner in which an energetic mind will press forward in pursuit of knowledge in spite of disadvantages and difficulties. Mr. Payne Knight did not begin the study of the Greek language till he was eighteen, and his attention was then chiefly directed to those subjects which illustrate Greek sculptures and coins, viz., Mythology and the Archaic Greek language, and the earliest productions of his pen were devoted to elucidate some obscure points of Greek mythology. He visited Italy before he was of age, and there acquired that taste for the fine arts, and especially for the productions of the Greek sculptor, which led to his forming the magnificent collection of ancient bronzes and coins bequeathed by his will to the British Museum*. The only one of Mr. P. Knight's works which has much interest for the general reader is " An Analytical Inquiry into the Principles of Taste," first published in 1805, and which has passed through several editions. In 1809 the Dilettanti Society published a splendid work entitled

* The value of this collection was estimated at 50,000l.

" Specimens of Ancient Sculpture selected from different collections in Great Britain," the subjects for which were chosen by Mr. P. Knight, and he wrote the preface and the description of the plates. He was also the author of several poems : " The Landscape," " Progress of Civil Society," "Monody of the Death of Mr. Fox," and " Alfred, a Romance in Rhyme," and of some articles in the Edinburgh Review. In 1820, he published an edition of the Iliad and Odyssey. His object in this edition was to restore the text of Homer to its original state. His " Inquiry into the Principles of Taste " was reviewed in the " Edinburgh Review " for January 1806. Mr. P. Knight was elected to serve in parliament for the borough of Leominster in 1780 ; and in 1784 he was chosen one of the representatives of Ludlow, for which place he continued to sit until 1806, when he retired from parliament.

Mr. Andrew Knight received his early education at Ludlow, from whence he was removed to a school of considerable reputation at Chiswick, then kept by Dr. Crawford. He was afterwards entered of Baliol College, Oxford, where the late eminent physician Dr. Baillie was his contemporary : who used to say of him, " that he managed to acquire as much Latin and Greek as most of his fellow-students, though he spent less time about it, and much less than he devoted to field sports." He was at this period and continued for many years afterwards to be an eager sportsman, and an excellent shot ; but with him, even in his boyhood, killing the game was only a secondary consideration to the opportunities which his long rambles with his gun afforded him for studying nature ; and from the facts and incidents collected at this early period he laid in a fund of information which formed the basis of many of his subsequent investigations.

He was at this time painfully shy, and it was difficult to draw him out ; but he was remarkable for the steadiness with which he resisted all attempts, whether by persuasion or raillery, to join in the intemperate habits then so common in the Universities.

His school holidays, and afterwards his college vacations, were spent either with his brother in London or with his mother, who had continued to reside at Wormesley Grange for some years after her husband's death ; but having sustained the loss of both her daughters (one in her 16th, the other in her 19th year), she removed to Maryknowle, a small house near Ludlow, which Mr. Payne Knight had fitted up as a temporary residence for himself during the time he was building Downton Castle. Some account of Mr. Andrew Knight's occupations and pursuits at this period has been furnished by the pen of his early friend the late Dean of Exeter, Dr. Landon. The intercourse of which the commencement is here described was continued till Mr. Knight's death, which was followed, after the lapse of only a few months, by that of the dean.

" My acquaintance with Mr. Thomas Andrew Knight commenced at Oxford, when he was a member of Baliol College, in 1778 or 1779, I cannot name the exact time. When at college in our leisure hours we often met, and frequently took walks together. Close application was not one among the characteristics of his college life. A little reading, with his extraordinary memory and great natural talents, went very far in improving the powers of his mind. His classical reading in Greek and Latin was not extensive, but whatever he once gave his mind to made impressions which he never lost. One line in Virgil, particularly of the Georgics, if quoted in our familiar conversation, would generally be followed by a recital of pages ; and the same faculty eminently displayed itself if an accidental reference were made to Milton's Paradise Lost, or Thomson's Seasons, when the mention of a single passage would draw from him an accurate repetition of a whole book, with scarcely a pause for recollection. In vacations from the University I frequently visited him when he resided with his aged mother at Maryknowle ; and his filial attention to the comfort and domestic happiness of that most excellent old lady it was always delightful to witness, and most strikingly evinced an affectionate and amiable disposition of heart. When amusements were not to be sought out of doors, we were by no means idle on a rainy day; and the manufacture of tackle for a day's fishing did not altogether preclude attention from subjects of a more important nature. In the evenings a desultory discussion on philosophical topics, on which neither of us were very deeply informed, served at least to awaken an inclination to be better acquainted with them; and in those early days inquiry was first made as to the authors who could best throw light upon mineralogy, chemistry, botany, agriculture, and the various branches of natural and experimental philosophy. The flame once kindled, excited great ardour in the pursuit of intelligence upon most of these sciences; and the quickness of perception and comprehension which marked the course of my friend's investigations soon outran the ordinary course of study, and led him to commence a course of discoveries upon intricate and novel subjects on which his precursors had rarely bestowed a passing thought."

Mr. Knight's mind, from the earliest dawn of his understanding, seemed peculiarly formed for the enjoyment of a country life; and the part of England on which his lot had fallen, was eminently calculated to draw forth and exercise the latent faculties of his mind. Its hills, its valleys, its rivers, its vegetable productions, its geological structure, and its meteorological changes, were to him objects of philosophical investigation; while the study of what Goldsmith so well denominates "Animal Biography," afforded him constant delight and amusement.

In this manner Mr. Knight passed some years, occasionally quitting his favourite pursuits to visit his brother in London, at whose house he never failed to meet a society calculated to exert the most beneficial influence on his mind and manners.

In 1790 he accompanied his brother, and his friend Mr. Townley, to Paris; but the symptoms of the approaching Revolution were becoming so fearfully manifest, that at the end of six weeks they returned to London, and Mr. Andrew Knight never again quitted England.

The following year Mr. Knight married Frances, the youngest daughter of the late Humphrey Felton, Esq., of Woodhall, near Shrewsbury. The gentleness of her disposition and her unceasing endeavours to promote his comfort and happiness during the forty-six years they were permitted to spend together, secured to her the affections of a heart so calculated for the reception of the endearing ties of domestic life, as that of Mr. Knight; and the pain of separation is now softened to her by a recollection of the uninterrupted harmony in which this long interval was passed.

On his marriage, Mr. Knight established himself at Elton, in the immediate vicinity of his mother's and brother's residences; the acquisition of a hothouse and a farm now enabled him to prosecute his experiments in horticulture and agriculture with more advantage than heretofore. His income, as a younger brother, was at this time limited, and it was astonishing how much he did to advance the science of horticulture with a garden and an establishment of the least expensive description; but one of his peculiarities was, the readiness by which, with his own hands and the assistance of a common carpenter or black-smith, he would construct all the machinery he required for conducting his most elaborate experiments.

About this time Mr. Knight became acquainted with Sir Joseph Banks; and this introduction had so important an influence on his future proceedings, that it should not pass unnoticed. It occurred in the following manner:—The Board of Agriculture had drawn up a set of queries, to which they desired to obtain answers from different districts; and an application had been made to Sir Joseph Banks to recommend persons properly qualified, to whom the queries should be addressed. Sir Joseph referred to Mr. Payne Knight to recommend some one for this purpose in Herefordshire; who mentioned his brother, as more likely than any one he knew to fulfil the object in view, from his practical knowledge of the agricultural operations of that part of England, as well as from the attention he had given to its natural history.

Mr. Andrew Knight was accordingly made known to Sir Joseph, who soon found that he was not only eminently qualified to effect the immediate object in view, but that he had made observations, and deduced theories from them, calculated to throw much light on the more abstruse subject of vegetable physiology; and he strongly urged him to lay the result of his researches before the public. Mr. Knight had not mixed a great deal in general society; he had not had access to many modern scientific works, and his information had been almost wholly derived from the study of nature; and it was not until he was, by Sir Joseph Banks, brought into contact with many of the most distinguished men in science and literature, who assembled at the evening converzatione in Soho Square, that he was himself aware that he had observed anything which had escaped the scrutiny of other naturalists.

In Sir Joseph, Mr. Knight had a friend always anxious to draw him forth, and zealously alive to his success; ever ready to obtain information for him on any subject, or to give his advice and assistance; and his suggestions were always received with the consideration they deserved and acknowledged with gratitude. At Sir Joseph's house he had occasionally opportunities of comparing his own observations and theories with those of many of the most celebrated naturalists of all countries; and it would probably have been advantageous to him had those interchanges of information and opportunities for discussion been more frequent, for it would have saved him trouble in working out facts which cost all the labour and time of original discoveries, and which labour would have been more profitably employed in building on the sub-structure already laid by other hands. He for some years purposely avoided to read the works of his precursors in the field of vegetable physiology, from an idea that, by the study of nature, unbiassed by the opinions of others, he should be most likely to arrive at truth; but he was at length induced to deviate from this course by the advice of his friend Sir Joseph.

In the latter years of Mr. Knight's life, age and other causes had conspired to make him less and less inclined to enter into general society, and he saw little of any one besides the members of his own family, excepting during his visits to London. But these visits became each year more curtailed; and though to the last his mind retained all its freshness and activity, it was evident to those about him that he wanted more frequent collision with minds similarly constituted to his own; which is always more requisite to powerful and original intellects than to those of humbler capacities.

Mr. Knight's first communication to the Royal Society was a paper "Upon the inheritance of decay among fruit-trees, and the propagation of debility by grafting," read April 30, 1795; and, in 1797, he published a "Treatise on the culture of the apple and pear, and on the manufacture of cyder and perry." In this work he repeated the same opinions which he had advanced in his paper, viz., that vegetable, like animal life, has its fixed periods of duration; and that however the existence of a variety of a fruit-tree may be protracted beyond the natural life of the original seedling plant, by grafting, or by unusually favourable circumstances of soil or situation, still there is a period beyond which the debility incident to old age cannot be stimulated; and to this he attributed the cankered and diseased state of most of the trees of the old varieties of cyder apples in the orchards of Herefordshire.

This hypothesis was so contrary to generally received opinions, that at first it met with considerable opposition; but the increasing decay of the old fruits, even where grafted on the most vigorous stocks, and the superior healthiness of the new varieties produced from seed, has caused Mr. Knight's theory to be now almost universally adopted. To remedy the ill-consequences that would have followed the decay of the old fruits, he set about raising new varieties of apples and pears from seed; but instead of following the old method of merely selecting seeds from good kinds*, it occurred to him, that by artificially

* So long ago as 1626, a Treatise on Orchards was published by William Lawson, in which he recommends for forming an orchard, that "the ground be

impregnating blossoms with the pollen of a different variety, possessing qualities of a contrary nature, but calculated, if combined with those of the kind operated upon, to produce excellence, and by then raising plants from the seeds so produced, the chances of obtaining valuable varieties would be considerably increased; and though many of the apples at first raised from seed in this manner did not answer his expectations, he eventually succeeded in creating new varieties of many fruits and excellent vegetables, which have long been cultivated and highly prized by the horticulturists of England, and probably by those of most civilised countries to whose climate they are suited.

The idea of improving fruits by crossing seems to have been entertained by Lord Bacon, though he was ignorant of the method of accomplishing it. After stating the effects of this course in producing mules in the animal world, he thus proceeds · " The compounding and mixture of plants is not found out, which, nevertheless, if it be possible, is more at command than that of living creatures; wherefore it were one of the most noble experiments touching plants to find this art; for so you may have a great variety of new plants and flowers yet unknown. Grafting doth it not: that mendeth the fruit, or doubleth the flower, but it hath not the power to make a new kind—for the scion ever overruleth the stock*."

If to Lord Bacon must be assigned the merit of having first suggested the possibility of producing new fruits in this manner, it was reserved for Mr. Knight to discover the means by which those "most noble experiments" were to be rendered successful; and to his discoveries we undoubtedly owe the innumerable varieties of excellent fruits that supply our tables, as

sown with kernels of the best and soundest apples and pears, and to leave the likeliest plants only in the natural place, removing others, as time and occasion may require;" but this practice does not appear to have been general, for Evelyn in his " Sylva," published some years afterwards, says—"Nothing is more facile than to raise new kinds of apples, *ad infinitum*, from kernels; yet in that apple county (Hereford), so much addicted to orchards, we could never encounter more than two or three persons that did believe it."

* Quarto edit. 1790, p. 97.

well as the almost endless profusion of beautiful flowers with which the process of hybridization has adorned our green-houses and flower-gardens.

The following extract from a letter from Sir Joseph Banks shows how new to the horticulturists of the year 1798 was this system, and how important he foresaw the results would be.

" I have, some time ago, read your work on the culture of cyder fruits with much pleasure. Your experiments on apples and grapes must be very tedious, but surely the success of those on annual plants will induce you to persevere. The chances of a valuable offspring must be materially multiplied by the stimulus of a different male ; who can tell but that this, through the medium of bees, or of the wind, is the only real origin of new varieties ? When you consider your experiments upon the fecundation of plants, and improving the kinds of them by coupling the best males and females of each sort, as unimport-ant matters, you really act very differently from what I feel myself disposed to do on the occasion. I am loth to speak in a dictatorial style, if my opinion differs from yours ; but I do confess, I think no experiments promise more public utility than those for improving the breeds of vegetables."

From this time Mr. Knight continued to contribute to the Transactions of the Royal Society the results of numerous expe-riments on the " Fecundation of plants," the cause of the " Rise of the sap in trees," the " Vessels through which it ascends and descends," the " Causes which influence the direction of the root," and a variety of similar subjects. In all these researches, the ingenuity and originality of the experi-ments, and the care with which the results were given, were so great, that the most captious of subsequent writers have admitted the correctness and value of the facts established by him : though the inferences he drew from them have, in some instances, been disputed. The great object he always had in view, and which he pursued through his long life with unde-viating steadiness of purpose, was utility ; and it was only when facts had some great practical bearing that he applied himself seriously to investigate the phenomena connected with them. His experiments on the descent of the radicle excited great attention among scientific horticulturists, and have perhaps been more generally known than any of his other researches[*]. The machinery by which he subjected seeds to rotary motion during the process of germination was constructed by his own hands, with no other assistance than that of an old carpenter, who was not remarkable for his intelligence. A representation of this machinery is given with the paper describing the experiment in the present work, and also in Sir H. Davy's Agricultural Chemistry, in which Sir H. adopts Mr. Knight's hypothesis, that plants probably owe the peculiar direction of their roots and branches almost entirely to the force of gravitation. Mr. Knight however, in a paper published a few years later, details some experiments which show, that certain other natural causes may occasionally so far act in opposition to gravitation, as to divert the radicle, as well as the fibrous roots, from the direction which gravitation would have impelled them to follow [†].

The experiments on the effects of rotary motion in counteracting the effects of gravitation were repeated by M. Dutrochet and other foreign physiologists, with various modifications, but always followed by the same results. On this subject a correspondence commenced between M. Dutrochet and Mr. Knight, which was continued during the remainder of his life.

Among other facts established by Mr. Knight's experiments is, that the ascending sap undergoes a change in its progress through the leaves, somewhat analogous to that which takes place in the blood of animals in its passage through the lungs; and that this elaborated sap afterwards descends through the bark, depositing in its course an inner layer of bark, and a new layer of wood, while the old external bark cracks and peels off as the stem or branch of the tree increases its dimensions, by the annual deposition of a layer of fresh wood. His views as to the vessels through which the sap ascends to the leaf have not

* See below, Paper No. VII.   † See below, Paper No. XIII.

been so generally adopted ; but this is a point on which some diversity of opinion still exists among physiologists. Mr. Knight considered the rays which are seen to diverge from the centre of a horizontal section of the trunk of a tree, and which in longitudinal sections is known as the silver grain, to be the vessels through which the descending sap is conveyed from the bark into the cellular cavities of the wood, there to remain till it combines with the ascending fluid in the following spring.

The following letter from Mons. Mirbel (though bearing date a few years later) refers to the papers written by Mr. Knight at this time ; and the candid and liberal spirit in which it is dictated is so honourable to the writer, that the insertion of it here reflects even more credit on him than on Mr. Knight. The feelings with which it was received by the latter will be seen by the following note, found among his papers :—

" M. Mirbel has changed his opinions respecting the transmutation of bark into alburnum ; and in a private letter conceded the point to me, in so manly and honourable a way, that I really felt much more sorry that M. Mirbel should have found himself called upon to make such a concession, than joy at my own triumph, which I may be supposed to have felt. The conduct of M. Mirbel greatly raises him in my esteem, and I should feel proud to follow his example."

" *Paris, ce* 20 *Mai,* 1816.

" Monsieur :

" J'ai reçu la lettre dont vous m'avez honoré, et je prie mon ami, Mons. le Comte de Mosbourg, qui part pour l'Angleterre, de vous porter ma réponse. Mon ouvrage étoit déjà imprimé quand j'ai eu connaissance de vos opinions ; elles m'ont d'abord paru spécieuses, et ensuite elles m'ont paru très-bien fondées Vous m'avez ouvert les yeux, et je vous en remercie, car de même que vous je ne cherche que la vérité. Se refuser à l'évidence est une folie dont un savant est tôt ou tard puni par la perte de sa réputation. Il vaut mieux changer de route que d'en suivre une qui nous égare. Je reconnois aujourd'hui que le liber ne se **change** point en bois ; qu'il est constamment repoussé à la circonférence ; et qu'il se forme annuellement, entre le corps ligneux et l'écorce, une couche de cambium, laquelle régénère le liber et le bois. Je crois que c'est à-peu-près là votre doctrine ; c'est en étudiant la nature que j'ai appris à apprécier vos travaux. J'ai répété les expériences de Duhamel ; il m'a semblé que s'il avoit fait des observations microscopiques, il seroit arrivé aux mêmes résultats que vous, et n'auroit pas laissé ses lecteurs dans une doute désespérante.

"M. Aubert du Petit-Thouars a combattu mes opinions, mais ce qu'il a mis à la place n'a pu me satisfaire ; c'est la raison pourquoi j'ai fait peu d'attention à sa critique ; la votre m'a ouvert les yeux sur mon erreur. J'ai fait une longue suite de recherches sur l'Orme, le Pommier, le Cérisier, et sur beaucoup d'autres arbres encore ; je crois cette fois avoir saisi la nature sur le fait, et je saisirai la première occasion de réfuter moi-même ma première doctrine. Je l'ai sérieusement examinée, et je pense qu'il me seroit facile de réordonner tous les faits, et de rendre cette partie de mon travail beaucoup plus exacte. . . . . . .

"Daignez agréer l'hommage de la haute considération avec laquelle j'ai l'honneur d'être, Monsieur,

"Votre très-humble serviteur,

"B. MIRBEL*."

* TRANSLATION OF M. MIRBEL'S LETTER.

"SIR :

"I received the letter with which you have honoured me on the 11th of last February, and I have requested my friend M. Le Comte de Mosbourg, who is going to England, to convey my answer to you. My work was already printed when I became acquainted with your opinions ; at first they appeared to me specious, but now they seem to me to be well founded.

"You have opened my eyes, and I thank you for it ; for, like yourself, I seek only for truth. To refuse to receive evidence is a folly for which a *savant* is sooner or later punished by loss of reputation.

"It is better to change a route than to follow one that leads us astray. I acknowledge now that the bark does not change into wood ; but that there is continually deposited between the wood and the bark a layer of cambium which generates new wood and bark. I believe this is nearly your doctrine. It is in studying nature that I have learned to appreciate your works.

"I have repeated the experiments of Duhamel, and it appears to me, that if he had made microscopic observations, he would have arrived at the same results as

In the year 1805 Mr. Knight was elected a Fellow of the Royal Society, and on the 4th of November, 1806, the Copley Medal was voted to him for his papers on vegetable physiology, and presented at the anniversary meeting on the 1st of December following, when Sir Joseph Banks delivered an address expressive of the sense the society entertained of the value of his discoveries.

But the time and attention he devoted to scientific pursuits did not divert him from the prosecution of objects which, though less calculated to secure him an eminent rank among philosophers, were gaining him the still more enviable distinction of a benefactor of his country.

He had by this time become well known as a practical agriculturist, and an improver of the breed of Herefordshire cattle. The stock of this county had been long distinguished for its superior quality ; the origin of this superiority he had taken some pains to discover, and the result of his inquiries led him to attribute it to the introduction from Flanders* of a breed of

yourself, and would not have left his reader in a perplexing state of doubt. M. Aubert du Petit Thouars has combated my opinions, but those he has substituted have not satisfied me, for which reason I have paid little attention to his criticisms. You have opened my eyes to my errors. I have made a long course of experiments on the elm, the apple-tree, the cherry, and on many other trees, and I believe I have this time detected nature in her operations, and I shall myself seize the first occasion to refute my original doctrine. I have seriously examined it, and I think it will be easy to rearrange the facts, and to make this part of my work much more correct.

" I have not yet, sir, received the work that you announce. What I know of you gives me beforehand a high opinion of your new researches. It is very important that we should clear up the chaos of vegetable physiology : this branch of general science is overloaded with error, and with fanciful theories ; we shall only succeed in clearing it up by substituting strict observation instead of vain hypothesis, and severe logic for frivolous reasoning : it is for you, above all others, to do us this service. Accept the testimony of the high consideration with which I have the honour to be, Sir,

" Your very humble Servant,
" MIRBEL."

* In Cuyp's pictures the cattle are usually represented of the Herefordshire colour, with white faces.

cattle by Lord Scudamore, who died in 1671, to whom the orchards of Herefordshire were also indebted for the introduction of many of their best apples.

An agricultural society was established in Herefordshire in 1797, in the formation of which Mr. Knight took an active part; and to the end of his life he was almost invariably present at its annual show. Both here and at Smithfield his cattle frequently obtained prizes, and, in his usual liberal spirit, he, on several occasions, offered premiums at Hereford for objects that he considered of importance to the farming interests.

In the year 1802, a Mr. Davidson was sent to England, by order of the Emperor of Russia, to procure some of the improved breed of cattle and sheep for the Imperial farm; and he was recommended by Sir Joseph Banks to Mr. Knight, through whose means a selection was made from the stock of other celebrated breeders, as well as his own, but for which he would not allow more than the market price to be charged. This commission was executed so much to the Emperor's satisfaction, that on Mr. Davidson's return to St. Petersburgh, the following letter was received by Mr. Knight:—

" *St. Petersburgh, 4th January*, 1803.

" SIR :

" On his arrival here, Mr. Davidson having represented the many civilities and attentions he had received from you while purchasing sheep and cattle for the Emperor, and also the very liberal and handsome manner in which you had parted with a heifer and several of your valuable flock of sheep, and procured others for him from your neighbours, I am directed by his imperial majesty to thank you, in the warmest manner, for the favours thus conferred upon him : he, at the same time, requests you will have the goodness to thank Mr. Martin and Mr. Steward for their kindness. Should there be anything of the same nature in his dominions which you might imagine could be of the least service to you, he will think himself happy in any opportunity you may afford him of returning the obligation.

I shall esteem myself favoured by any application to me upon
the subject, and will immediately upon receiving it lay it before
his imperial majesty.

"I have the honour to be,

"Your obedient humble servant,

"N. NOVOSSILZOFF."

Mr. Knight had, in 1799, received a gift from George III. of a
Merino ram, some of which had been imported for the purpose of
improving the wool of the native breeds of sheep; and he had
obtained a mixed breed, between the Merino and the Ryeland,
to which he for some years paid much attention, and had regularly reported the result of these experiments to Sir Joseph Banks.
Not many years before his death, he imported some Norwegian
ponies, which, though neither particularly handsome nor active,
he thought, from their great strength and hardy habits, were
likely to effect an improvement in the breed of horses adapted to
agricultural and other uses, where strength and hardihood are
more valuable qualities than spirit or beauty. A cross with the
London dray-horse produced some animals combining many of
the good qualities of both parents.

It was during his annual visit to the metropolis, in the
spring of 1803, that he was introduced, by their mutual
friend Sir Joseph Banks, to Sir Humphrey (then Mr.) Davy,
who was about to deliver a course of lectures on "the Chemistry
of Agriculture," before the Board of Agriculture, and who was
anxious to avail himself of Mr. Knight's experience and enlightened views on some of the points on which he had to treat.

The acquaintance thus begun soon ripened into a warm
friendship, and a correspondence commenced which was continued, with few interruptions, till the lamented death of Sir
Humphrey in 1829.

Mr. Davy visited Mr. Knight at Elton in the summer of
1803, in company with Mr. Greenough, with whom he was
proceeding to make a tour in Wales and Ireland; and for many
years afterwards he rarely failed to spend some days either

at Elton or Downton Castle, for the purpose of enjoying his favourite amusement of fly-fishing. These days he commemorates in the following passage of his " Salmonia" :—

Day 8th. Scene, Downton.

" *Halietus :* I do not think, as the day advances, there will be any deficiency of light, and I shall not be sorry for this, as it will enable you to see the grounds of Downton, and the distances in the landscape, to more advantage.

" *Poietes :* This spot is really very fine ;—the fall of water— the picturesque mill—the abrupt cliff, and the bank covered with noble oaks above the river, compose a scene such as I have rarely beheld in this island.

" *Halietus :* We will wander a little longer through the walks. There you will enter a subterraneous passage in the rock beyond the mossy grotto. Behold the castle or mansion-house, clothed in beautiful vegetation of which the red creeper is most distinct, rises above on the hill ! After we have finished our walk and our fishing, I will, if you please, take you to the house, and introduce you to the worthy master, whom to know is to love, and to whom all good anglers should be grateful, and who has a stronger claim to a more extensive gratitude—that of his country and of society—by his scientific researches on vegetable nature, which are not merely curious but useful, and which have already led to great improvements in our fruits and plants, and generally extended the popularity of horticulture."

The following letters contain allusions to some of these visits, which were a source of so much gratification to Mr. Knight, that it is hoped no apology will be required for thus preserving a memorial of them :

" Royal Institution, August 25, 1808.

" My Dear Sir :

" I have just sent your excellent paper on the functions of the alburnum to press. Do you wish for any extra copies ? Our society will expect with anxiety a continuation of your important researches ; and I trust we shall have a paper from

you the beginning of the next session.   I shall ask permission
to witness the results of some of your experiments in the course
of the next month.   I think of leaving London for a fortnight,
and there is no place that I have so great a desire to visit as
your delightful scenery.   The hope of the pleasure of your
society, the banks of the Teme, and the grayling fishing, are an
assemblage of temptations which will induce me to bend my
course towards Herefordshire.   Two philosophical friends, Mr.
Children and Mr. Pepys, have promised to be my companions
in this little journey, and we propose to establish our head-
quarters at Leominster and Leintwardine, from which last place
I shall have the opportunity of paying you a visit, and I hope
you will permit us all to join in a fly-fishing party.

"I have been much engaged in experiments since I had the
pleasure of seeing you, and I have succeeded in decomposing
all the earths*, which turn out to be highly combustible metals
united to oxygen.   I am, my dear sir, with respectful compli-
ments to Mrs. Knight,

" Very sincerely your obliged,

" H. Davy."

" Cobham, Kent, November 3, 1810.

"My Dear Sir:

"I cannot yet profit by the kind permission you have given
me to submit my ideas upon vegetable chemistry to your observations and corrections, for I have only just commenced that
part of my labours, and I do not hope to be able to get through
it till the beginning of the spring.   In considering the physiology
of the subject, I shall have little to do but to record your
labours, for you have *created* almost all the *science* we possess
on that interesting subject; my aim will be to throw out some
chemical hints upon the nature of vegetable nutrition, and the
conversion of dead into living matter, and which may at length

---

* The experiments thus simply reported form the subject of his second
Bakerian Lecture; and, " since Newton's first discoveries in Optics, it may be
questioned, whether so successful an instance of philosophical induction has ever
been afforded."—See Paris's Life of Sir Humphrey Davy.

produce new investigations. I have often, since I was at Downton, had occasion to check *feelings* which certainly were *too selfish* to be indulged in for more than a moment. When I have seen a fine day, and the flies sporting in the sunshine, I have sighed and said, ' What would such a day be worth at Downton ! ' In the first week after I returned, I rejoiced when the wind blew from the east.

" We were unfortunate in our weather : but to have had a week of fine days, and good fishing, added to our general stock of pleasures whilst we were with you, would have been above the common balance of human enjoyment, and we might have considered ourselves, in the superstitious spirit of the ancients, *nimis fortunati.*

" I have been much employed, since my return, in pursuing investigations upon the nature of air and water, and their conversion into each other*. The inquiry becomes more difficult as it becomes more refined ; but I hope to be able to give some decided views upon the subject. Many thanks for the interest you express in my experiments. I am little anxious about speculative opinions, yet I shall omit no explanation that may assist research : facts are what we ought to value, and they must be permanent even among the revolutions of opinion.

" When the weight of the atmosphere was first proved by the Torricellian experiment, the Italian philosopher was abused, and a thousand false explanations of the barometer given by monks and jesuits. One never hears now of Father Linus's invisible threads of suspension for the mercury ! the fact belongs to the immutable in natural philosophy.

" I beg to be remembered very respectfully and kindly to Mrs. Knight, and all the family at Downton. Believe me your goodness and hospitality have not been thrown away upon an ungrateful man.

" I am, my dear Sir,

" Very truly, always your obliged,

" H. DAVY.

* The results of these experiments will be seen in the Bakerian Lectures of 1810.

"This is almost the first hour of leisure that I have had since I received your letter. I am here at grass for two or three days, in the midst of fine woods, but without a Teme or a Downton."

The paper " On the Functions of the Alburnum," alluded to in Sir H. Davy's second letter, was published by the Royal Society, but has not been included in the selection of papers for the present work, because the theory it presents has not been established, and it will be seen, that at a subsequent period, Mr. Knight himself was disposed to adopt the opinions of M. Dutrochet, who ascribes the ascent of the sap to electrical agency. A letter addressed to Sir Joseph Banks on this subject will show what were his views at that time, and the observations on which they were founded.

" *August* 14, 1799.

"MY DEAR SIR :

" I am very much obliged to you for my ram, which arrived very safe, and in perfectly good condition. I shall try different crosses with him this autumn, and I shall have great pleasure in sending you the results of such trials as I shall make. You may depend on the statements I shall send being perfectly accurate, if without any other merit.

" I will take this opportunity of mentioning the observations and opinions I spoke of in my last letter, relative to the ascent of the sap in trees, though I fear it will occasion me to trouble you with an epistle of immoderate length. If I become a troublesome scribbler to you, I must claim your pardon on the ground that you have made me such ; for without the attention I have been honoured with from you, I am certain I should never (in print) have scribbled at all. In the observations I am going to state, there will probably be little, perhaps nothing, new to you; but as I do not know how much will be new, I will state the whole as if I supposed it such.

" It is, I think, easy to prove every theory I have seen on the subject false; that of capillary attraction is surely without foundation, not being any way equal to propel the sap in the manner described by Hales. Dr. Hunter's opinion, that the sap is raised by the expanded air-vessels pressing on the sap-vessels, does not agree with the fact, that the sap flows with great force when the temperature of the surrounding air is declining; nor do I see a force here adequate to the effect produced. Dr. Darwin's imagination is generally too strong for his judgment; and it has, I suspect, created more in this case than nature has done. My theory may perhaps be more absurd than either; but such as it is, I will profit by the permission you have given me to lay it before you.

" There are two kinds of grain in wood; the one usually called the false or bastard, the other the true or silver grain. The former consists of those concentric circles which mark the annual increase of the tree; and the latter is formed of polished laminæ diverging in every direction, from the centre towards the bark of the tree, slightly adhering to each other at all times, and scarcely at all during the spring and summer, whence the increased brittleness of wood at these seasons. If you will examine a piece of English oak, you will find the laminæ I describe, and that every sap tube is touched by it at short distances, and is slightly diverted by it from its course. If these laminæ be expansible by increase of temperature, I conceive that they are placed as well as possible to impel the sap to the extremities; and that they are expansible by change of temperature I am led to suspect, by their being much affected and put in motion by the state of the atmosphere long after the tree has ceased to live. I shall at present confine my observations to the English oak, though the same observations are applicable in a greater or less degree to every other kind of tree, and even to the cabbage-stalk. In sawing oak into boards, it is usual to cut it, as much as is possible, into what are called quarter boards; being so named from the tree being first cut into quarters. In a true quarter board the laminæ of the silver grain lie exactly parallel with the surface of the board, and a board thus sawn is never seen to deviate from its horizontal position when laid in a floor. If, on the contrary, a board be sawn across the silver grain, it will during many years be incapable of bearing changes of temperature and moisture without being warped, nor will the strength of very strong nails be able to prevent the inconvenience thence arising. On this account quarter boards are always sold at a much higher price than others, which are here called bastard boards. If a board of the latter kind be laid in the floor with that surface uppermost which grew nearest the centre of the tree, it will show a disposition to become convex; if with the other surface uppermost, concave. The latter being much more inconvenient, this circumstance ought to be attended to by workmen; but it is, I believe, wholly unknown to them. I do not suppose this property in wood to have been attended to by the makers of harpsichords or pianofortes; if it has not, it is probably the cause why some instruments keep in tune better than others.

"You have, perhaps, remarked that when an oak has been stript of its bark and exposed to the sun and air, its surface becomes full of small clefts, which continue for a long time to contract and expand with the changes in the weather—you will find that these are always formed by the laminæ of the silver grain having parted from each other. This restless temper in it (of which I could point out other instances) has convinced me that it was not made to be idle; and as no other power appears to me to have been discovered capable of propelling the sap to the height described by Dr. Hales, I am much disposed to believe that this is the office which nature has assigned it, and that the following may possibly be the mode of acting. All bodies being more or less expansible by heat, and the silver grain appearing to be of a very irritable temper, I infer that it will expand and press on the sap vessels, whenever the temperature of the surrounding air is increasing by the presence of the sun or other causes; and that it will contract again during the cold of the night, or other adventitious decrease of heat. These effects will first take place in the smaller branches—later in the trunk (owing to its greater bulk and the temperature of the fluid it receives from the earth), and last at the root. If we suppose these laminæ to contract first in the smaller branches during the decreasing temperature of the evening, the resistance the rising fluid will meet with in those branches will be less than the pressure exerted in the trunk and large boughs, and the sap will, in consequence, flow with greater freedom during the evening and night (as my experience induces me to believe it does); and during this time plants ought, according to this theory, to grow most, as a few experiments I have made incline me to believe they do. In the morning the increasing temperature of the air would put the sap in the smaller branches in motion, and thus supply the progress of vegetation during the day. No kind of weather appears so well calculated to produce the expansion and contraction I have supposed to exist, as that in which there are frequent hot gleams of sun with intervening clouds and showers; and in such weather I think plants usually make the most rapid progress.

" When trees are burst by frost, it is, I believe, usually supposed to arise from the congelation and consequent expansion of the fluid remaining in the sap vessels; but this opinion I think must be erroneous, for the sap vessels (in the common kind of fracture) are not ruptured, nor does the fracture follow their direction—it follows that of the silver grain; and I believe that the internal part of the tree is cleft by the expansion of the external part, owing to the sudden change of temperature in the end of long and hard frosts, as frequently happens to other hard and brittle bodies. The silver grain is here extremely well placed to produce this effect, and I have little doubt does produce it. But there is another species of rupture, common in pollard trees, which follows the circular line of the sap vessels; and this is probably occasioned by the freezing of the sap.

" My letter has grown to a most immoderate length, and I therefore will not at present trouble you with further observations. If you think there is any prospect of my being right, I will endeavour, in the course of this autumn and next spring, to make further experiments and observations. My opinions on this subject have been the same during the last six or seven years, but I have lately been paying much attention to the cause of blights, and I have reason to believe that they depend much on an imperfect and irregular supply of sap. There is one species of blight, the mildew, of whose nature I have satisfied myself during the last month. It appears to me to be evidently a plant of the cryptogamous class, as you have probably long since known, with oval capsules and globular seeds.

You were so kind as to say you had taken some copies of the paper I had the honour to address to you in the spring. I will not trouble you to send them to me, but I shall be much obliged to you if some time in the autumn you will send a copy or two to Mr. Felton, who, I believe, has the honour of your acquaintance. He is Mrs. Knight's uncle, and requested me to send him a copy.

" I am, dear Sir,

" Your much obliged obedient servant,

" T. A. KNIGHT."

SIR JOSEPH BANKS IN REPLY.

" Soho Square, April 10, 1800.

" MY DEAR SIR :

" Your very interesting letter would not have remained so long unanswered, had I not been for the last month in a state of persecution from the multiplied duties of my new station in the Committee for Trade. I have seldom had a day to spare : and till the holydays relieved me, I thought I should never again be permitted to return to my favourite pursuits; and during my absence from London, the impossibility of consulting some of my friends, whose opinions upon the subject of the circulation of sap I have been used to rely upon, prevented my writing.

" Whether any of our predecessors may have been better qualified to investigate the physiology of plants than you are, I shall not decide upon ; but that you are eminently qualified for such undertakings I will most readily declare. Your observations and experiments are all new to me, and have given me infinite pleasure. I have only, therefore, to request a continuation of your friendly communications, either in manuscript or in print, as you may think fitting, and I promise you I shall receive them with no little avidity.

" I dare not venture to decide on the ingenious conclusions you have drawn from your experiments ; they are so wholly new, and so much beyond the usual range of opinions. I observe, however, that Dr. Darwin, who mixes truth and falsehood, ingenuity and perversity of opinions, exactly in the manner we mix the ingredients of punch, has gone beyond your speculation of a nervous system in plants, by suggesting that they may have a brain. I confess, also, that he does not follow up his assertion with half the force of reason which you adduce in support of yours.

" Nothing appears to me likely to develope the internal structure of plants so much as the analogy they bear to animals, whose structure is more easily examined : nature seems in organic bodies to have followed one uniform plan ; that is, she has arranged a certain number of parts necessary for the structure of the most perfect work of creation, and varied her works, principally by subtracting something from each, from the man to the mushroom, which is like a man furnished with lacteals in the form of roots, but has no occasion for a stomach, or for the powers of digestion.

" Plants have no digestive powers : and putrefaction appears to me to do the office for plants which digestion performs for animals, by assimilating the parts of substances that have been animal or vegetable ; both feed alike on what has at some former period been organized, and on nothing else.

" I hope you will not disappoint us after the hopes you have given us of a visit this spring. We shall be in high beauty very soon. Kew gardens will be beautiful in a fortnight's time.

" Believe me, my dear Sir,

" With sincere esteem and regard, most faithfully yours,

" JOSEPH BANKS."

Among the numerous societies to which the present age has given birth, none, perhaps, have been followed by more beneficial results to the community at large than the Horticultural Society. The proposed establishment of this society was first communicated to Mr. Knight by Sir Joseph Banks, as follows:

" *Soho Square, March* 29, 1804.

   " My Dear Sir :

" It having occurred to some of us here, that a Horticultural Society might be formed, upon a principle not very dissimilar from that of the numerous Agricultural Societies, which, if they have done no other service, have certainly wakened a taste for agriculture, and guided the judgments of those who wished to encourage it; two meetings have been held in order to commence the establishment, the proceedings of which I enclose to you. You will see that I have taken the liberty of naming you as an original member."

John Wedgewood, Esq., was the first projector, and on the society being constituted on the 14th of March, 1804, the rules and regulations which had been suggested by Mr. Wedgewood were adopted.*

On the 30th of March, a meeting was held for the appointment of an annual council and officers, when the Earl of Dartmouth was elected President, Mr. Wedgewood, Secretary, &c.

The first part of the Transactions was published in 1807. It opens with an introductory paper written by Mr. Knight, and also contains another paper from his pen, " On Raising New and Early Fruits ;" read November 4, 1806. From this time every succeeding part of the Society's Transactions contain several communications from him.

In order to put the Society upon a more firm foundation, and to give it a higher character, both in this and foreign countries, it was determined to obtain a charter, which was

---

* The account here given of the origin and progress of the Horticultural Society is extracted from a communication from the Secretary, Mr. Bentham.

granted in April 1808, and on Lord Dartmouth dying, about the end of the year 1810, Mr. Knight was elected President on the 1st of January, 1811, and continued to fill that office during the remainder of his life. His residence in the country prevented, indeed, his usually taking a part in the deliberations of the council; but it enabled him more effectually to promote the objects of the Society, by the prosecution of his investigations; and on every occasion where his time or his purse could be made available to its interests, his assistance was always most liberally given. With one or two exceptions, he was present at the anniversary meetings on the 1st of May, till the last year of his life.

At the period when Mr. Knight became President, the Society had made little progress; and its rapid increase afterwards is, in a great measure, to be attributed to Mr. Sabine, who became a member about the same time, and afterwards accepted the office of secretary, and whose zeal and activity, supported by the reputation of the President, gave a new impulse to its exertions, and enlisted among its supporters not only men of science and practical gardeners, but nearly all the rank and wealth of the kingdom. With the ample means thus placed at the disposal of the Society, information and produce were collected from all parts of the world, and were distributed with unsparing liberality; and by the sound physiological principles taught by the President, and the unceasing activity of the Secretary, a complete revolution was effected in the science and practice of gardening, and a great public benefit was conferred throughout the kingdom, by inducing many in every class of life to employ their leisure hours in an innocent and healthy pursuit.

The Society first established a small experimental garden at Kensington in the commencement of the year 1818; but this being found too limited, and too much within the influence of the London atmosphere, it was determined to select another site, and the present garden of thirty-three acres was taken a few years afterwards, and the stock finally removed there in the early part of the year 1822. The great expense attending the establishment, and keeping up of so large a garden, together with the failure of the parliamentary grant and the royal subscription, both of which the Society had been led to expect, but which it never received, added to some losses which it sustained a few years afterwards, gave a temporary check to its means; but the active support of its many zealous friends enabled it to recover its position, without contracting for a moment the field of its usefulness, and long before his death, Mr. Knight could safely contemplate this society as a permanent means of applying to the benefit of the community those physiological principles which he had laboured through life to establish.

One of the earliest means adopted by the council for promoting the improvement of horticulture, was the establishment of medals as a reward for merit; these were first given in the year 1808, and on the 1st of May, 1814, the gold medal was voted by the Society to Mr. Knight, " For his various and important communications to the Society, not only of papers printed in their Transactions, but of grafts and buds of his valuable new fruits."

A few years later, the council thought it desirable to establish a class of medals of a smaller size than the original ones; and soon after the death of Sir Joseph Banks, in 1819, on carrying this resolution into effect, they embraced this opportunity of recording their sense of the benefits the Society had derived from his support and influence, by calling it the Banksian Medal, and placing Sir Joseph's profile on the obverse of the medal.

In the year 1835, in consequence of the extensive distribution of these medals, the dies had become worn out; at the same time, the encouragement to horticulturists which they had given had been so manifest, that it was determined to have three dies prepared by one of the first artists of this country. An emblematic representation of Flora, attended by the four Seasons, was selected as the design for the large medal; the head of Sir Joseph Banks was again adopted for the smaller one; and for the intermediate one, the council determined that no device could be more appropriate, and at the same time more acceptable to those whom it was intended to encourage, than a similar profile of Mr. Knight. The die of the Knightian medal was accordingly executed, together with the two others, by Mr. Wyon, and was first distributed to those to whom it had been awarded in the course of the year 1837. At a meeting of the Society held on the 4th of May, 1836, it was resolved, " That the first impression of the Society's new large medal be struck in gold, and presented to Thomas Andrew Knight, Esq., for the signal services he has rendered to horticulture by his physiological researches." This resolution having been transmitted to Mr. Knight, he signified his acceptation of it in the following letter, characterised by that liberality, which he showed in all his transactions with the Society :—

" May 6, 1836.

"MY DEAR SIR,

" I feel highly honoured and flattered by the wishes of the members of the Horticultural Society of London, that the first impression of their new gold medal should be presented to me, and I shall receive it with very great pleasure, provided I be permitted to subscribe a sum equivalent to its cost, to be employed in liquidation of the debt of the Society, but not upon any other conditions.

" I remain, &c.,
" T. A. KNIGHT.

" George Bentham, Esq., Secretary."

From the preceding details, relating to the establishment and progress of a Society so intimately connected for many years with Mr. Knight, we now resume the thread of the narrative in noticing his pursuits and occupations in the country.

In the spring of 1809, Mr. Andrew Knight and his family quitted Elton and removed to Downton Castle, which Mr. Payne Knight had given up to his brother, having built himself a cottage in the grounds, in which he passed his mornings during the summer and autumn months ; the rest of the year he spent

in London.   He still received his visitors at the castle, and frequently joined the family party at dinner, or in the evening,
and the arrangement probably contributed to the comfort of all
parties ; for while it relieved the elder brother from the trouble
unavoidably attendant on a large country establishment to a
bachelor, it afforded many advantages to Mr. Andrew Knight
and his family.

Different as were the characters and dispositions of the
brothers, the most perfect good understanding and kind feelings
invariably subsisted between them ; and on the death of Mr.
Payne Knight, in May, 1824, his loss was acutely felt by his
brother.

The subjects to which Mr. A. Knight chiefly devoted his
attention at this period will be seen by a reference to his writings. It is a source of regret that not many of his private
letters to his friends have been preserved which would have any
interest for the general reader, but a few will be given in this
place.

To John Williams, Esq., Pitmaston.

"*Elton*, 1807.

"My dear Sir,

" I had sooner written to thank you for the information with
which you have provided me, respecting your improved method
of managing vines, but that I was from home till some days
after your letter arrived here, and I have subsequently been
every day necessarily engaged much more than suits my eyes,
which do not bear very close application.

" We have long known that pears can live on branches from
which a portion of bark is taken in a ircle ; but this operation
has always been injudiciously performed, and the improvement
you mention is certainly your own.   The effect of taking off a
circle of bark is to occasion a stagnation of the descending sap,
which is probably repelled back into the buds and fruit, and
occasions the one to turn into blossom-buds, and supplies the

other, in your experiments, with a more abundant portion of food than it would otherwise obtain; and I have shown that the wood of a fir-tree above such a decorticated space, was one-fifth larger than the wood of the same tree below : the specific gravity of the one being 0.590, and that of the other only 0.491. Your experiment, like the preceding, which is in the Philosophical Transactions of 1806, affords, I think, strong evidence in support of my theory, in which the sap is supposed to descend down the bark; and on that, as well as other accounts, is very acceptable to me.

"I had occasion to write to Sir Joseph Banks the day after I came home, and I sent him an account of your experiments on the vine, with which I am sure he will be much pleased. I think an account of them would be very well received by the Horticultural Society*. I feel greatly interested in them, I assure you, both as a gardener and as they afford strong evidence in support of my opinions respecting the circulation of the sap.

" I remain," &c.

The two letters that follow relate to some experiments on the effects of voltaic electricity on vegetable life, which Mr. Williams had undertaken at the suggestion of Sir Joseph Banks and Mr. Knight, whose attention had been directed to this subject, by experiments made by Dr. Wilson Philip, proving the powerful influence of a current of electrical fluid when applied to the digestive organs of animals; while by some other writers it had been denied that any effect was produced by similar application of electricity.

Some seeds of the *Vicia faba* were subjected by Mr. Williams during the process of germination to a current of voltaic electricity, and the result was, that vitality was quickly destroyed by a strong charge; and that even the slightest that could be given produced a manifestly injurious effect on the plant, and destroyed it when long persisted in. One remarkable effect

---

* A paper on " A method of hastening the maturation of grapes," was communicated to the Horticultural Society by Mr. Williams, May 3, 1808.

apparent was, that when the radicle ceased to vegetate, it did not change colour like a decaying root from end to end, but in alternate rings of black and white. Mr. Williams thought this effect probably indicated that some parts of the organization of the root were more susceptible of the electrical influence than others, but considered that it deserved further investigation.

"*January* 3, 1818.

"My dear Sir,

"I have had so much writing on my hands, that I had not at once eyes and time to write to you sooner, which must plead my excuse for my apparent inattention.

"The metallic oxydes still stand, I think, too prominent; and I, if I had not seen the experiment, should, as a member of the Royal Society, on hearing the paper read, be more disposed to attribute the death or sickness of the plants to the operation of metallic poison, than to the voltaic battery. No one can possibly rob you of the discovery you have made, as my correspondence with Sir Joseph Banks upon the subject will prove. You will, I hope, still appear in the same part of the Transactions*.

"I think it expedient that a few more experiments be made, as soon as you begin to warm your vinery. There can be little difficulty in proving whether oxyde of iron, that is, the red oxyde, which I suppose to be the kind produced by the voltaic battery, is poisonous to plants, by putting a dozen rusty nails into a tumbler of water with plants. I am of opinion that it would not produce the least injury; for red soils (which are much more fertile than pale yellow and white) contain about thirteen per cent. of red oxyde, and plants are well known not to be destroyed by strong chalybeate springs.

"Many beans may be placed over a tumbler of water, with their radicles descending into the water, through which a strong stream of galvanic electricity may be made to pass, as in your

* Mr. Williams' paper was read before the Royal Society, but was not printed in the Transactions.

experiment with sprigs of mint, when I had the pleasure of being at Pitmaston last.

" I think some of the facts you mention, not being important, had better be omitted; for short papers, like short sermons to most congregations, are more agreeable to the members of the Royal Society, some of whom come there with rather a strong propensity to fall asleep: there is also, among philosophers of the present day, a belief that the electric fluid produced little or no effect upon animal or vegetable life; and to oppose this belief strong facts, and those few in number, being most easily remembered and weighed, act most powerfully; and every fact which is not really strongly in favour operates injuriously.

" The winter has set in rather severely, notwithstanding the dispersion of arctic ice; but we must not decide, till we have seen a few springs, upon its operation in chilling our climate.

" Sincerely yours,

" THOMAS A. KNIGHT."

TO THE SAME.

" Downton, January 21, 1818.

" MY DEAR SIR,

" I have been some time from home or I should have written to you sooner upon the subject of your paper. I am myself perfectly satisfied that your conclusions respecting the influence of the voltaic battery upon plants are correct; but the opposite opinion that the electric fluid produces no effect upon vegetable being, has got possession of the public mind, owing to erroneous conclusions having been drawn by former writers, who had imagined themselves to have witnessed the influence of electricity to be great in promoting the growth of plants. Your paper must therefore come before judges who have already drawn conclusions in direct opposition to yours. My wishes relative to your paper can point to one object only, which is that of doing you credit; and I fear that unless strong evidence can be made to meet the possibility of the operation of metallic oxydes, Sir Joseph Banks would request you to delay its presentation to the Royal Society. I have had a correspondence with him upon the subject; and I see that the erroneous conclusions which have been drawn relative to the influence of common electricity upon plants, have made a strong impression upon his mind.

" Relative to climate : the public attention is at the present moment pointedly directed to the important fact of the dispersion of the enormous collection of ice in the North, to which Sir Joseph Banks, and probably almost every philosopher, who has directed his attention to the subject, has attributed an intimate connexion with our cold weather in the spring ; and consequently this is the precise moment in which an amelioration of our spring weather from these causes is anticipated ; therefore it is a bad moment for a paper, attributing the change in our climate to local causes, to appear before the Royal Society.

" France, you know, has made very little good wine for several years, yet no change of culture has taken place in France likely to influence the temperature of its climate within the last seventy years ; and the fact that Europe has grown milder by the destruction of its forests, appears to be universally admitted. I, however, make these remarks merely for your consideration, and am ready to act just as you wish me to do.

" Mrs. Knight and my daughters beg to be kindly remembered to Mrs. Williams and family.

" Yours sincerely."

To one of his Daughters.

" May 2, 1826.

" My dearest F——,

" Our meeting passed off as usual yesterday, and the apparent feeling of the members was so friendly that I could almost call it affectionate. I said a few words to them respecting the magnitude and increased importance of the Society, and suggested the consideration whether the office of president ought not to be held by a person of higher rank and consequence than myself, and requested that whenever it appeared to them that a beneficial change of president might take place, no tenderness of feeling towards me ought to influence them ; and that whether I continued my office of president, or became an ordinary member of the Society, my best exertions should never be wanting to support its prosperity. I was much cheered, and I believe the wish that I should continue my office is generally entertained by the Society ; so I suppose I am likely to continue P. H. S.

"We have just received from the North-west coast of America from one of our collectors, named Douglas, a large collection of seeds of plants, amongst them some of a new species of raspberry, of much merit, of which I had before heard ; and of a most beautiful plant of the genus *Gualteria*, which is allied to the Vacciniums, but bears a very close resemblance to an Arbutus, and flourishes in the deepest shade, even that of a dense pine-forest ; and is perfectly hardy. Its fruit is also sweet and palatable, and Mr. Douglas told me that he had lived wholly upon it for three days and a half ; but as a shrub it is thought a great acquisition indeed. I shall bring down some seeds of it, and I hope to raise many plants, some for you.

"Our collector proposes, when he has sent all he can home by a ship, to march across the continent of America to the country of the United States on this side, and to collect what plants and seeds he can in his journey ; but it is but too probable that he will perish in the attempt. Mr. Sabine says, that if he escapes, he will soon perish in some other hardy enterprise or other*. It is really lamentable that so fine a fellow should be sacrificed. He is the shyest being almost that I ever saw ; and upon my requesting, the year before the last, to ask him some questions respecting a part of America through which he had travelled, Mr. Sabine said ' Now Douglas will be terribly frightened ;' and so, with all his daring personal courage as to

---

* This prophecy was unhappily fulfilled only a few years afterwards by the death of Mr. Douglas, who, while exploring the mountains in the interior of Owhyee, one of the Sandwich Islands, fell into a pit constructed for taking wild cattle, and was gored to death by a bull, which had previously been captured.

actual danger, he appeared to be, till I had talked to him for some time in a friendly and familiar way. With very kind remembrances to all,

"Ever your affectionate father."

In the summer of 1827, Mr. Knight had the gratification of receiving a visit from Monsieur Dutrochet, with whom he had long held an intercourse by letter, though they had not previously met. The extract given below from a letter to Mr. Williams, relates to this visit.

"We came here (Downton) from London in a single day, or we should have had great pleasure in spending a day at Pitmaston. I brought with me my French correspondent, Mons. Dutrochet, who I mentioned to you as the discoverer of the extraordinary circumstance that animal and vegetable membrane, which under ordinary circumstances are impervious to water, readily admit that fluid to pass through them when their opposite sides are in contact with a fluid of greater density, or in some instances possessing different chemical powers ; and the facts he had advanced render it doubtful whether any mechanical agent is at work in raising the sap in trees, except the membrane, which separates the cells from each other, which are excited to act by some power, probably chemical, in the sap. M. Dutrochet spent a fortnight here, during which we made some experiments together, and investigated the hypotheses of different writers. He travelled 550 miles, and back again, with no further object than to have an opportunity of conversing upon the subject of vegetable physiology. I found him a very intelligent and generally well informed man, and he returned a very zealous horticulturist. The inhabitants of his vicinity, the neighbourhood of Tours, appear to be extremely ignorant of horticulture, and to know nothing of varieties of fruit of any kind beyond those described by Duhamel."

Mr. Knight's time was divided between philosophical and horticultural investigations, and the fulfilment of the duties of a country gentleman. He had ceased to occupy any land himself, but he paid great attention to the cultivation of his estate by his tenantry ; and though he was on all occasions a most liberal and indulgent landlord, and ever ready to afford encouragement and assistance to active and intelligent tenants, he was firm in insisting on the adoption of a proper course of management.

He was happy in his home, and beloved by all about him ; and his healthful and peaceful occupations, while they supplied never-ceasing employment for his active mind, kept him free from the mortifications and disappointments which are too frequently attendant on a life of public service, or a course of ambition.

On the 29th of November 1827, Mr. Knight was unhappily called upon to sustain the heaviest affliction that can fall on a father, in the death of his only son, by a blow as unexpected as it was overwhelming.

The following account of this singularly promising young man, extracted from a memoir written by one of his friends* soon after his death, will show as far as words can do, how irreparable was the loss of such a son and brother, to a family whose hearts were only too strongly fixed upon him.

" The dreadful accident which cut off in the prime of life an only son, and one who was even less the object of the admiration of his family for his talents than he was of their affection for his amiable qualities, took place at his father's house on the 29th of November, 1827. Mr. Knight was shooting, in the company of two gentlemen, in the woods at Downton Castle, when a casual shot struck him in the eye, and passed into the brain. He met the blow with fortitude and resignation—not a reproach escaped him. He was immediately conveyed to an adjoining cottage, where he soon fell into a state of insensibility, having exerted himself, as long as his faculties remained to him, in endeavouring to alleviate the misery of his unfortunate companion who had inflicted the blow. Medical

* The Rev. Thomas Salwey, Vicar of Oswestry.

aid was soon procured; but it was a case that no human art could reach. He lingered until about ten o'clock on the following morning, when he expired, apparently without pain—the only circumstance which could shed a gleam of consolation over the agony of those hours during which his afflicted relatives watched over him.

" In drawing a brief sketch of this lamented young man, we feel that we cannot better describe him than by saying, that he combined in a remarkable manner the talents of his uncle and his father, whose names have long been familiar to the literary and scientific world, both at home and abroad; the former having been justly regarded as one of the most distinguished scholars, the latter as one of the first physiologists of his age.

"The reputation of his uncle, and his own education at Eton, had led him to become intimately acquainted with the classics; and one of the highest gratifications which his intimate friends derived from his society arose from that keen perception of their beauties which, with the aid of a powerful memory, enabled him so happily to apply them to passing scenes.

" From Eton he removed to Trinity College, Cambridge, where he made a considerable progress in mathematics. He became also well acquainted with metaphysics, a branch of knowledge in which he took much pleasure. It has been objected to metaphysics that they lead to scepticism; but they whose originality of mind leads them to seek for truth in new and unbeaten tracks, where few are capable of following them, are perhaps too hastily accused of disregarding the important truths of revelation. Whatever danger, however, may arise from the study of metaphysics to less powerful minds, the subject of this memoir was possessed of qualities which prevented his being long misled by them. To a patient investigation of truth, and that jealousy in its admission which, whilst it is the mark of a superior mind, is at the same time the ground of that confidence we place in its decisions, he united an openness to conviction, and a candour in acknowledging it, that few are possessed of. Whilst he delighted them by following our

deepest metaphysicians through all the subtleties of their ingenious disquisitions, his intimate friends can bear testimony that
the evidences of revealed religion had latterly occupied much
of his attention, which he discussed in that spirit of candour,
and with that fair mode of argument, which can alone make
our faith a rational one.

"There were few branches of knowledge into which the
acute mind of this gifted individual had not led him; but those
in which he took most delight were the different branches of
natural history, particularly zoology, ornithology, and botany.
Few, indeed, at his age have possessed a mind stored with such
deep and varied information; for a quickness of perception,
carrying him at once through all the ordinary paths of knowledge, made him appear to start from the point at which others
rested as their goal. The energy of a powerful mind led him
at once to cope with difficulties, which others need the discipline
of habit to enable them to encounter with success; hence arose
the acquisition of a deep and varied store of information, apparently without effort or application.

" The same originality of mind, which made him delight in
pursuing some of the least beaten tracks of knowledge, guided
him also in the choice of his travels. It was to those countries
on the Continent of Europe, where man has done the least in sub-
duing nature, that he bent his steps—Norway, Sweden, Lapland,
and Finmark, became the field of his researches. Here, in the
company of his friend, George Chichester Oxenden, Esq., he
encountered difficulties and hardships which the less hardy
frame of the enterprising Clarke prevented him from attempt-
ing. Blessed with stronger constitutions, they traversed $2\frac{1}{2}^{\circ}$
of latitude between Tornea and the Icy Sea, principally on
foot, carrying their own provisions, occasionally exposed to
imminent danger from the half-frozen state of the lakes and
rivers they had to pass over, and sleeping for many nights
together on the snow. They at length reached the North Cape,
and afterwards, from the little village of Hammerfest, embarked
on board a Russian trader for Archangel, with the intention of

wintering at Soroke, in the Gulf of Kandalax, but the vessel
having been disabled in a storm, in want of provisions, and the
crew in a state of insubordination, they were compelled to leave
her, and to return in an English vessel they fortunately fell in
with in the White Sea.   A second storm obliged them to run
into a harbour near the island of Hitteroen, on the coast of
Norway.   Here our travellers separated, Mr. Oxenden returning home, and Mr. Knight proceeding to St. Petersburg, by the
way of Drontheim and Stockholm.

" Upon his return to his native country, Mr. Knight sedu-
lously devoted himself to those duties which have raised so high
the character of the English country gentleman.   As an impartial and enlightened magistrate ; as a zealous and liberal patron
of public improvements ; as the friend and protector of the
poor ; as one who from his talents was destined to take a lead
in that station in which his large property would have placed
him ; his country, and the county of Hereford in particular,
will long lament him.   A refined and highly principled mind,
and a natural modesty of character, had already gained him the
esteem of a large circle of acquaintance ; while his amiable disposition, and his attachment to his relations, which indeed was
one of the most striking features of his character, had secured
to him in an eminent degree the affections of his own family
and of his friends.

" His remains were interred at Wormsley, in the county of
Hereford, near those of his uncle ; and though, in compliance
with the wishes of his family, his funeral was strictly private,
the regrets of a whole county and the tears of the poor followed
him to his early grave.

" Mr. Andrew Knight was born on the 23d of June, 1796,
and was therefore in his 32d year."

A belief in the unerring wisdom by which the affairs of this
world are guided and directed, was so firmly impressed in Mr.
Knight's mind, that no murmur escaped him, at the mysterious
dispensation that had blasted all his fondest hopes.   He soon

resumed his usual occupations, but in a manner from which it was evident, that his chief object was to endeavour to withdraw his mind from a contemplation of his bereavement; but the pleasure they had once afforded to him was gone, and the interest he had hitherto taken in all around him, was now converted into a painful source of recollection.

In a letter to a friend written in the course of the following year he says :—

" I am at present, as I have been for some months, not in a state of mind to attend to, or interest myself about anything. I endeavour all I can to rouse myself into action, and I trust I shall in time succeed; for I know that I cannot long survive in a state of idleness.

" I cannot but feel consoled and gratified by the interest taken in the calamity of my family by all classes. My son, if his life had been spared, I am confident would have fully justified the favourable opinion generally entertained of him. As a father, he never gave me pain, except when the ardour of his character, and I may say his absolute love of danger, excited very painful apprehensions in my mind. The ways of Providence are hid from our sight, but the rule by which all is guided is just, and life is at best but an uncertain blessing, and it is perhaps weakness to mourn for the dead."

To a casual observer a slight appearance of nervous excitement was soon the only symptom that indicated the change this blow had made—but to those who lived with him, and were anxiously watching the workings of his mind, the fearful struggle that was going on within, was painfully apparent: disappointment, nevertheless, never, for one moment, had power to sour the sweetness of his temper, and he seemed to be always trying to fill the blank in his heart, by bestowing, if possible, redoubled kindness and affection upon those who were still spared to him.

It was long before he was like himself again ; and even to the close of his life, though time had done much by its softening influence to restore his mind to a healthy tone, there had been impressions made under the first overwhelming influence of this blow, which no effort of reason, nor the persuasions of those around him, could ever entirely eradicate. The following letter from Sir Humphry Davy shows how warmly he sympathised in Mr. Knight's affliction.

" Park Street, January 17, 1828.

" MY DEAR SIR,

" I have three or four times within the last six weeks taken up the pen and begun to write to you ; but I have always laid it down again, fearing to trust myself with a subject on which I could not write without feeling deeply, and great mental agitation.

" I have grieved with you, but in such the most awful visitation of evil belonging to human nature, it is almost vain to attempt to offer consolation : yet, considering life as a great system in which all is for good, and believing that the intellectual and moral part of our nature is as indestructible, as the atoms that compose our frame, I feel the conviction that where a mind so highly gifted, and so little selfish, is removed from this scene of being, apparently so prematurely, it is to act in a better and nobler state of existence.

" The noblest spirits often return the soonest to the source of intellectual life, from which they sprung : and they are surely the happiest ; whilst we are to wait the trials of sorrow, sickness, and age.

" I offer my most ardent wishes for your recovery, and that of Mrs. Knight. I know the agony of *spes fracta,* but even in this case, time, the great soother, creates a new source of hope.

" I wish I could give you a more satisfactory answer to your kind inquiries respecting my health. Dr. Philip has been very kind to me, but 'my body does me sorely wrong.' I sometimes hope, and sometimes despair, of ultimate recovery. My paralytic symptoms are much diminished : but still I cannot get rid of stiffness in my right arm and leg. I am now amusing myself with inquiries in natural history, and I hope, in the spring, to make some inquiries respecting the transmigrations of some of the anglers' water-flies.

" The Gardens of the Zoological Society are flourishing, and there are a good many animals collected there.

" The political bark, left by Mr. Canning without a pilot, seems quite wrecked; and I believe there will be some difficulty in building another. The country is in a very critical state; there certainly never was a moment in which less political talent appeared; but I am writing on a subject which every body seems to be alike ignorant of, and the business is, I fear, in hands weak in talent though strong in influence.— I am, my dear Sir,

" Very sincerely your obliged friend,

" H. DAVY."

The full measure of distress brought on Mr. Knight by his bereavement, cannot be known unless it be mentioned, that in consequence of expressions open to ambiguous interpretation used by Mr. Payne Knight in his will, an amicable law-suit had already been commenced with all parties interested, in order that the right succession to his estates might be determined.

The death therefore of his son, to whom the property would unquestionably have descended, made this already painful position tenfold more distressing to him; and though the happy disposition of his mind to look on the bright, rather than the dark side of the prospect, supported him through the remainder of his life, the uncertainty in which he continued even to his death, as to the power of disposing of his estates, was often a source of anxiety and grief to him.

Before Sir Humphry Davy quitted England for the last time, he published a fourth edition of his Lectures on Agricultural Chemistry, which he dedicated to Mr. Knight; and thus announced it in a letter written on the eve of his departure, May 20th, 1828.

" It was my ardent wish to pay you a visit before I left

England; but I do not feel myself sufficiently strong. I must defer it till another, and a better season. The extremely severe course of diet and regimen keeps my spirits very low, and, my physicians tell me, this is absolutely necessary; and whether I live or die, I am resolved to live according to rule, and to give my constitution a fair chance.

"I have sent a copy of my Agricultural Chemistry to the Horticultural Society, addressed to you. If any thing it contains relating to Vegetable Physiology is of value, it is owing to you, and in my dedication I perform at once an act of public duty and of private friendship. Should I recover my health, I have various plans of scientific labour, principally on natural history: and in the wintry state of my mind, I live principally on hope. I beg my kindest remembrances to Mrs. Knight, and all your family; and I am, my dear Sir,

"Most sincerely yours,

"H. DAVY."

Sir H. Davy died at Genoa, on the 28th of May, 1829. Of all Mr. Knight's friends, there was not one in whose society he so much delighted, and whom he could so ill at this time have spared: there were many points in which the feelings of both were peculiarly in accordance. They were both impelled by the same ardour in the investigation of truth, and the same desire to render their talents and their labour beneficial to their fellow-creatures.

The investigation of nature in all the various forms of creation, was a source of delight to both; and the keen perception of the charm of poetry, which Mr. Knight possessed in no common degree, caused him to derive the highest gratification from the singular combination of poetic imagery with deep philosophic discussion, which often characterized Sir H. Davy's conversation.

In his will, Sir H. Davy left Mr. Knight a seal ring, bearing the impression of a fish, in remembrance of the days passed together on the banks of the Teme.

Mr. Knight had constantly been urged by his horticultural friends in England, and on the Continent, to collect all he had written together, to publish it as a single work; and at one time he entertained serious thoughts of commencing a labour of this kind; but a habit of procrastinating, or perhaps it may be more properly said, the ardour of pursuit which constantly impelled him forward to seek untrodden ground, made the task of working over the old irksome, and hence he unfortunately never accomplished this desirable object: and from the same causes he declined to write articles on vegetable physiology for the Edinburgh Encyclopædia, and the Society for the Promotion of Useful Knowledge, to do which he was strongly urged. The subjoined extract from a letter from Monsieur De Candolle, shows that an intention of translating and collecting Mr. Knight's papers was entertained at Geneva, but it was abandoned in consequence of Mr. Knight expressing his intention of undertaking the task himself.

" Genève, 5 Juin, 1829.

"MON CHER MONSIEUR:

"Un homme que vous connoissez peut-être de réputation, M. Le Baron Creed, (qui a traduit en François l'ouvrage de Thaer,) passant l'hiver à la campagne, et voulant employer les longues soirées d'hiver à quelque chose d'utile, est venu me demander de lui désigner quelque ouvrage à traduire: je lui ai proposé de traduire et de réunir en un volume tous vos divers mémoires sur la physiologie végétale et l'horticulture. Ce plan lui a souri, mais avant de le mettre à exécution je me suis chargé de vous écrire pour vous demander—1°. Si cette réunion de vos divers mémoires en un corps d'ouvrage ne vous seroit pas désagréable, et si vous y donnez votre consentement. 2°. Si vous vouliez m'envoyer la liste complette de vos mémoires, afin que nous ne risquions pas d'omettre quelqu'un. 3°. Si dans le cas où vous aviez quelques additions ou corrections à faire à l'un d'eux, vous voudriez l'adresser à M. Creed, pour qu'il l'ajoutât en votre nom; et enfin si vous vouliez lui permettre de correspondre avec vous pendant la durée de son travail. Il

E

attendra votre réponse pour entreprendre son travail.   Je serois
de ma part charmé de voir réunis des mémoires d'un si grand
intérêt, et qui sont à-present dans des collections si volumi-
neuses, et qu'on n'a pas toujours sous la main.

" Recevez, mon cher Monsieur et collègue, l'expression de
la haute et sincère considération avec laquelle je suis votre
humble et dévoué serviteur,

" DE CANDOLLE*."

The letters which have been hitherto introduced into this
memoir, have shown the kind and friendly intercourse which
existed between Mr. Knight and some of the first philosophers
of his age.   In order to do full justice, however, to the kindness of his heart, we trust to be excused for exhibiting his
character as a parent, by introducing the following specimens
of his correspondence with his children ; which will show how
vivid even in advanced age was his sympathy in the sufferings
or happiness of others, and how unabated the warmth of his
affections.

---

* " My dear Sir,

" A man whom you perhaps know by reputation, M. le Baron Creed, (who
has translated into French the work of Thaer), is going to pass the winter in
the country, and being desirous to employ the long winter evenings in some
useful occupation, he has come to me to request me to point out to him some
work to translate.   I have proposed to him to translate, and to unite in one
volume, your various papers on vegetable physiology and horticulture.   He ap-
proves this plan ; but before executing it, I have undertaken to write to you, to
ask, first, if this combination of your different papers into one work would be
disagreeable to you, and if you give your consent ; secondly, if you would
send a complete list of your papers, that we may not risk omitting any ; and
thirdly, that if you have any additions or corrections to make to any of them,
he may add them in your name ; and fourthly, if you will allow him to
correspond with you during his labours, in case he should require any explanations.   He will wait your answer to begin his work.   For my own part, I shall
be charmed to see combined together papers of such great interest, and which
at present are scattered among works so voluminous, that one has them not
always at hand.   Receive, dear sir and colleague, the expression of the high
and sincere consideration with which I am your humble and devoted servant,

" DE CANDOLLE."

" *Sept.* 2, 1830.

" MY DEAREST F——,

" I read your very kind letter with some degree of melancholy pleasure, though mingled with much pain. The certainty that the dear object for whom we all mourn must be happy, must be to you, as it is to us all, the chief solace and comfort. She is much happier than she could ever have been in this troubled world; she loses nothing; for a few short years probably of more painful than pleasing existence must have brought her to the end of this present life.

" The opinion that persons quitting this life have felt the glowing happiness you describe, is not new. The following lines are quoted in the Spectator, but by whom written I do not know :

> ' Leaving the old, both worlds they view,
> Who stand upon the confines of the new.'

" I wish to repeat to you again, what I said in my last, that time will render your feelings less acutely painful than you can now imagine, and that we may look back upon such scenes of past distress with some degree of melancholy pleasure, particularly when we can look forward, as you can with confidence, to meeting the dear object of your past solicitude in a better world. I need not tell you, if words would tell, what I feel for your sufferings, but you have still some blessings left, to which a large portion of the human race are strangers, and I hope you will look forward with hope to the remaining portion of your life, and to our all meeting again in a happier world. Remember me most kindly to Mr. S.

" Your ever affectionate Father."

" *Downton, May* 30, 1833.

" MY DEAREST CHARLEY,

" It was with very painful feelings that I interfered to persuade you not to go to Paris, upon which I thought you had set your heart: and I felt great pain at the thoughts of robbing you of pleasure ; but I thought I foresaw danger, much danger, in your going, in the state of health I thought you in ; and I should have been most miserable had you gone. It is a generally-received opinion that age blunts the feelings ; but I could never at any period of my life have felt more acutely than I now feel everything in which your health and comforts are involved. My own life I value at little ; I have only to look forward to increasing debility and decay of power of body and mind : but to your health and life I look forward with very different feelings, and I am much more anxious to see you in health than to retain my own life.

"You must spend the enclosed cheque, which you were to have spent at Paris, in any way that may give you most pleasure ; and I insist on your keeping it. I shall bear the expense of a journey to Paris, whenever you choose to go.

"Thy ever affectionate Father, T. A. K."

"*Downton, July* 19, 1834.

"Dearest E——,

"I have sent you a draft on my banker, which I hope will enable you to send poor Horace to the Lee without inconvenience. You have both had a severe struggle, but I trust your constitution has not been permanently injured, and I venture to hope that his (as not unfrequently occurs) has been favourably changed and improved.

"The termination of hot dry weather, and the abundant rain of yesterday, of which I hope and conclude you have had a share, will, I trust, be favourable to poor Horace.—I have been in some degree confined by one of my little attacks of gout, and my foot continues slightly swelled, but I have never been prevented going to my garden.

"Your mother is pretty well, but I am sorry to say not so strong as she was last year. She, however, walks to church and back without suffering from too much fatigue ; and unless after sleeping ill, her health is tolerable, and her appetite not defective ; and although she is not so strong as I could wish to

see her, she is upon the whole well for her time of life.   I beg
to be kindly remembered to Mr. Walpole, and pray tell him that
I shall be happy to see him here for as long a period as will suit
his engagements.

" Ever your affectionate Father."

Mr. Knight continued occasionally to communicate the
results of his observations and investigations to the Royal and
Horticultural Societies.   His last paper in the Philosophical
Transactions was " On the hereditary Instincts of Animals,"
which was read on the 25th of May, 1837.   He took much
pleasure in cultivating the attachment of the brute creation,
and it was sometimes a subject of doubt whether his children's
pet birds and animals shared most largely in their affection or
in his; but besides the indulgence of the kindness of his disposition, he was thus afforded opportunities of observing many
peculiarities in the habits of creatures thus brought immediately under his eye, and relieved from the restraint which
the fear of man, by long continuance converted into an instinct, usually throws in the way of the naturalist.   His fond-
ness for animals was not of that senseless kind which is shown
by lavishing unreasonable indulgences on them; but it was
dictated by a true benevolence, which would have led him to
suffer pain himself, rather than have been the cause of it to a
worm or a fly.   He was very particular as to the manner in
which the game and poultry were killed for the supply of his
table; and he sometimes even superintended the operation
himself, that he might be sure it was done in the manner
calculated to cause least pain.   At the time when he was an
eager sportsman, he has often been known to spend half the
day, and remain out long after his dinner-hour, in hunting for
a wounded bird; and if unsuccessful in his search, the idea of
the sufferings of the poor creature seemed to weigh upon his
mind, and he would not unfrequently resume his search early
on the following morning.

Among domesticated animals, Mr. Knight particularly delighted to trace the hereditary direction which cultivation through successive generations had given to natural instinct; and in the course of his experiments on the improvement of fruit and animals, he had made many curious observations as to the qualities which are transmitted by one or the other parent; and he sometimes amused himself with endeavouring to trace in human subjects the same analogy, by which certain moral and physical peculiarities were derived, some from one parent, and some from the other, and which he was disposed to imagine might be reduced to something like rule. His opinions on this subject are glanced at in the subjoined letter to Sir George Mackenzie, as well as his view on the tendency of modern education, both immediately and prospectively; and a few extracts from letters to other of his friends touching on similar points will follow.

" *Downton, Sept.* 29, 1836.

" My dear Sir,

" I have delayed troubling you with a letter, till I had read with attention both your little publications, and that of Dr. Caldwell. Both have given me very great pleasure ; and though I cannot say that I am so much a phrenologist as either of you, yet I perfectly agree with you in the conclusion which you have drawn in a great extent of cases, that certain forms of skull are favourable, as indicating powers of thought; and I have long believed that exertion of mind through successive generations, and proper selection of males and females, might give not only greatly enlarged powers of mind, but also better organised brain, and skulls of better forms. Upon the ill effects of modern education we are entirely of the same opinion ; and I perfectly agree with Dr. Caldwell respecting the ill effects of subjecting the brain of young subjects to any degree of painful labour. I have seen, during the course of a very long life, many very clever over-educated children ; but I have never seen any instance in which the brain-worn child of twelve years old dis- played at a later period much powers of mind. Talents which have been early visible, have, in a great variety of instances, continued to improve; but the possessors of these were not early subjected to more labour than they could bear; and the ordinary labours of education were not in any degree oppressive to them.

" I also believe Dr. Caldwell's opinion, that dyspeptic cases are to a great extent brain cases, to be well founded. He has not mentioned the singular discovery of Dr. Wilson Philip, that if the eighth pair of nerves be divided, and the divided ends be made to point in some what different directions, and the nervous communication be thus intercepted, digestion is immediately suspended totally; but that it may be made to go on perfectly well by causing a current of galvanic fluid to pass down from the neck to the stomach. Your late illustrious countryman, and my fellow-collegian and friend, Dr. Baillie, entertained previously, I believe, somewhat similar opinions respecting the influence of the operation of the brain upon the stomach. Soon after my opinions respecting the creations and motions of the fluids of plants and other matters connected with vegetable physiology were made public, and when, with the exception of Sir Joseph Banks, I had no supporter, my time and mind were laboriously occupied in a great variety of experiments, I became unwell, my stomach ceased to act, and I thought myself fast approaching to the termination of my labours. I then consulted Dr. Baillie, who gave me an extraordinary prescription : —'Take no more medicine ; walk more, and think less.'

" I entertain very nearly as exalted an opinion of the ignorance of a large portion of our legislators as you do : either they cannot, or they will not think. The Mayor of Worcester some years ago, when George III. addressed him, said, ' Please your majesty, Lord Coventry speaks for me.' Many of our legislators might say, ' and thinks for me.' As sagacity in the brute creation certainly becomes hereditary when exercised through successive generations, stupidity, I believe, becomes hereditary also; and, according to Dr. Caldwell's theory, the injurious effects of too early labour of the infant brain, must operate hereditarily. The early and excessive labour to which girls are subjected in acquiring skill in music, has long appeared to me to operate very injuriously upon their constitution and form. The roses in young ladies' cheeks, if unchanged, would, I do not doubt, appear much less bright to my eyes now than they did half a century ago : indeed, I am sure that it is so ; for I recollect perfectly well, that when I was a child, the plumage of the breasts of the male chaffinches appeared to me nearly as bright as those of the male bullfinch now do ; but I can distinguish straight from crooked now, as well as I could do at any period of my life ; and I am quite certain that the hollow, sunken chests presented by many of the young ladies of the present day of the affluent and highly-cultivated classes were not as common, or nearly so, sixty years ago, as they now are ; and I have heard on good authority, that such flat and sunken chests are not seen among the less educated girls of Ireland. With us the ears and fingers of girls are exercised, not their minds rationally exerted and amused; and I cannot avoid believing that the offspring of such parents are often born without the power of thinking deeply. I have heard it remarked by a very sensible countryman of yours, that among families which have long lived in affluence and been highly educated, a hundred men of quick parts would be found for one deep reasoner.

" I beg to assure you, that I felt very highly gratified by the belief that your very short visit to Downton proved agreeable to you ; and Mrs. Knight, and the other parts of my family who had the pleasure to meet you here, have begged me strongly to express their hopes and wishes that you will soon repeat it, and for a longer period. I cannot but feel highly flattered and gratified by the published account you were so kind as to send me of your visit to me, but I fear that your friendship has led you to speak much more favourably of me than I deserve.

" I am much inclined to doubt whether any phrenologist, by examining the exterior form of our heads, would be able to decide that our minds resembled each other, as I think they do, and as my family all thought. I have heard several people remark that my head and Davies Gilbert's are alike, and my family made the same observation. Mathematics have been the favourite study of my ancestors, so that nature perhaps made me for a mathematician, and accident a naturalist.

" I shall have great pleasure in sending you anything which my garden affords, and you wish to receive, either in the autumn or in spring, as you will direct, with models of my traps. Mrs. Knight and my family beg to join in kind remembrances, and I remain, my dear Sir, very sincerely yours,

" T. A. K."

The subjoined is an extract from the account alluded to of a visit to the President of the Horticultural Society of London, from the pen of Sir George Stewart Mackenzie, Bart., which appeared in the Edinburgh Chronicle of September 1838, describing the impression made on his mind by a day spent with Mr. Knight, at Downton.

" The venerable and talented proprietor of Downton, surrounded by a princely domain of ten thousand acres of rich and beautiful country, thinks of nothing but of what may be useful to his fellow-creatures. He received us with that unostentatious but kindly welcome which displayed the true spirit of hospitality ; regarding a visit as a favour conferred on the host, and not on the guest; and which at once excites mutual benevolence, that operates like magic in giving birth to friendship. It is true, we had seen our excellent host once before, and enjoyed occasional correspondence with him during many years. But notwithstanding, on entering a house for the first time, we felt a little awkward, as Scotchmen generally do in such circumstances. In a short time, however, this was brushed off by attention from every side ; and we experienced with much delight the ease, grace, and kindliness of English hospitality.

" Our venerable host, active and energetic in his 78th year as a man of 40, is one of those rarities among men, that know

everything—who can put their hand to everything, and give
a sound philosophical reason for what they do. He is one who
can discern rottenness in church and state, as well as canker
in a fruit-tree, and can fathom both. He can see the traps
set for the people, as they are closely analogous to those ingenious ones he sets for the blackbirds that come to devour his
fruit. He soon introduced us to his garden, which we were
most anxious to see. We found no display—nothing for show
—all was perfectly simple and business-like, and full of experiment. Various modes of culture were in progress with everything; and reasons were given for commencing every experiment.

" Were we to attempt describing all that we noticed in a
garden at which, on account of its plainness, those who regard
show and display would turn up their noses, it would be proper to think of writing a volume. We will therefore conclude
by stating that Mr. Knight has not yet subscribed to the theory
of the rotation of crops derived from the experiments which
showed that plants deposited excrementitious matter; the
theory being that, while such matter is useless to the plants
that reject it, other plants are nourished by it. Further experiments are wanted to elucidate this curious subject; and no
one has better means to confirm or overset the theory than
Mr. Knight."

The simple means by which Mr. Knight effected his earliest
and most important discoveries have been already mentioned;
and Sir George M'Kenzie correctly describes the appearance
of the garden at Downton, at a period when Mr. Knight had
for many years possessed the power of obtaining whatever
would have facilitated the most extensive application of his own
theories to practice; but it was still characterised by the same
simplicity. In his own mind were combined, probably more
than in that of any other person who ever lived, the qualities
of a physiologist and a practical gardener; and whatever success attended his horticultural operations resulted from his

sound knowledge of the vital actions of plants, founded on philosophical investigation, and his skilful adaptation of the external forces by which they are regulated.

The following extracts from various letters of Mr. Knight's
will serve further to illustrate his views on various questions
connected with the habits of domesticated animals.

" The observations of your sporting friend, that dogs which
have not been regularly and well fed will bury their superfluous
food, is well founded; but perhaps it had been more correctly
applied, if he had extended it to families of dogs; for I do not
think that the descendant of a long succession of parlour and
lap dogs would do this, though he were not well fed; and I
entertain very little doubt that the offspring of a breed which
through successive generations had been ill fed, would hide his
superfluous food, though he had been well and regularly fed.

" I have been struck with astonishment to see to what an
extent the offspring of a breed of Norfolk water-dogs, which
they there call Retriever, would do spontaneously, what their
parents had been taught to do, of which I could give many
instances.

" If you contrast the various actions of the different families of
dogs,—the truffle-hunter—the fox-hunter—the pointer—setter
— springing-spaniel — shepherd's dog — bull-dog — the silent
South-sea dog, with the native manners of the wild type, the
wolf, we shall not wonder at many irregularities in the actions
of different families of domesticated animals of the same
species."

" I think if the habits of any two families of the same species
of domesticated animals were attentively watched and compared,
great diversities of action would be observable. If we were to
draw our conclusions respecting the sagacity of the horse from
observations of the actions of a Welsh mountaineer pony, we
should pronounce that species of animal to be singularly saga-
cious in distinguishing a bog from sound ground. He knows

it perfectly by the smell; but the blood-horse shows no such sagacity—he is a perfect idiot in that respect.

" If a botanist who had only seen that variety of the *Brassica oleracea* which we call a cauliflower, described it, how little would his account agree with the observations others would have made who had seen the Scotch kale and ox-cabbage Bees have been stated to fortify their hives against the ingress of enemies in those countries where such enemies are found ; while we see no indication of such precautions here, where, through many successive generations, no such enemies have presented themselves."

" We find abundant facts to prove that not only animals, but plants also, adapt their habits to incidental external circumstances.   The crab, the pear, and the plum are produced in a state of nature only upon trees covered with sharp thorns ; and wheat, in anything approaching its natural state, is always strongly bearded.

" The wild duck sagaciously conceals its nest, and covers its eggs when it leaves them ; but the same bird domesticated often drops them at random ; and if it makes a nest, it is in so open a place, that the crows destroy them.   The tame goose cannot be trusted with its own eggs ; for it will sit so long, when it lays one, that it will spoil those previously laid."

An anecdote is given in the first volume of Mr. Jesse's Gleanings in Natural History, on the authority of Mr. Knight, of a fly-catcher, which he used often to mention as one of many instances that had come under his observation, of the exercise of a degree of intelligence, apparently surpassing the limits of the instinct given to animals to guide and direct their proceedings in the ordinary mode of existence appointed to them, and to indicate a power of adapting the habits of an individual, on whom cultivation had not exerted any influence, to exigencies which could rarely, if ever, occur.

This bird, for several successive years, built its nest in a stove in the kitchen garden at Downton Castle, into which it had free access through an aperture made to admit air. Mr. Knight observed that during the process of incubation the old bird was absent much more often from its nest than is usual during that process, and yet that it had evidently not abandoned its eggs; he therefore watched its motions closely, and soon discovered the curious fact, that the bird quitted its nest when the thermometer rose to about 71° or 72°, and returned to it when the temperature sunk again; thus seeming to have a knowledge that only a certain degree of heat was necessary to the eggs, and that, being furnished from another source, its own labours might be dispensed with. The ostrich in the torrid regions of Africa leaves her eggs, in like manner, to the influence of the sun's rays during the day; but Buffon and other naturalists deny that there is any foundation for the vulgar belief of her abandoning them altogether, and state that she constantly returns to sit upon them during the night."

The two following were addressed to Dr. Bevan, author of a work on the honey-bee :—

" In the course of my experiments I have had many opportunities of observing the peaceful and patient disposition of bees, as individuals, which Mr. John Hunter has also in some measure noticed. When one bee had collected its load and was just prepared to take flight, another often came behind it, and despoiled it of all it had collected. A second, and even a third load was collected, and lost in the same manner, and still the patient insect pursued its labour without betraying any symptom of impatience or resentment* : when, however, the hive is approached, the bee appears often to be the most irritable of animals. They are probably by nature little disposed to fight,

---

* The author of Insect Architecture in the Library of Entertaining Knowledge, after quoting the above from Mr. Knight, adds, —" Probably the latter circumstance at which Mr. Knight seems to have been surprised, was nothing more than an instance of the division of labour, so strikingly exemplified in every part of the economy of bees."

when they have nothing to fight for, as when they have first swarmed; but they appear to become acquainted, and to place confidence in persons who are much with them, and from whom they have never received injury. A labourer who looked after my bees at the time I was making experiments upon them, would put his fingers into the mouth of the hive, and push away the bees to show me the newly-formed comb, without apparently giving any offence."

"*Downton, July* 1829.

"I believe that I have been to an unjust extent sceptical respecting the accuracy of M. Hubert's statements. I have found so much inaccuracy in the writings of vegetable physiologists, that I am often probably somewhat unreasonably difficult to convince; and I recollect one of my friends having told me that when he had said to Sir Joseph Banks that I believed some statement, Sir Joseph jestingly remarked, 'he (meaning me) is an excellent person to believe after.' The evidence of bee's-wax being an animal secretion is so strong, that I cannot question it, and I think you have satisfactorily explained why it may be made into thinner combs in the autumn than in the spring*. I think you will also find it more brittle and white than the spring combs are; though possibly the spring combs may have received some colouring matter after their first formation. Whatever may be the cause of the difference of colour, I believe you will find such difference to exist; and a Polish friend of mine, whose acquaintance with the management of bees in that country, where the wax forms an article of considerable value comparatively with the price of other articles, was extensive and accurate, informed me that the autumnal combs are always

---

* Dr. Bevan had, in answer to some arguments of Mr. Knight's in favour of wax being a vegetable production, detailed experiments to prove that it was secreted by the membrane which lines the sacklets of the working bee; and he accounted for the more liberal use of it in spring by the supposition that the comb was made thicker in that season for the purpose of resisting the struggles of the nymphs, and that its tenuity in autumn might be attributable to the cells being at this season chiefly intended for repositories for honey.

separated from the others on account of their superior white-ness, and the consequent diminished labour of bleaching*.

" I have been and am still engaged in some experiments upon the potato, which plant has given me more physiological infor-mation than all the remainder of the vegetable world; and where it has not given me the information I wanted, it has directed me where to find it. I think it is capable of much improvement as an article of human food, and that varieties may be formed which as food to animals will cause a larger supply of animal food to be brought to the market than can be obtained from all the varieties of the turnip.

" If business or pursuit of pleasure should bring you into this vicinity, I shall be happy to see you at Downton. I have not much to show that is likely to interest you, and my habits of activity are necessarily sinking under the weight of seventy years ; though I am grateful that I still retain my health, and my powers of memory and of mind little changed, I believe.

" I remain," &c.

" I wrote down, some days ago, a few observations upon the screech-owl, which was formerly supposed often to visit the win-dows of the chambers of the sick and dying. Lady Macbeth says, ' It was the owl that shrieked, the fatal bellman, which gives the stern'st good-night.' I happened once to have heard this shriek, which was uncommonly loud, and most hideous, bearing no resemblance whatever to any of the ordinary cries of the owl. I saw the bird at the moment when it was uttering its horrid shriek at the window of a person who was lying ill of a fever. The owl was at that time a greatly more abundant bird than it now is ; and it was, I do not doubt, led by its nice sense of smell, and, like the raven, was the announcer of present, not the prophet of future ill, the patient having in this case recovered."

" Having retired under the shade of an oak in a very hot

---

* Mr. Knight has suspected this to have been caused by the bleaching effect of the atmosphere during the summer.

morning of September 1st, 1835, I observed a shower of honey-dew to descend in innumerable small globules (which become visible when seen in one light) from the leaves of the tree, upon which I found a very large number of aphides, from whose bodies the honey-dew appeared to be ejected with considerable force. I, in consequence, brought home a branch of the tree, which I so placed, that the light, in an otherwise dark room, should shine only upon such branches; and I then obtained clear evidence that the aphis can discharge its honey with considerable force. It is consequently often found in situations at which it could not have arrived by the mere influence of gravitation. I suspect this circumstance has led to the belief of the existence of two kinds of honey-dew, one being immediately ejected by plants; I doubt the existence of more than one kind, for I have often found a minute aphis, by the aid of a lens, in the small globules, apparently emitted by a leaf."

These specimens may serve to show in some degree how Mr. Knight's mind was always at work, and with what alacrity it seized on whatever contributions of knowledge nature threw in his way, and also the manner in which he extracted out of every-day experience facts which illustrated or confirmed former speculations.

He carried on a very extensive correspondence, not only with many of the men most distinguished for their attainments in science in Great Britain, but with most of the writers on vegetable physiology and horticulture on the continents of Europe and America. A large collection of interesting letters were preserved by him; but the limits of this memoir precludes the insertion of more than the few that have been given. He was also a corresponding member of numerous Societies for the encouragement of horticulture and agriculture in Europe, America, and Australia.

The readiness with which he communicated the results of his investigations, and the practical objects to which they led, caused incessant application to be made to him by horticulturists of all grades; and, as he never withheld information or assistance from the most humble of his applicants, his time was much occupied in answering letters, and in sending off packages of plants, &c. of the new varieties of fruits and vegetables he had raised, which he distributed with an unsparing hand; still from the time the new poor laws came into operation, notwithstanding his advanced age and his numerous avocations, he took an active part in their administration; and no cause but indisposition ever prevented his attending the weekly meetings of the board of guardians at Ludlow; for he considered that the benefits to be derived from this law would be materially diminished, if not annihilated, unless the country gentlemen lent their assistance in enforcing the proper fulfilment of its provisions.

Another subject in which he latterly took much interest was the commutation of tithes; and in 1834 he published a pamphlet, suggesting the adoption of meat as the basis on which to found the calculations of the value of tithes, instead of corn.

Though early in life Mr. Knight had been considered delicate, he had, for a long course of years, enjoyed almost uninterrupted good health, which his mode of life was well calculated to confirm: he spent many hours of every day in the open air, in his garden, or in walking about his estate: he had always been remarkable for his abstemious habits; he rarely tasted wine or any fermented liquor, and ate little animal food; which it is to be feared he persevered in to an injurious extent, for, when the powers of the stomach became diminished by the decay incidental to old age, a more generous diet would probably have had a beneficial effect on his constitution. For the last three years of his life, occasional symptoms of dyspepsia appeared, and, during the winter of 1837-8, he suffered a good deal from derangement of the digestive organs, which at times produced a very distressing sense of suffocation. He had a

F

severe attack of this kind in April, but as he was anxious to have the advice of Dr. Wilson Philip, he proceeded to London at the usual time.

He spent a day with his friend Mr. Williams on the road, and though much enfeebled by his illness, he bore the journey without apparent fatigue, and expressed his hopes that he should soon be restored to his usual state of health. On the 1st of May, he did not feel equal to taking the chair at the anniversary meeting of the Horticultural Society, nor did he ever leave the house after his arrival in London; but he saw several of his friends, conversed cheerfully, and seemed to enjoy their society.

The medicines prescribed by Dr. Philip had relieved several of the most unfavourable symptoms ; and the state of his pulse, which was as regular as that of a person in perfect health, for some time led his family to hope that he was going on well, notwithstanding that his amendment was less decided than they wished; and even when some degree of anxiety had begun to be felt as to the final issue of his illness, no symptom indicated immediate danger ; though it was apparent, from the subjects on which he conversed, that he thought it probable he should not recover, and that tenderness for the feelings of Mrs. Knight and his eldest daughter, who were with him, alone prevented his declaring this opinion in more direct terms.

He spoke with affection of the absent members * of his family, and of the arrangement he had made of his affairs ; while, to those who had the happiness of being present, he expressed in most affecting terms all that was most grateful and consoling to them to dwell upon, of his feelings to them, and of his deep thankfulness for the many blessings he had enjoyed in the course of his long life, and of his readiness to leave the world whenever he was summoned to do so.

* Sir William and Lady Rouse Boughton were detained in the country by the serious illness of the former, and Mr. and Mrs. Francis Walpole were abroad.

Illness and suffering never elicited from him one expression of impatience; they only drew forth fresh proofs of the kindness and unselfishness of his nature. At times the sense of suffocation he experienced was exceedingly distressing, but the moment that a diminution of the symptoms allowed him to speak, he never failed to tell those about him that he was better, knowing the comfort it would afford them.

After passing a tolerably tranquil night, early on the morning of Friday, May 11th, 1838, he suddenly fell back on his pillow, and drew his last breath without a sigh or a struggle.

His end was as peaceful as had been the pursuits and occupations of his long and useful life; and few men have descended to the grave more beloved, or more sincerely regretted by all ranks of society.

His remains were interred at Wormesley on Tuesday, May 22nd, near to those of his brother and his lamented son.

Many of Mr. Knight's friends were desirous to have shown the last proof of regard for his memory by attending his remains to the grave; but such offers were declined, with a very few exceptions, from a conviction that a simple unostentatious funeral would best have accorded with his own feelings when living. Every mark of respect was shown in the towns and villages through which the procession passed; and the large body of his tenantry, of whom it was chiefly composed, as they followed him to his last resting-place, evinced how strongly they felt, that in him they had lost their best friend, and the kindest and most indulgent of landlords.

A monument has since been erected to his memory by his widow, with the following inscription from the pen of the Rev. — Lee.

THOMAE . ANDREAE . KNIGHT . A.M. R.S.S.

HORTULANORUM . SOCIETATIS . APUD . LONDINENSES . PRAESIDI

QUEM . SUMMO . INGENII . ACUMINE . ET . VI . PRAEDITUM

CERTAM . PERFECTAM . QUE . RERUM . SCIENTIAM . IMPENSE . PERQUIRENTEM

IMPRIMIS.HORTORUM . CULTURAE.PROVEHENDAE. OPERAM . ET . STUDIUM . NAVANTEM

PIETAS . ERGA . DEUM

QUEM . EX . TOTIUS . NATURAE . MENTE . ATQUE . ANIMO . BENEVOLUM . AGNOVIT

COMITAS . ERGA . SUOS

QUORUM . COMMODIS . STUDIO . ACERRIMO . INSERVIEBAT

PARITER . EXORNAVERUNT

RERUM . NATURAE . COGNITORI . DILIGENTISSIMO . ET . LOCUPLETI

SCIENTIAE . ATQUE . DOCTRINAE . FAUTORI . STRENUO . ET . BENEFICO

VIRO . OPTIMO . HOMINI . AEQUISSIMO

CONJUX . CONJUGI . AMANTISSIMO

H. M. P. C. L. M.

ANNO . SACRO . MDCCCXXXVIII.

At a meeting of the Horticultural Society, held on the 19th of June, for the purpose of electing a President in the room of Mr. Knight, it was resolved—" That this meeting deeply deplore the loss the Society has sustained by the death of their late President, T. A. Knight, Esq., an individual not less distinguished for his private worth than for his public usefulness; whose memory, from the urbanity of his manners, the kindness of his disposition, his attachment to science generally, more especially to that branch patronised by this Society, will be long cherished, as his decease will be sincerely lamented."

In December following, the Duke of Sussex resigned the Chair of the Royal Society, and in a farewell address, delivered on this occasion, his Royal Highness alluded to those distinguished members whose loss the Society had sustained since the last anniversary meeting, and when speaking of Mr. Andrew Knight, described him as " having possessed very great activity of body and mind, with singular perseverance and energy in the pursuit of his favourite science ; a lucid and agreeable writer, who had by his labours developed views of the greatest value and interest in vegetable physiology, as well as in practical horticulture." After giving a sketch of Mr. Knight's labours, his Royal Highness concluded by saying, " It would be difficult to find any other contemporary author, in this or other countries, who had made such important additions to the knowledge of horticulture and the economy of vegetation."

Before closing this brief and imperfect, though it is hoped not unfaithful memoir of Mr. Knight's life, a portion of the task which is, perhaps, the most difficult remains to be accomplished.

If those by whom this memoir has been drawn up have felt themselves unequal to exhibit the workings of his mind in the investigation of the truths of philosophy; if they have not ventured to point out what are the errors he has exposed, and the difficulties he has cleared up; or what are the new facts that he has added to science, it is satisfactory to them to reflect, that his own works, which have received the approbation of most of the naturalists of Europe, have done this more fully than could have been effected by any one less qualified than himself to write on the subject.

But the acquisition of philosophic truth, and the study of the works of creation—which we have the highest authority for believing to be not merely a noble and legitimate exercise of man's powers of mind, but one acceptable to his Creator, and for the comprehension and investigation of which his mind seems to have been expressly adapted—is not the great object of life.

It is in the cultivation of man's moral powers, and in his reception and acting upon those truths which the highest exercise of reason would not have discovered, that the end of creation is to be looked for: and the memoir of Mr. Knight would be incomplete, without an attempt at least to delineate those deeper and more hidden principles which stamp the moral and religious character of an individual. It will be felt that this in all cases is a delicate and difficult task; and if any part of what is said should be thought to have been dictated rather by affection than by unbiassed judgment, it will, it is hoped, be conceded that Mr. Knight's own family had far more frequent and intimate opportunities of knowing his feelings and principles, than any other persons could possess; and in finishing what they are well aware is a feeble delineation of his character, by touching on a few points not already noticed, they trust it will be believed that they say no more than is the result of their sincere convictions.

Like other persons of ardent temperament, Mr. Knight felt strongly on all occasions; and his sense of honour was of a nature, perhaps, almost too chivalrous for the every-day concerns of life. He was slow to discover evil in others, but when he had been once led to suspect a want of integrity and fairness, he too hastily expressed such opinion; and hence he sometimes might have appeared to those who did not know the working of his mind, to have been guided by feelings very opposite to the true ones, for no heart ever more overflowed with kindness and charity to all mankind than his, and no one was more sincerely disposed to judge of others, "as he would himself be judged." A more extended intercourse with mankind would probably have had a beneficial influence on his mind on this point, but it would perhaps have robbed it of somewhat of that guilelessness and simplicity, which were among the most engaging peculiarities of his character.

In politics, the same apparent bitterness, but originating in the same high feeling, was sometimes displayed. He was a Whig of the old school, and though a strong advocate for reformation of abuses, and an admirer of liberal measures, he was decidedly opposed to more extended suffrage, vote by ballot, triennial parliaments, and other schemes of the ultra Liberal party: but, from having lived through the days when, owing to the long continuance of one party in power, abuses had crept into the administration of government, his prejudices had been excited against the Tories; and truth demands the admission, that he sometimes expressed himself of persons and measures in terms which his best friends regretted; but if convinced that he had formed an erroneous judgment, no one was more ready than himself to admit he had been wrong ; and his forgiveness of similar offences against himself, and his forgetfulness of injuries, have more than once been manifested in instances where others thought the provocation received might have justified a lasting estrangement.

The warmth of his feelings, it cannot be denied, sometimes warped his judgment ; and the faculty of fairly balancing opposite contingencies, and giving to each its due weight, and thus arriving at a cool and impartial estimate, was not one of the qualities in which his understanding most excelled. He was too much disposed to act on the impulse of the moment, and this often exposed him to subsequent inconvenience and annoyance ; though the ill consequences that might have arisen from this failing were generally averted by the kindness of his heart, and the strict integrity and sense of justice by which all his actions were controlled.

It must always be difficult for children to speak of the failings of a father ; and this difficulty is tenfold increased, when these were so overbalanced by what is great and good, as was the case in Mr. Knight's character, and when the kindness and affection by which every act of his domestic life was guided, prevented his little faults from being perceptible to his family, except at a distance; but in touching on the evil as well as the good, they feel sure they are only doing what his own upright and manly mind would have approved.

The unguarded expressions in which it has before been mentioned that Mr. Knight occasionally spoke of men and measures, was also sometimes the cause of misconception as to the nature of his religious opinions. It was very far from true that he disbelieved the fundamental truths of Christianity ; on the contrary, he often referred to them both as a test of truth, and a rule of conduct. He was not attached to any particular party or sect, but always declared his belief that all would be objects of Divine mercy, whose actions and conversation were controlled and directed by the influence of Christian principle. He entered life at a time when, as the warmest supporters of

the Church of England admit, a lamentable laxity prevailed in
her discipline; and unfortunately several strong cases of derelic-
tion of duty in her ministers came under his observation; but
he rejoiced in the progressive improvement that has since that
period been gradually accomplished in the habits of the clergy;
and in discussing his favourite subject, a modification of the
tithe laws, he never failed to mention, as the great object, the
advantage that would ensue from an alteration of these laws
to the cause of religion: and he always expressed himself desirous that the parochial clergy should generally be better provided than at present, with the means of living in comfort
themselves, and of affording temporal assistance to their flocks.
He had himself originally been intended by his family to enter
the church; but he declined to accede to their wishes, from the
deep sense he entertained of the responsibility of the duties
that such a course would entail; and when he saw men, who
he believed had taken upon themselves the solemn vows of the
ordination service from mercenary motives, and whose conduct
would have been offensive in a layman, he was in the habit of
delivering in very strong terms his opinion of the injurious
fluence which such persons were likely to exercise on the
spiritual interests of those committed to their charge: but the
zealous and hard-working clergyman was sure to receive from
him, not only the warmest expression of approbation and respect, but every proof of esteem and kindness; and those who
had the charge of the parishes in the neighbourhood of Downton Castle, can testify how readily he afforded his co-operation
and assistance to every plan for the relief and benefit of their
parishioners.

By many persons, who do not think themselves deficient in
religious principles, the evangelical party are made an object of
ridicule; but in this Mr. Knight never joined; and though he
might sometimes think the zeal of some of its members mis-
directed, he was willing to give them the full credit due to good
intentions; and he never would allow that the adoption of a
higher rule of duty should be a cause for reproach.

His charities were very extensive, and it was only by chance that those who most shared his confidence became acquainted with the large sums he distributed. It was the spontaneous feeling of his heart, that it is more blessed to give than to receive; and when he bestowed money or did an act of kindness that caused him some personal inconvenience, he always endeavoured to make it appear, that for some reason or other, it happened to be an accommodation to himself, and that he was the party on whom the favour was conferred.

The indulgence and patience he evinced in conversing with the ignorant and the dull was pre-eminent; no arrogance of manner ever displayed itself while arguing with an inferior disputant. He himself knew too much, not to make ample allowance in others for a want of acquaintance with any subjects which he had more particularly studied; and with the greatest readiness he avowed his own ignorance when questioned as to any point on which he did not feel himself competent to afford the desired information. When his children were young, he was always ready to lay aside his book to answer their questions, or to assist in their amusements; he was anxious to cultivate in them a taste for horticulture, natural history, and other rational pursuits; and his daughters now look back to the hours spent with him in his study, or in his garden, as among the happiest recollections of their childhood.

Even after he had entered his eightieth year, it was delightful to watch the spirit with which he shared in the sports of his grandchildren, and the trouble he took to provide occupation and amusement for them, and the pleasure which he derived from the success of his labours.

What is said by his sons of Mr. Knight's favourite poet, Crabbe, may be most appropriately applied to himself, "that as the chief characteristic of his heart was benevolence, so that of his mind was a buoyant exuberance of thought, and a perpetual exercise of intellect, a youthful tenderness of feeling, and a smile of indescribable benevolence." Like Crabbe, too, he had no great "love for painting, or music, or architecture, and little for what a painter's eye considers the beauty of landscape," but he had the strongest perception and enjoyment of the charms of poetry. Pope, Johnson, Gray, and Crabbe ranked first in his estimation among the English poets ; and for the writings of Byron, Rogers, Campbell, and Mrs. Hemans, he had a high admiration. His memory was wonderfully retentive, and no one who was much in his society could fail to remark the peculiar readiness and aptitude of his quotations. Whether the subject of conversation were grave or lively, he had always at command some strikingly apposite illustration of ideas casually expressed ; and the deep feeling of its beauties which characterised his manner of reciting poetry, added much to the effect of the passages so happily selected ; and if encouraged to go on, he would repeat page after page of all his favourite authors.

The singular powers of memory he possessed were combined with a very uncommon facility for retaining even the words in which ideas were conveyed to his mind. On one occasion, at the house of his friend, the late Sir Uvedale Price, a gentleman present quoted a passage from Gibbon's Roman History ; Mr. Knight expressed a doubt whether he had used the exact words of Gibbon ; and in confirmation of his opinion repeated a page and a half from the work, including the passage in question. On the book itself being referred to, the accuracy of his quotation was established. This was not a singular instance, for had it been Hume, or Robertson, or almost any other standard work of history or philosophy, that had been referred to, he would probably have been equally master of any striking passage.

At another time Dr. Cornewall, then Bishop of Hereford, repeated to Mr. Knight an epitaph on Douglas, eighth Duke of Hamilton, containing twenty-two lines, with the merit of which he was much struck, and some discussion on its beauties followed. When Mr. Knight came down to breakfast the next morning, he had recalled the whole of the lines to his recollection, and on their being written down from his dictation, they were found to be perfectly correct.

To the end of his life this power of memory, which is usually one of the first that fails, remained almost unimpaired. All that he read or heard his mind retained with the same distinctness that it would have done in former days ; and when he was in his seventy-seventh year, he acquired by heart nearly the whole of Campbell's poem of " The Last Man," which he then for the first time met with, with nearly the same ease that he had done the epitaph more than thirty years before.

Mr. Knight's form was muscular and powerful, and till his last illness, and notwithstanding his advanced age, his step was as firm, and his figure as erect, as it had ever been, though his height was nearly six feet : his complexion was fair, and his eyes blue ; his hair was light brown, but at an early age he became bald, and the fine intellectual form of his head was very striking. His countenance, though not handsome, beamed with intelligence and benevolence, and was a type of the qualities of his mind and heart.

The limits of this work will not allow that a more detailed view of Mr. Knight's character should be given ; but if, among those who knew him, a good-natured smile may sometimes have been called forth by any little peculiarities, arising from the origi- nality of his mind, his friends will agree that few lives ever abounded more in works of kindness and charity than his, and that the object foremost in his thoughts was that of making his investigations into the more abstruse branches of natural history, the basis of designs for the improvement and benefit of his fellow-creatures.

Had the task of delineating his character fallen into other hands, his family would have rejoiced ; and it will be a source of deep and lasting regret, if the inability they strongly feel to do justice to his noble nature, should have caused him to appear to those who had not the happiness of knowing him less wise and good than he was.

LIST OF MEDALS PRESENTED TO MR. KNIGHT.

1806—Royal Society Gold Copley Medal.

1814—Horticultural Society Gold Medal.

1815—Large Silver Medal for Black Eagle Cherry.

1817—Do. do., for Waterloo Cherry.

1818—Do. do., for Elton Cherry.

1822—Silver Banksian Medal for new Pears.

1836—New large Gold Medal.

1801—Society of Arts Silver Medal for Turnip Drill.

1815—Caledonian Horticultural Society Gold Medal, " In testimony of their gratitude for his valuable discoveries, the result of patient and laborious research in Vegetable Physiology—science having been his guide."

1826—Massachusetts Agricultural Society Medal. (A circular plate of silver, two inches and a quarter in diameter, inscribed) " The Massachusetts Society for Promoting Agriculture, to Thomas Andrew Knight, Esq. of Downton Castle, England, as a tribute to an eminent Physiologist, and a benefactor to the new world."

1830—Swedish Academy of Agriculture, " Grand Silver Medal."

LIST OF SOCIETIES OF WHICH MR. KNIGHT WAS A MEMBER.

1804—Royal Society of London.

1804—Royal Society of Edinburgh.

1804—Horticultural Society of London.

1825—Hon. Member of Royal Botanical Society of Glasgow. Hon. Member of Medico-Botanical Society, London. Hon. Associate of Verulam Society of London.

1818—Hon. Member of Society of Naturalists of Berlin.

1820—Hon. Member of Horticultural Society of Potsdam.

1822—Hon. Member of Literary and Philosophical Society of New York.

1822—Hon. Member of New York County Agricultural Society.

1822—Hon. Member of New York Historical Society.

1822—Hon. Member of Massachusetts Agricultural Society.

1823—Corresponding Member of Prussian Horticultural Society.

1824—Corresponding Member of Columbian Horticultural Society.

1824—Corresponding Member of the Pomological Society of Guben.

1828—Hon. Member of Swedish Agricultural Society.

1829—Hon. Member of Imperial Natural History Society of Moscow.

1829—Hon. Member of Horticultural Society of the Département du Nord (France).

1832—Hon. Member of Agricultural Society of Western Australia.

1832—Hon. Corresponding Member of ' La Sociedad Patriotica de la Habana.

1833—Hon. Member of Horticultural Society of Charleston.

1833—Hon. Member of Lower Canada Horticultural Society.

1834—Hon. Member of South Carolina Horticultural Society.

The following list contains most of the new varieties of fruits raised by Mr. Knight, which he considered worth preserving :

Apples.—Spring-grove Codling.   Downton Lemon Pippin.   Herefordshire Gillyflower.   Grange Apple, &c.

Cherries.—Elton, Waterloo and Black Eagle.

Strawberries.—Elton and Downton.

A large and long-keeping red Currant.

Plums.—Ickworth Impératrice.—A large purple Plum not named, and two improved Damsons.

Nectarines.—Impératrice, Ickworth, Downton, and Althorp.

Pears—Monarch, Althorp Cressane, Rouse Level, Winter Cressane, Belmont, and many others.

Many excellent and productive varieties of Potatoes, of which the only one named is the Downton Yam.

The Knight Pea, and improved varieties of Cabbage.

In making the following selection from the numerous communications addressed by Mr. Knight to the Royal and Horticultural Societies, the object kept in view has been to embody the whole of those which give an account of the important physiological experiments carried on, or facts observed by him, or in which are consigned the theoretical or practical results deduced from these experiments and observations. Those relating to temporary, controversial or other matters, now deprived of the interest they possessed at the time when read, are here omitted.

In the arrangement of the papers, it has been thought best to adopt the chronological order : thus showing the gradual steps attained by Mr. Knight in the pursuit of his inquiries, and simplifying the references he was in the habit of making to previous memoirs. This order has only been departed from in as far as was necessary to separate the communications made to the Royal Society from those made to the Horticultural Society ; for it has appeared as if Mr. Knight's object, in determining to which body he should address himself, was to place on record, in the Philosophical Transactions, the general physiological principles he laid down, and in the Horticultural Transactions to detail the practical application of those principles.

Three papers on questions of Animal Economy, of considerable importance, but not immediately connected with Horticulture, or Vegetable Physiology, are given in an Appendix.

PART I.

PAPERS ON VEGETABLE PHYSIOLOGY,

READ BEFORE

THE ROYAL SOCIETY, IN THE YEARS 1795 TO 1816.

REPRINTED FROM THE PHILOSOPHICAL TRANSACTIONS.

I.—OBSERVATIONS ON THE GRAFTING OF TREES.

[Read before the ROYAL SOCIETY *April 30th, 1795.]*

THE disease from whose ravages apple and pear trees suffer most is the canker; the effects of which are generally first seen in the winter, or when the sap is first rising in the spring. The bark becomes discoloured in spots, under which the wood, in the annual shoots, is dead to the centre; and, in the older branches, to the depth of the last summer's growth. Previous to making any experiments, I had conversed with several planters, who entertained an opinion, that it was impossible to obtain healthy trees of those varieties which flourished in the beginning and middle of the present century, and which now form the largest orchards in this county (Herefordshire). The appearance of the young trees, which I had seen, justified the conclusion they had drawn; but the silence of every writer on the subject of planting, which had come in my way, convinced me it was a vulgar error, and the following experiments were undertaken to prove it so.

I suspected that the appearance of decay in the trees I had seen lately grafted, arose from the diseased state of the grafts, and concluded that if I took scions or buds from trees grafted in the year preceding, I should succeed in propagating any kind I chose. With this view, I inserted some cuttings of the best wood I could find in the old trees, on young stocks raised from seed. I, again, inserted grafts and buds taken from these on other young stocks, and, wishing to get rid of all connexion with the old trees, I repeated this six years ; each year taking the young

shoots from the trees last grafted.   Stocks of different kinds were tried;
some were double-grafted, others obtained from apple-trees which grew
from cuttings, and others from the seed of each kind of fruit afterwards
inserted on them.   I was surprised to find that many of these stocks
inherited all the diseases of the parent trees.

The wood appearing perfect and healthy in many of my last-grafted trees,
I flattered myself that I had succeeded ; but my old enemies, the moss
and canker, in three years convinced me of my mistake.   Some of them,
however, trained to a south wall, escaped all their diseases, and seemed
(like invalids) to enjoy the benefit of a better climate.   I had before
frequently observed, that all the old fruits suffered least in warm situations, where the soil was not unfavourable.   I tried the effects of laying
one kind, but the canker destroyed it at the ground.   Indeed I had no
hope of success from this method ; as I had observed that several sorts,
which had always been propagated from cuttings, were as much diseased
as any others.   The wood of all the old fruits has long appeared to me
to possess less elasticity and hardness, and to feel more soft and spongy
under the knife, than that of the new varieties which I have obtained
from seed.   This defect may, I think, be the immediate cause of the
canker and moss, though it is probably itself the effect of old age, and
therefore incurable.

Being at length convinced that all efforts to make grafts from worn-
out trees were ineffectual, I thought it probable that those taken from
very young trees raised from seed could not be made to bear fruit.
The event here answered my expectation.   Cuttings from seedling apple-
trees of two years old were inserted on stocks of twenty, and in a bearing
state.   These have now been grafted nine years ; and though they have
been frequently transplanted to check their growth, they have not yet
produced a single blossom.   I have since grafted some very old trees
with cuttings from seedling apple-trees of five years old : their growth
has been extremely rapid, and there appears no probability that their
time of producing fruit will be accelerated, or that their health will be
injured, by the great age of the stocks.   A seedling apple-tree usually
bears fruit in thirteen or fourteen years ; and I therefore conclude, that
I have to wait for a blossom till the trees from which the grafts were
taken attain that age; though I have reason to believe, from the form of
their buds, that they will all be extremely productive.   Every cutting,
therefore, taken from the apple (and probably from every other) tree,
will be affected by the state of the parent stock.   If that be too young
to produce fruit, it will grow with vigour, but will not blossom ; and if

it be too old, it will immediately produce fruit, but will never make a healthy tree, and consequently never answer the intention of the planter. The root, however, and the part of the stock adjoining it, are greatly more durable than the bearing branches; and I have no doubt but that scions obtained from either would grow with vigour, when those taken from the bearing branches would not. The following experiment will, at least, evince the probability of this in the pear-tree :—I took cuttings from the extremities of the bearing branches of some old ungrafted pear-trees, and others from scions which sprung out of the trunks near the ground, and inserted some of each on the same stocks. The former grew without thorns, as in the cultivated varieties, and produced blossoms the second year; whilst the latter assumed the appearance of stocks just raised from seeds, were covered with thorns, and have not yet produced any blossoms.

The extremities of those branches, which produce seeds in every tree, probably show the first indication of decay ; and we frequently see (particularly in the oak) young branches produced from the trunk, when the old ones have been dead. The same tree when cropped will produce an almost eternal succession of branches. The durability of the apple and pear I have long suspected to be different in different varieties ; but that none of either would vegetate with vigour much, if at all, beyond the life of the parent stock, provided that died from mere old age. I am confirmed in this opinion by the books you did me the honour to send me: of the apples mentioned and described by Parkinson, the names only remain ; but many of Evelyn's are still well known, particularly the red-streak. This apple, he informs me, was raised from seed by Lord Scudamore in the beginning of the last century*. We have many trees of it, but they appear to have been in a state of decay during the last forty years. Some others mentioned by him are in a much better state of vegetation, but they have all ceased to deserve the attention of the planter. The durability of the pear is probably something more than double that of the apple.

It has been remarked by Evelyn, and by almost every writer since, on the subject of planting, that the growth of plants raised from seeds was more rapid, and that they produced better trees than those obtained from layers or cuttings. This seems to point out some kind of decay attending the latter modes of propagation; though the custom in the public nurseries of taking layers from stools (trees cropped annually close to the ground) probably retards its effects, as each plant rises immediately from the root of the parent stock.

* Probably about the year 1634.

Were a tree capable of affording an eternal succession of healthy plants from its roots, I think our woods must have been wholly overrun with those species of trees which propagate in this manner, as those scions from the roots always grow in the first three or four years with much greater rapidity than seedling plants. An aspen is seldom seen without a thousand suckers rising from its roots; yet this tree is thinly, though universally, scattered over the woodlands of this country. I can speak from experience, that the luxuriance and excessive disposition to extend itself in another plant, which propagates itself from the root (the raspberry), decline in twenty years from the seed. The common elm being always propagated from scions or layers, and growing with luxuriance, seems to form an exception ; but, as some varieties grow much better than others, it appears not improbable that the most healthy are those which have last been obtained from seed. The different degrees of health in our peach and nectarine trees may, I think, arise from the same source. The oak is much more long-lived in the north of Europe than here; though its timber is less durable, from the number of pores attending its slow growth. The climate of this country being colder than its native, may, in the same way, add to the durability of the elm; which may possibly be further increased by its not producing seeds in this climate,—as the life of many animals may be increased to twice its natural period, if not more, by preventing their seeding.

II.—ACCOUNT OF SOME EXPERIMENTS ON THE ASCENT OF THE SAP IN TREES.

[*Read before the* ROYAL SOCIETY, *May* 14th, 1801.]

THESE experiments were made on different kinds of trees ; but I shall confine myself to those I have made on the crab-tree, the horse-chestnut, the vine, and the oak ; and shall begin with those made on the crab-tree.

Choosing several young trees of this species in my nursery, of something more than half an inch diameter, and of equal vigour, I made two circular incisions through the bark, round one half the number of them, about half an inch distant from each other, early in the spring of 1799 ; and I totally removed the bark between these incisions, scraping off the external coat of the wood. The other half I left in their natural state.

At the usual season, the sap rose in equal abundance in all; and their branches shot, during the whole spring, with equal luxuriance. But that part of the stems (of the trees whose bark had been taken off) which was below my incisions, scarcely grew at all, whilst all the parts above the incisions increased as rapidly as in the trees whose bark remained in the natural state ; the upper lips of the wounds also made considerable advances towards a union, but the lower ones made scarcely any.

Soon after midsummer, those parts of the wood which had been deprived of bark became dry and lifeless, to some depth ; and the sap, in consequence, meeting obstruction in its ascent, some latent buds shot forth, in some of the plants, below the incisions. When one of the shoots which these buds produced was suffered to remain, the part of the stem below it began immediately to increase in size ; but if it was at any distance below the incision above, the part between it and that incision still remained very nearly stationary, so as to be, in the autumn, almost a whole year's growth less than the stem above the incisions.

Choosing other stocks, which had each a strong lateral branch, I removed the bark, in the manner described, in two places ; the one above, and the other below, each lateral branch. The sap here passed both my incisions as freely as in the former experiment ; the lateral branches between them grew with the greatest vigour, and the part of the stem between those branches and the lower incisions increased much in size. I varied these experiments in every way that occurred to me ; and the result uniformly was, that those parts of the stems and branches which were above the incisions, and had a communication with the leaves, through the bark, increased rapidly ; whilst those below the incisions scarcely grew at all, till a new communication with the leaves through the bark was obtained, by means of a lateral shoot below the incisions. It now appeared to me to be probable that the current of sap which adds the annual layer of wood to the stem must descend through the bark, from the young branches and leaves ; and to these my attention was in consequence directed.

Towards the end of the summer, when some young luxuriant shoots of my apple-trees had attained a proper degree of firmness, I made four circular incisions through the bark of each, as in the preceding instances ; and I removed the bark in two places, leaving a leaf between the places where the bark was taken off. Examining them frequently during the autumn, I found that the insulated leaf acted just as the lateral branch had done ; the part of the bark and stem between it and the lower incision being apparently as well fed as any other part of the tree ; and it grew as much. Making similar incisions on other branches of the same age, I left similar portions of insulated bark, without a leaf between the incisions ; but in these no apparent increase in the size of the wood was discoverable.

I was still unacquainted with the channel through which the sap was conveyed into the leaf; and therefore, having obtained a deeply-tinged infusion, by macerating the skins of a very black grape in water, I prepared some annual shoots of the apple and of the horse-chestnut in the manner above mentioned; then, cutting them off a few inches below the incisions of the bark, I placed them for some hours in the coloured infusion. Making transverse sections of them afterwards, I found that the infusion had passed up the pores of the wood, beyond both my incisions, and into the insulated leaves; but it had neither coloured the bark, nor the sap between it and the wood; and the medulla was not affected, or at most was very slightly tinged at its edges.

My attention was now turned to the leaves: these in the apple-tree are attached to the wood by three strong fibres or tubes, (or rather bundles of tubes,) one of which enters the middle of the leaf-stalk, and the others are on each side of it. In the horse-chestnut there are seven or eight bundles of a similar kind of tubes in each leaf; through these the infusion had passed, and had communicated its colour to them, through almost the whole length of each leaf-stalk. Examining these tubes more minutely, I found that they were surrounded with others, which were free from colour, and appeared to be conveying, in one direction or the other, a different fluid. On tracing these downwards, I discovered that they entered the inner bark, and had no immediate communication with the tubes of the wood. I now endeavoured, in the same manner, to trace back those vessels which had carried the infusions into the leaves, and I readily found them to be perfectly distinct from the common tubes of the alburnum. They commence a few inches below the leaf to which they belong, and they become more numerous as they approach it; everywhere surrounding the medulla in bundles, as represented in plate i. To these vessels the spiral tubes are everywhere appendages. I do not know that any specific name has been given to these vessels; and, therefore, as they constitute a centre, round which the future alburnum is formed in the succulent annual shoot, I will call them the central vessels, to distinguish them from the spiral tubes and the common tubes of the wood. In plates ii. and iii. the direction of these vessels, with the spiral tubes, in their passage from the sides of the medulla to the leaf-stalk, is delineated in a transverse and longitudinal section; they extend to the extremities of the leaf, where I believe they terminate. Plate iv. presents two sections of the leaf-stalk of the horse-chestnut; the first being taken from the middle of the stalk, and the second from its base. Lying parallel with, and surrounding the above-mentioned vessels, appear other vessels, which, I conclude, return the sap to the tree : for when a leaf was cut off which had imbibed a coloured infusion, I found that the native juices of the plant flowed from these vessels, apparently unaltered, as has been remarked by Dr. Darwin. These vessels descend through the inner bark, (as delineated in plates I., II., and III.) and appear to extend from the extremities of the leaves to the points of the roots.

The whole of the fluid, which passed from the wood to the leaf, seems to me evidently to be conveyed through a single kind of vessel; for the spiral tubes will neither carry coloured infusions, nor in the smallest degree retard the withering of the leaf, when the central vessels are divided. But the annexed figures appear to point out at least two kinds of returning vessels. And I think it by no means improbable that two kinds exist with distinct offices; for there is a new layer of alburnum and a new internal bark to be formed. I have, however, seen it asserted somewhere, in the writings of Linnæus and other naturalists, that the internal bark is annually converted into alburnum. But this is totally erroneous ; and a vigorous shoot of the apple-tree often presents in its transverse sections, when three or four years old, as many layers in its bark, each of which once formed its internal vascular lining.

As the bark appeared to me now to receive its nutrition through the leaf, I wished to see what effect would be produced by gradually reducing the quantity of the leaves. I had a luxuriant shoot of the vine in my vinery, exactly in the stage of growth I wanted ; and this branch therefore was towards its point every day deprived of a small portion of its leaf. The bark, in consequence, became shrivelled and dry ; and at length the buds below vegetated, and the point of the shoot died, apparently from the want of nourishment. I here observed, as I had frequently done before, that almost the whole action of each leaf lies between itself and the root ; for the branch, in this case, was perfectly well fed below the uppermost unmutilated leaf, but failed immediately above it.

Every branch in which I had yet attempted to trace the progress of the sap having contained its medulla uninjured, the action of that substance next engaged my attention, and I made the following experiments on the vine :—Having made a passage about half an inch long, and a line wide, into a strong succulent shoot of this plant, I totally extracted its medulla, as far as the orifice I had made would permit me. But the shoot grew nearly as well as the others whose medulla had remained uninjured, and the wound soon healed. Making a similar passage, but of greater length, so that part extended above, and part below, a leaf and bud, I again extracted the medulla. The leaf and bud with the lateral shoot annexed (in the vine) continued to live, and did not appear to suffer much inconvenience, but faded a little when the sun shone strongly on them.

I was now thoroughly satisfied that the medulla was not necessary to the progression of the sap; but I wished to see whether the wood and leaf could execute their office when deprived at once of the bark and medulla. With this view, I made two circular incisions through the bark, above and below a leaf; and I took off the whole of the bark between them except a small portion round the base of the leaf. Having then perforated the wood, where I made each of my incisions through the bark, I destroyed the medulla in each place, as in the preceding experiments. The leaf, however, continued fresh and vigorous; and a thin layer of new wood was formed round its base, as far as the bark had been suffered to remain.

Whilst I was waiting the result of the preceding experiments, I made a few efforts to discover another branch of circulation, namely, that which takes place within the fruit, and conveys nourishment to the future offspring. My experiments were here, however, confined almost entirely to two species of fruit, the apple and the pear; and therefore, as the organization of different fruits is evidently different, I do not consider my observations such as can throw much general light on the subject. Examining the fruit-stalks of the apple, the pear, the vine, and some other fruit-trees, I found their organization to be nearly similar to that of the branch from which they sprang, and to consist of the medulla, the central tubes, a very small portion of wood, the spiral tubes and those of the bark, and the two external skins. Tracing the progress of these in the full-grown fruits of the apple and pear, I found, as Linnæus has described, that the medulla appeared to end in the pistilla. The central vessels diverged round the core, and, approaching each other again in the eye of the fruit, seemed to end in ten points at the base of the stamina, to which, I believe, they give existence. The spiral tubes, which are, in all other parts appendages to these vessels, I could not trace beyond the commencement of the core; but as the vessels themselves extend through the whole fruit, it is probable that the spiral tubes may have escaped my observation. Linnæus supposes the stamina to arise from the wood. I should not venture to state an opinion in opposition to his; but I believe he has not anywhere distingushed those I call the central vessels, from the common tubes of the wood.

Having hitherto found that all advancing fluids appeared to pass either along the tubes of the alburnum, or along the central vessels, I had little doubt that the fruit was fed through the latter; but my efforts to ascertain this, in the autumn of 1799, were not successful. In the last spring I was more fortunate. Placing small branches of the apple, the pear, and the vine, with blossoms not yet expanded, in a decoction of logwood, I found that the colouring matter readily passed up the central tubes of the fruit-stalks of all; and in the apple and pear, I easily traced it, through the future fruit, to the base of the stamina. The office of the tubes in the bark did not appear in this experiment; but as I have reason to believe the motion of the sap in the bark to be always retrograde, I am disposed to conclude that it is so here, and that, through the bark of the stalk, any superfluous humours existing in the fruit, from excessive humidity of weather, or other cause, are carried back, and absorbed by the tree. I have, however, very frequently repeated an experiment on the vine, which, I think, evidently proves that the fluid returned (if any), is essentially different from that which is derived from the leaf. In the culture of this fruit, I have frequently pinched off the young shoot, immediately above a branch, as soon as the latter became visible in the spring, letting the leaf opposite the bunch remain. In this case, the wood below the upper leaf acquired nearly its proper length and substance. But when I have taken off that leaf, the wood between the bunch and the next leaf below, has ceased to elongate; and has remained, in form and substance, similar to the small fruit-stalk attached to it.

I was long at a loss to conjecture by what means nutrition was conveyed to the seeds of the apple and pear; for I had reason to believe that it was not done by the medulla; and I had previously ascertained that the seeds would derive nourishment from the pulp, when the fruit was taken prematurely from the tree. At length, in a large apple, which was just beginning to decay, I found a number of minute vessels, leading from the pulp to the tubes which originally constituted the lower parts of the pistilla, and to which the seeds are attached. These now appeared to me evidently to be the channels of nutrition to the seeds; and since I have known what I have to look for, I find these vessels sufficiently visible in every apple: there are, however, five other tubes, which pass along the external edges of the cells of the core, to which I do not venture to assign an office. It appears to me not very improbable, that the internal organization of this fruit will be found to bear some resemblance to the placenta and umbilical cord of the animal economy. If transverse and longitudinal sections of young apples and pears be made, soon after the blossom has fallen, the pulp will appear to be of two kinds : one of which is included within the vessels which carry up coloured infusions ; and this seems to be formed by continuation of the vessels and fibres within the wood. The other part appears to belong, in a great measure, to the bark : it is in very small quantity in the very young fruit ; but, at its maturity, it constitutes much the greater part of the pulp. The vessels, however, which diverge into the external pulp, and probably convey nourishment to it, appear to be continuations of the central vessels, every where, I believe, accompanied, as in the leaf, with minute ramifications of the tubes of the bark. The substance of the core is similar to that of the silver grain of the wood, of which it may possibly be a continuation.

The force with which the sap has been proved to ascend, by Hales, banishes every idea of mere capillary attraction. The action of the spiral tubes appears much more adequate to the effects produced, and I readily admit the supposed action of these, wherever they are found ; but I have so often attentively searched in vain for them, with glasses of different powers, in the root, in the alburnum, and in the bark, that I cannot but question their existence in those parts. Attached to the central vessels, in the annual shoot, in the fruit-stalk of different trees, in the tendril of the vine, in the leaf, and in the seed, the spiral tubes certainly exist, and are in most cases visible without the aid of a lens. But as I have not been able to discover them in other parts of the tree, and as the different authors I have looked into have not distinguished those I call the central vessels from the common tubes of the alburnum, nor marked the difference in the organization of the annual branch and annual root, I must venture to call their accuracy here in question, though with great deference for their opinions.

Linnæus and others have attempted to account for the ascent of the sap, by the expansion of the fluids within the vessels of the plant, by the agency of heat. But the sap rises under a decreasing, as well as under an increasing temperature, during the evening and night (if it be not excessively cold), as well as in the morning and at noon ; and it is sufficiently evident, that the heat applied to the branches of a vine within the stove, cannot expand the fluids in the stems and roots, which grow on the outside. It is also well known, that the degree of heat required to put the sap into motion, in this plant, is not definite, but depends on that to which the plant has been previously accustomed. Thus a vine, which has grown all the summer in the heat of a stove, will not be made to vegetate during the winter by the heat of that stove ; but, if another plant of the same variety, which has grown in the open air, be at any time introduced, after it has dropped its leaves in the autumn, it will instantly vegetate. This effect appears to me to arise from the latter plant's possessing a degree of irritability, which has been exhausted in the former, by the heat of the stove, but which it will acquire again during the winter, or by being drawn out, and exposed for a short time to the autumnal frost. On the same principle, we may point out the cause why seedling plants always thrive better in the spring than in the autumn, though the weather be apparently less favourable. In the former season, the stimulus of heat and light is gradually becoming greater than that to which the plant has been accustomed ; in the latter season, it becomes gradually less.

There is another circumstance attending trees that have been made to blossom early in the preceding spring, which has always appeared to me an extremely interesting one. If a peach-tree, for example, be brought into blossom in one season in the beginning of February, by artificial heat, it will spontaneously show strong marks of vegetation at the approach of that season in the succeeding year ; and, if it be not well protected, it will expose its blossoms to almost inevitable destruction. I do not see any cause to which this effect can be attributed, except to the accumulated irritability of the plant.

That heat is the remote cause of the ascent of the sap cannot, I think, be doubted ; and, perhaps, frequent variations of it are, in some degree, requisite ; (for plants have always appeared to me to thrive best with moderate variations of temperature ;) but the immediate cause will, I think, be found in an intrinsic power of producing motion, inherent in vegetable life ; and I hope to be able to point out an agent, by which the mechanical force required may possibly be given.

There is, you know, in every kind of wood, what workmen call its grain, consisting of two kinds, the false or bastard, and the true or silver grain. The former consists of those concentric circles which mark the annual increase of the tree ; and the latter is composed of thin laminæ, diverging in every direction from the medulla to the bark, having little adhesion to each other at any time, and less during the spring and summer than in the autumn and winter ; whence the greater brittleness of the wood in the former seasons. These laminæ (which are of different width in different kinds of wood) lie between, and press on, the sap vessels of the alburnum ; they are visible in every wood that I have had an opportunity to examine, except some of the palm tribe ; and these appear to me to have peculiar organs, to answer a similar purpose. If you will examine a piece of oak, you will find the laminæ I describe ; and that every tube is touched by them at short distances, and slightly diverted from its course.

If these are expansible under changes of temperature, or from any cause arising from the powers of vegetable life, I conceive that they are as well placed as is possible to propel the sap to the extremities of the branches; and their restless temper, after the tree has ceased to live, inclines me to believe that they are not made to be idle whilst it continues alive.

I shall at present confine my observations to the English oak, though the same are applicable, in a greater or less degree, to every other kind of wood.   In sawing this tree into boards, it is usual to cut it, as much as possible, into what are called quarter-boards ; which are so named because the tree is first cut into quarters.   In a perfect board of this kind, the saw exactly follows the direction in which the tree most readily divides when cloven : in this case, the laminæ of the silver grain lie parallel with the surface of the board ; and a board thus cut, when properly laid on the floor, is rarely or never seen to deviate from its true horizontal position.   If, on the contrary, one be sawed across the silver grain, it will, during many years, be incapable of bearing changes of temperature and of moisture without being warped ; nor will the strength of numerous nails be sufficient entirely to prevent the inconvenience thence arising. That surface, of a board of this kind, which grew nearest the centre of the tree, will always show a tendency to become convex, and the opposite one concave, if placed in a situation where both sides are equally exposed to heat and moisture.   You may probably have observed, that when an oak has been deprived of its bark, and exposed to the sun and air, its surface has been everywhere covered with small clefts.   These are always formed by the laminæ of the silver grain having parted from each other ; and they will long continue to open and close again with the changes of the weather.   In the last summer, I very frequently placed pieces of oak, recently deprived of its bark, in a situation where it was fully exposed to the sun, but defended from rain.   The surface of the tree, in a few hours, presented a great number of small clefts, into which I put, in the middle of the day, the points of small iron pins.   Examining these late in the evening, I found that the wood closed so much as to hold them firmly, and, early in the next morning, they were not easily withdrawn ; but as the influence of the sun increased, the clefts again gradually opened, as in the preceding day, and the pins always dropped out.   I could never discover that any weight was gained by the wood during the night ; but I was not provided with a balance of proper sensibility to ascertain this point.   This experiment was frequently repeated, and always with precisely the same result.   After long exposure to the air and light, the wood loses this property.

If the motion I have supposed the silver grain to possess, in the living tree, be more than you think can be properly admitted to belong to vegetable life, I will request your attention to the power of moving in the vine leaf, on which I have made many experiments. It is well known that this organ always places itself so that the light falls on its upper surface ; and that, if moved from that position, it will immediately endeavour to regain it : but the extent of the efforts it will make, I have not anywhere seen noticed. I have very frequently placed the leaf of a vine in such a position, that the sun has shone strongly on its under surface ; and I have afterwards put obstacles in its way, on whichever side it attempted to escape. In this position the leaf has tried almost every method possible to turn its proper surface to the light ; and I have several times seen one which, having tried during several days to approach the light in one direction, and having nearly covered its under surface, by bending its angular points almost to touch each other, has unfolded itself again, and receded farther from the glass, to approach the light in an opposite direction. As the whole effect here produced appears to arise merely from the light falling on the under surface of the leaf, I cannot conceive how the contortions of its stalk, in every direction, can be accounted for, without admitting, not only that the plant possesses an intrinsic power of moving, but that it also possesses some vehicle of irritation ; and, without this, it will I think be difficult to explain how the heat applied to the branch of the vine, within the stove, can put the sap in the roots and external stem into motion. It may be objected, that these are always ready when the branch calls for nourishment, and that they are no way affected by the internal heat. But this I cannot admit to be the case ; because I have found that the stem suddenly becomes extremely susceptible of injury from cold, as soon as the branch begins to vegetate ; and that its whole powers will be paralysed for some days, by exposure for a few hours to a freezing temperature.

I have had very frequent opportunities of observing a remarkable power in trees, of transferring their sap from one tube to another ; for I have often intersected, in the trunk, every tube which led to a lateral branch, and still this branch has derived a considerable portion of nourishment from the trunk. And if the tubes of an annual shoot of the oak be traced downwards in the autumn, they will be found to pass along the layer of wood of the preceding summer, without any apparent communication between them and the tubes of any former year's growth. Yet the sap rises through the whole of the white wood ; and it must be transferred from the internal tubes to those near the surface, which alone appear to communicate with the central tubes of the young shoots and leaves.   Indeed, we have frequent evidence that trees possess this power; for we see that the whole sap of the stock is carried into an inserted bud or graft.

I at one time suspected that a small portion of sap, in its descent from the leaves, had been carried down by the wood, through my incisions, in the preceding experiments on the crab-tree, because I observed a very small increase in size, in the lower part of the stocks ; which, I think, could not have taken place without some matter derived from the leaves. But subsequent observation induces me to believe, that the small quantity of additional matter found in the lower part of the stock came from a different source.   In those experiments I paid little attention to any small shoots which sprang from the trunk at some distance below the incisions ; and the buds, which usually began to vegetate about mid-summer, were not always rubbed off till some minute leaves appeared. Through these I now believe that a small quantity of sap was thrown into the bark, and carried up through its tubes by capillary attraction, when the current from above was intercepted ; for the increase of size in the stock always diminished, as it ascended towards the incision ; which, I think, would not have been the case, had it been produced by nourishment descending from the upper parts of the tree.

Nothing has occurred in the preceding experiments to throw much light on the office of the medulla, to which Linnæus and subsequent writers have annexed so much importance ; but I will now endeavour to point out one of its offices.   In the young and succulent shoot this substance is extremely full of moisture ; and, as there is an immediate communication between it and the leaf, through the central tubes, I conclude it forms a reservoir, to supply the leaf with moisture, whenever an excess of perspiration puts that in a state to require it.   Some reservoir of this kind appears to me to be necessary to plants, for their young leaves are excessively tender, and they perspire much; and cannot, like animals, fly to the shade and the brook.   In the mature annual branches, and in those of more than one year old, the medulla is dry, and, I think, it is evidently lifeless ; but the space it occupies is never filled with wood, as some naturalists have imagined.

The heart or coloured wood, distinguished from the alburnum, seems to execute an office somewhat similar to the bone in the animal economy. The rigid texture of the vegetable fibre, has rendered this substance unnecessary in the young subject ; but, as the powers of destruction, both from winds and gravity, increase in a compound ratio with the growth of the tree, some stronger substance than the alburnum may be supposed to be wanting to support the additional weight of fruit and seeds. In the root this substance cannot be wanted, and there it is not found; but if the mould be taken away from the roots round the trunk, so that they are exposed to the air, and made to support the weight of the tree, they become as full of coloured wood as the trunk and large branches. Having cut through the alburnum of an oak all round, not the slightest mark of vegetation appeared in the succeed-ing spring; and, having been unable to impel either air or water through its tubes, I conclude that the coloured wood of the oak is without circulation:—I see very little reason, however, to admit that it is with-out life in a young or middle-aged tree. The new matter which enters into the internal part of the alburnum, on its conversion into heart or coloured wood, seems to be of a nature different from the alburnum itself; for it not only changes its colour, which is nearly white, to a dark brown, but it renders it at least ten times more durable. Some portion of this increased durability may, perhaps, be attributable to the superior solidity of the coloured wood; but a little attention to the common kinds of English timber, (omitting the resinous tribe,) will convince us that these qualities, though frequently found together, have very little connexion with each other. If a number of oaks of the same age be examined, it will be found that, in some individuals, the alburnum consists of a greater number of annual layers than in others, and that the coloured wood will have approached nearer the bark on one side than on the other, in the same tree; the termination also of the coloured wood, and the commencement of the alburnum, are often found in the middle of an annual layer of wood; and each substance, at the points of contact, possesses all its characteristic properties. The alburnum, I think, evidently extends itself laterally, without any radicles descend-ing from the leaves or buds above. I have often procured a union, by grafting, between trees of different kinds, and have sometimes found mere varieties of the same species of tree, whose wood was sufficiently distinguishable, in every stage of future growth, to allow me readily to trace their line of union. The wood of the graft does not at all descend below its original place of junction with that of the stock; which, immediately below, wholly retains its native character; and, in the part where both are spliced together, each constantly extends itself in the direction of the divergent laminæ of its silver grain. The heart wood also appears to increase by lateral extension; but I am ignorant of the channels through which the additional matter is conveyed to it.

I will now take the liberty of stating a few of the conclusions that I have ventured to draw from the foregoing, and many similar experiments. As I have not been able to find the spiral tubes anywhere, except immediately surrounding the medulla in different parts, in the seed, and in the leaf, and as they everywhere terminate at short distances, I conclude that the sap is not raised by their agency; nor by the central vessels, to which they are appendages: for these extend no greater length downwards than the spiral tubes, and terminate with them at the external surface of that annual layer of wood to which they belong; and they have not any apparent communication with the similar vessels of the succeeding year. In the lower parts of hollow trees they must long have ceased to exist at all: and, in all trees, except very young ones, they are (as it were) ossified within the heart wood; and those in the annual shoots and buds are often a hundred-and-fifty feet distant from the roots, from which they are supposed to raise the sap.

The common tubes of the alburnum, (which do not appear to me to have been properly distinguished from the central vessels by the authors that I have read,) extend from the points of the annual shoots to the extremities of the roots; and up these tubes the sap most certainly ascends, impelled, I believe, by the agency of the silver grain. At the base of the buds, and in the soft and succulent part of the annual shoot, the alburnum, with the silver grain, ceases to act and to exist; and here, I believe, commences the action of the central vessels, with their appendages, the spiral tubes. By these the sap is carried into the leaves, and exposed to the air and light; and here it seems to acquire (by what means I shall not attempt to decide) the power to generate the various inflammable substances that are found in the plant. It appears to be then brought back again, through the vessels of the leaf stalk, to the bark, and by that to be conveyed to every part of the tree, to add new matter, and to compose its various organs for the succeeding season. When I have intentionally shaded the leaves, I have found that the quantity of alburnum deposited has been extremely small.

In speaking of the circulation within the apple and pear, I wish to express myself with much less decision, as I have not seen the effects of taking up any of those vessels into which the coloured infusions did not enter. The internal organization of the leaf, and of the wood, of those trees which have a central medulla, seems to admit but of little variation, and (as far as I have had opportunities to examine) of no essential difference; whilst that of different fruits is extremely various. The external vascular parts of the apple and pear, abstracted from those which seem to carry nourishment to the seeds, appear to me to resemble, in some respects, those of the leaf; and, relative to the offspring, I suspect that they perform a somewhat similar office.

III.—ACCOUNT OF SOME EXPERIMENTS ON THE DESCENT OF THE SAP IN TREES.

[Read before the ROYAL SOCIETY, *April* 21, 1803.]

IN a memoir which I had the honour to present two years ago[*], I related some experiments on trees, from which I inferred, that their sap, having been absorbed by the bark of the root, is carried up by the alburnum, or white wood of the root, the trunk, and the branches; that it passes through what are there called central vessels, into the succulent part of the annual shoot, the leaf-stalk, and the leaf; and that it returns to the bark, through the returning vessels of the leaf-stalk. The principal object of this paper is to point out the causes of the descent of the sap through the bark, and the consequent formation of wood.

These causes appear to be gravitation, motion communicated by winds or other agents, capillary attraction, and probably something in the conformation of the vessels themselves, which renders them better calculated to carry fluids in one direction than in another. I shall begin with a few observations on the leaf, from which all the descending fluids in the tree appear to be derived. This organ has much engaged the attention of naturalists, particularly of M. Bonnet: but their experiments have chiefly been made on leaves severed from the tree; and, therefore, whatever conclusions have been drawn stand on very questionable ground. The efforts which plants always make to turn the upper surfaces of their leaves to the light, have with reason induced naturalists to conclude, that each surface has a totally distinct office; and the following experiments tend strongly to support that conclusion.

I placed a small piece of plate glass under a large vine leaf, with its surface nearly parallel with that of the leaf; and, as soon as the glass had acquired the temperature of the house in which the vine grew,

* See the preceding paper.

H

I brought the under surface of the leaf into contact with it, by means of a silk thread and a small wire adapted to its form and size. Having retained the leaf in this position one minute, I removed it, and found the surface of the glass covered with a strong dew, which had evidently exhaled from the leaf. I again brought the leaf into contact with the glass, and, at the end of half an hour, found so much water discharged from the leaf, that it ran off the glass when held obliquely.

I then inverted the position of the leaf, and placed its upper surface in contact with the glass : not the slightest portion of moisture now appeared, though the leaf was exposed to the full influence of the meridian sun. These experiments were repeated on many different leaves ; and, the result was, in every instance, precisely the same. It seems, there-fore, that, in the vine, the perspiratory vessels are confined to the under surface of the leaf ; and these, like the cutaneous lymphatics of the animal economy, are probably capable of absorbing moisture, when the plant is in a state to require it. The upper surface seems, from the position it always assumes, either formed to absorb light, or to operate by the influ-ence of that body : and if any thing exhale from it, it is probably vital air, or some other permanently elastic fluid. It nevertheless appears evident, in the experiments of Bonnet, that this surface of the leaves of many plants, when detached from the tree, readily absorbs moisture.

Selecting two young shoots of the vine, growing perpendicularly against the back wall of my vinery, I bent them downwards, nearly in a perpendicular line, and introduced their succulent ends, as layers, into two pots, without wounding the stems, or depriving them of any portion of their leaves. In this position, these shoots, which were about four feet long, and sprang out of the principal stem, about three feet from the ground, grew freely, and in the course of the summer reached the top of the house. As soon as their wood became sufficiently solid to allow me to perform the operation with safety, I made two circular incisions through the bark of the depending part of each shoot, at a small distance from each other, near the surface of the mould in the pots, and I wholly removed the bark between the incisions ; thus cutting off all communication through the bark between the layers and the parent stems. Had the subjects of this experiment now retained their natural position, much new wood and bark would have been formed at the upper lip of the wounds, and none at all at the lower, as I have ascertained by frequent experiment. The case was now different : much new bark and wood was generated on the lower lip of the wounds, because uppermost by the inverted position of the branches ; and I have no doubt, but that the new matter thus deposited owed its formation to a portion of sap which descended by gravitation from the leaves growing between the wounded parts and the principal stems.

The result of this experiment appears to point out one of the causes why perpendicular shoots grow with much greater vigour than others ; they have probably a more perfect and more rapid circulation.

The effects of motion on the circulation of the sap, and the consequent formation of wood, I was able to ascertain by the following expedient. Early in the spring of 1801 I selected a number of young seedling apple-trees, whose stems were about an inch in diameter, and whose height, between the roots and first branches, was between six and seven feet. These trees stood about eight feet from each other ; and, of course, a free passage for the wind to act on each tree was afforded. By means of stakes and bandages of hay, not so tightly bound as to impede the progress of any fluid within the trees, I nearly deprived the roots and lower parts of the stems of several trees of all motion, to the height of three feet from the ground, leaving the upper parts of the stems and branches in their natural state. In the succeeding summer, much new wood accumulated in the parts which were kept in motion by the wind ; but the lower parts of the stems and roots increased very little in size. Removing the bandages from one of these trees in the following winter, I fixed a stake in the ground, about ten feet distant from the tree, on the east side of it ; and I attached the tree to the stake, at the height of six feet, by means of a slender pole about twelve feet long ; thus leaving the tree at liberty to move towards the north and south, or more properly, in the segment of a circle of which the pole formed a radius ; but in no other direction. Thus circumstanced, the diameter of the tree from north to south, in that part of its stem which was most exercised by the wind, exceeded that in the opposite direction in the following autumn, in the proportion of thirteen to e even.

These results appear to open an extensive and interesting field to our observation, where we shall find much to admire, in the means which nature employs to adapt the forms of its vegetable productions to every situation in which art or accident may deposit them. If a tree be placed in a high and exposed situation, where it is much kept in motion by winds, the new matter which it generates will be deposited chiefly in the roots and lower parts of the trunk ; and the diameter of the latter will diminish rapidly in its ascent. The progress of the ascending sap will of course be impeded ; and it will thence cause lateral branches to be produced, or will pass into those already existing. The forms of such branches will be similar to that of the trunk ; and the growth of the insulated tree on the mountain will be, as we always find it, low and sturdy, and well calculated to resist the heavy gales to which its situation constantly exposes it.

Let another tree of the same kind be surrounded, whilst young, by others, and it will assume a very different form. It will now be deprived of a part of its motion, and another cause will operate :—the leaves on the lateral branches will be partly deprived of light, and, as I have remarked in the last paper I had the honour to address to you, little alburnum will then be generated in those branches. Their vigour, of course, becomes impaired, and less sap is required to support their diminished growth ; more, in consequence, remains for the leading shoots ; these, therefore, exert themselves with increased energy ; and the trees seem to vie with each other for superiority, as if endued with all the passions and propensities of animal life.

An insulated tree in a sheltered valley will assume, from the foregoing causes, a form distinct from either of the preceding* ; and its growth will be more or less aspiring, in proportion to the degree of protection it receives from winds, and its contiguity to elevated objects, by which its lower branches, during any part of the day, are shaded.

When a tree is wholly deprived of motion, by being trained to a wall, or when a large tree has been deprived of its branches, to be regrafted, it often becomes unhealthy, and not unfrequently perishes, apparently owing to the stagnation of the descending sap, under the rigid cincture of the lifeless external bark. I have, in the last two years, pared off this bark from some very old pear and apple trees, which had been regrafted with cuttings from young seedling trees, and the effect produced has been very extraordinary. More new wood has been generated in the old trunks, within the last two years, than in the preceding twenty years ; and I attribute this to the facility of communication which has been restored between the leaves and the roots, through the inner bark. I have had frequent occasion to observe, that wherever

* Not only the external form of the tree, but the internal character of the wood, will be affected by the situation in which the tree grows ; and hence, oak timber which grew in crowded forests appears to have been mistaken, in old buildings, for Spanish chesnut. But I have found the internal organization of the oak and Spanish chesnut to be very essentially different. (See a magnified view of each in plate 5.)

The silver grain and general character of the oak and Spanish chesnut are also so extremely dissimilar, that the two kinds of wood can only be mistaken for each other by very careless observers. Many pieces of wood found in the old buildings of London, and supposed to be Spanish chesnut, have been put into my hands ; but they were all most certainly forest oak.

the bark has been most reduced, the greatest quantity of wood has been deposited.

Other causes of the descent of the sap towards the root I have supposed to be capillary attraction, and something in the conformation of the vessels of the bark. The alburnum also appears, in my former experiments, to expand and contract very freely under changes of temperature and of moisture ; and the motion thus produced must be in some degree communicated to the bark, should the latter substance be in itself wholly inactive. I however consider gravitation as the most extensive and active cause of motion in the descending fluids of trees ; and I believe that, from this agent, vegetable bodies, like unorganized matter, generally derive, in a greater or less degree, the forms they assume : and probably it is necessary to the existence of trees that it should be so. For if the sap passed and returned as freely in the horizontal and pendent, as in the perpendicular branch, the growth of each would be equally rapid, or nearly so : the horizontal branch would then soon extend too far from its point of suspension at the trunk of the tree, and would inevitably perish, by the increase in a compound ratio of the powers of destruction, as compared with those of preservation.

The principal office of the horizontal branch, in the greatest number of trees, is to nourish and support the blossoms, and the fruit, or seed ; and as these give back little or nothing to the parent tree, very feeble powers alone are wanted in the returning system. No power at all had been fatal ; and power sufficiently strong wholly to counteract the effects of gravitation had probably been in a high degree destructive. And it appears to me by no means improbable, that the formation of blossoms may, in many instances, arise from the diminished action of the returning system in the horizontal or pendent branch.

I have long been disposed to believe the ascending fluids in the alburnum and central vessels, wherever found, to be everywhere the same ; and that the leaf-stalk, the tendril of the vine, the fruit-stalk, and the succulent point of the annual shoot, might in some measure be substituted for each other ; and experiment has proved my conjecture, in many instances, to be well founded. Leaves succeeded and continued to perform their office when grafted on the leaf-stalk ; the tendril and the fruit-stalk alike, supplied a branch grafted upon them with nourishment. But I did not succeed in grafting a fruit-stalk of the vine on the leaf-stalk, the tendril, or succulent shoot. My ill success, however, I here attribute solely to want of proper management, and I have little doubt of succeeding in future.

The young shoots of the vine, when grafted on the leaf-stalk, often grew to the length of nine or ten feet ; and the leaf-stalk itself, to some distance below its juncture with the graft, was found in the autumn to contain a considerable portion of wood, in every respect similar to the alburnum in other parts of the tree.

The formation of alburnum in the leaf-stalk seemed to point out to me the means of ascertaining the manner in which it is generated in other instances; and to that point my attention was in consequence attracted. Having grafted leaf-stalks with shoots of the vine, I examined, in tranverse sections, the commencement and gradual formation of the wood. It appeared evidently to spring from the tubes which, in my last paper, (to which I must refer you,) I have called the returning vessels of the leaf-stalk ; and to be deposited on the external sides of what I have there named the central vessels, and on the medulla. The latter substance appeared wholly inactive; and I could not discover anything like the processes supposed to extend from it in all cases into the wood.

The organization of the young shoot is extremely similar to that of the leaf-stalk, previous to the formation of wood within it. The same vessels extend through both ; and therefore it appeared extremely probable, that the wood in each would be generated in the same manner ; and subsequent observation soon removed all ground of doubt.

It is well known that, in the operation of budding, the bark of a tree, being taken off, readily unites itself to another of the same, or of a kindred species. An examination of the manner in which this union takes place, promised some further information. In the last summer, therefore, I inserted a great number of buds, which I subsequently examined in every progressive stage of their union with the stock. A line of confused organization marks the place where the inserted bud first comes into contact with the wood of the stock ; between which line and the bark of the inserted bud new wood regularly organized is generated. This wood possesses all the characteristics of that from which the bud was taken, without any apparent mixture whatever with the character of the stock in which it is inserted. The substance which is called the medullary process, is clearly seen to spring from the bark, and to terminate at the line of its first union with the stock.

An examination of the manner in which wounds in trees become covered, (for, properly speaking, they never can be said to heal,) affords further proof, were it wanted, that the medullary processes, (as they are improperly named), like every other part of the wood, are generated by the bark.

Whenever the surface of the alburnum is exposed but for a few hours to the air, though no portion of it be destroyed, vegetation on that surface for ever ceases : but new bark is gradually protruded from the sides of the wound, and by this, new wood is generated. In this wood the medullary processes are distinctly seen to take their origin from the bark, and to terminate on the lifeless surface of the old wood within the wound. These facts incontestably prove, that the medullary processes, which in my former paper I call the silver grain, do not diverge from the medulla, but that they are formed in lines converging from the bark to the medulla, and that they have no connexion whatever with the latter substance. And surely nothing but the fascinating love of a favourite system, could have induced any naturalist to believe the hardest, the most solid, and most durable part of the wood, to be composed of the soft, cellular, and perishable substance of the medulla.

In my last paper, I have supposed that the sap acquired the power to generate wood in the leaf; and I have subsequently found no reason to retract that opinion. But the experiment in which wood was generated in the leaf-stalk, apparently by the sap descended from the bark of the graft, induces me to believe that the descending fluid undergoes some further changes in the bark, possibly by discharging some of its component parts through the pores described and figured by Malpighi.

I also suspected, since my former paper was written, that the young bark, in common with the leaf, possessed a power, in proportion to the surface it exposes to the air and light, of preparing the sap to generate new wood ; for I found that a very minute quantity of wood was deposited by the bark, where it had not any apparent connexion with the leaves. Having made two incisions through the bark round annual shoots of the apple-tree, I entirely removed the bark between the incisions, and I repeated the same operation at a little distance below, leaving a small portion of bark unconnected with that above and beneath it. By this bark a very minute quantity of wood in many instances appeared to be generated at its lower extremity. The buds in the insulated bark were sometimes suffered to remain, and in other instances were taken away ; but these, unless they vegetated, did not at all affect the result of the experiment. I could therefore account for the formation of wood in this case, only by supposing the bark to possess in some degree, in common with the leaf, the power to produce the necessary changes in the descending sap ; or, that some matter, originally derived from the leaves, was previously deposited in the bark ; or that a portion of sap had passed the narrow space above, from which the bark had been removed, through the wood. Repeating the experiment, I left a much greater length of bark between the intersections; but no more wood than in the former instance was generated. I therefore concluded that a small quantity of sap must have found its way through the wood from the leaves above; and I found that when the upper incisions were made at ten or twelve lines distance, instead of one or two, and the bark between them, as in former experiments, was removed, no wood was generated by the insulated bark.

I shall conclude my paper with a few remarks on the formation of buds in tuberous-rooted plants, beneath the ground. They must, if my theory be well founded, be formed of matter which has descended from the leaves through the bark. I shall confine my observations to the potato. Having raised some plants of this kind in a situation well adapted to my purpose, I waited till the tubers were about half grown ; and I then commenced my experiment, by carefully intersecting with a sharp knife the runners which connect the tubers with the parent plant, and immersing each end of the runners thus intersected in a decoction of logwood. At the end of twenty-four hours I examined the state of the experiment; and I found that the decoction had passed along the runners in each direction; but I could not discover that it had entered into any of the vessels of the parent plant. This result I had anticipated; because I concluded that the matter by which the growing tuber is fed must descend from the leaves through the bark; and experience had long before taught me that the bark would not absorb coloured infusions. I now endeavoured to trace the progress of the infusion in the opposite direction, and my success here much exceeded my hopes.

A section of potato presents four distinct substances; the internal part, which, from the mode of its formation and subsequent office, I conceive allied to the alburnum of ligneous plants; the bark which surrounds this substance; the true skin of the plant; and the epidermis. Making transverse sections of the tubers which had been the subjects of experiments, I found that the coloured infusion had passed through an elaborate series of vessels between the cortical and alburnous substances, and that many minute ramifications of these vessels approached the external skin at the base of the buds, to which, as to every other part of the growing tuber, I conclude they convey nourishment.

IV.—EXPERIMENTS AND OBSERVATIONS ON THE MOTION OF THE SAP IN TREES.

[Read before the ROYAL SOCIETY, *February* 16, 1804.]

IN the Observations on the Descent of the Sap in Trees, which I last year took the liberty to lay before the Royal Society, I offered a conjecture, that the vessels of the bark, which pass from the leaves to the extremities of the roots, were, in their organization, better calculated to carry the fluids they contain towards the roots than in the opposite direction. I had not, however, at that time, any experiment directly to support this supposition; but I thought the forms generally assumed by trees in their growth, evinced the compound and contending actions of gravitation, and of an intrinsic power in the vessels of the bark, to give motion to the fluid passing through them. In the account of the experiments which I have now the honour to address to you, I trust I shall be able to adduce some interesting facts in support of that inference.

Having selected, in the spring of 1802, four strong shoots of the vine, growing along the horizontal trellis of my vinery, I depressed a part of each shoot, whilst it was soft and succulent, about three inches deep, into the mould of a pot placed beneath it for that purpose; but without making any wound, or incision, in the young shoots thus employed as layers.

In this position they remained during the succeeding summer; and, in the autumn, had nearly filled the pots, which were ten inches in diameter, with their roots. As soon as the leaves had fallen, the layers were disengaged from the parent stocks; and about five inches of wood, containing one bud, were left, both at the proper and the inverted end of each layer. Every bud was also, by previous management, made to stand at an equal distance from the mould in the pots, and with an equal elevation, of about thirty-six degrees. About one inch of wood was likewise left at each end of every layer, beyond the buds.

In the succeeding spring, the buds vegetated strongly, both at the proper and at the inverted ends of the layers, as the experiments of Hales and Duhamel had given me reason to expect; and in one instance, the bud at the inverted end of the layer grew with greater vigour than that at its proper end: but the growth of these buds was not the object which I had in view.

I have already stated, that nearly an inch of wood was left at each end of every layer, beyond the bud; and to this wood, at the inverted ends of the layers, my attention was chiefly directed: for if the vessels of the bark possessed the powers I attributed to them, I concluded that the sap would be impelled to the inverted ends of the layers, and be there employed in the production of new wood and roots; and in this my expectations were not disappointed. At the proper end of the layers, the wood immediately beyond the buds became dry and lifeless early in the succeeding summer; the stems also, between the buds and the mould in the pots, increased in size as usual; and nothing peculiar occurred. But at the inverted end appearances were extremely different: new wood here accumulated rapidly beyond the buds, and numerous roots, of considerable length, were emitted, whilst no sensible growth took place between the base of the young shoots and the mould in the pots.

It having been proved by Duhamel that inverted parts of trees readily emit roots, I expected to derive further information from cuttings of this kind: I therefore planted, in the autumn of 1802, forty cuttings of the gooseberry-tree, and an equal number of the common currant-tree; one half of each being inverted. Of the former, not one of the inverted cuttings succeeded; whereas few of the latter failed; and in these I had an opportunity of observing the same accumulation of wood above the bases of the annual shoots, and the same mode of growth, in every respect, as in the inverted vines; except that no roots were emitted at their upper ends. The same thing occurred, without any variation, in inverted grafts of the apple-tree.

If it be admitted, according to the theory I have on a former occasion laid before you, that the sap descends from the leaves through the vessels of the bark; and that such vessels are, in their organisation, better calculated to carry their contents towards the original roots than in the opposite direction; it will be extremely easy to explain the cause of the accumulation of wood, and the emission of roots, above, instead of below, the base of the annual shoots. The vessels of the bark (the *vaisseaux propres* of Duhamel) commencing in the leaves, were formerly traced by M. Mariotte, and subsequently by myself, (being ignorant of his discovery,) to the extremities of the roots; and when a cutting, or tree, is planted in its natural position, the sap passes downwards through these to afford matter for new roots, and to increase the bulk of those already formed, having given proper nutriment to the branches and trunk in its descent. But, in the inverted cutting, or tree, these vessels become inverted; and, if their organisation be such as I have supposed it, a considerable part of that fluid, which naturally descends, will be carried upwards, and occasion the production of new wood, above, instead of below, the junction of the annual shoot with the older wood, as in the experiments I have described. The force of gravitation will, however, still be felt; and, by its agency, sufficient matter to form new roots may be conveyed to those parts of the inverted cutting, or tree, which are beneath the soil. Besides, if we suppose a variation to exist in the powers or organisation of the vessels which carry the sap towards the root, we may also attribute, in a great measure, to this cause, the different forms which different species or varieties of trees assume; for, if the fluid in these vessels be impelled with much force towards the roots, little matter will probably be deposited in the branches; which, in consequence, will be slender and feeble, as in the vine; and there is not any tree that has been the subject of my experiments, in which new wood accumulated so rapidly at the upper end of inverted plants. To an excess of this power, in the vessels of the bark, we may also ascribe the peculiar growth of what are called weeping trees; for, by this power, the effects of gravitation will be, in a great degree, suspended; and the pendent branch will continue healthy and vigorous, by retaining its due circulation. The perpendicular branch will, however, still possess some advantages; for, in this, gravitation will act on the fluid descending from the leaves; and these will, of course, absorb from the atmosphere with increased activity. A greater quantity of matter will therefore enter, within any given portion of time, into vessels of the same capacity; and this increased quantity may frequently exceed that which the vessels of the bark are immediately prepared to carry away. Much new wood will in consequence be generated, and increased vigour given; and, the same causes operating through successive seasons, will give the ascendancy we generally observe in the perpendicular branch.

In the preceding experiments none of the layers, or cuttings, exceeded a few inches in length; and, to the summit of these the sap appeared to rise, through the inverted tubes of the wood, nearly as well as in those which retained their natural position. But some further experiments had induced me to suspect that this would not be the case in longer cuttings; I therefore planted, in the autumn of 1802, twelve cuttings of the sallow, (*Salix caprea,*) inverting one-half of them. The whole readily emitted roots, and grew with luxuriance; but their modes of growth were extremely different. In the cuttings which stood in their natural position, vegetation proceeded with most vigour at the points most elevated; but, in the inverted cuttings, it grew more and more languid as it became distant from the ground, and nearly ceased, towards the conclusion of the summer, at the height of four feet. The new wood also, which was generated by these inverted cuttings, accumulated above the bases of the annual shoots, as in the preceding instances.

These facts appear to prove, that the vessels of plants are not equally well calculated to carry their contents in opposite directions ; and, I think, afford some grounds to suspect that the vessels of the bark, like those which constitute the venous system of animals, (to which they are in many respects analogous,) may be provided with valves, whose extreme minuteness has concealed them from observation.

The experiments, and still more the plates, of Hales, have induced naturalists to draw conclusions in direct opposition to the preceding. But the plates of that great naturalist are not always taken correctly from nature* ; and plates, under such circumstances, however fair and candid the intentions of an author may be, will too often be found somewhat better calculated to support his own hypothesis than to elucidate the facts he intends to state.

The preceding peculiarities in the growth of inverted cuttings, appear to have escaped the observation of Duhamel; and, as very few instances of error, or want of accurate observation, will ever be found in the works of that excellent naturalist, I must request permission to send you some of the subjects of my experiments, as vouchers for my own accuracy.

Of the inverted cuttings employed by Duhamel, a small portion only appears to have remained above the ground; and, under such circumstances, the different forms of those growing in their natural, or inverted, position would be scarcely observable. It appears also, from his experiments, that such inverted cuttings, in subsequent years, grow with as much vigour as others that are not inverted; whence we must conclude that the organisation of the internal bark becomes again inverted, and adapted to the position of the branch. The growth of some inverted plants of the gooseberry-tree, which I obtained, many years ago, from layers, gave me reason to draw a different conclusion ; for these always continued weak and dwarfish. I do not, however, entertain the slightest degree of doubt but that the assertion of Duhamel is perfectly correct.

I intended to have added some observations on the reproduction of buds and roots of trees ; but these would necessarily extend the present paper to an immoderate length ; I shall therefore reserve them for a future communication, and conclude with an account of an experiment which more properly belongs to the paper I had the honour to address to you last year, but which had not then succeeded.

I have stated in that paper, that the leaf-stalk, the fruit-stalk, and the

---

* The eleventh plate (Vegetable statics) is that to which, in this place, I particularly allude.

tendril, of the vine, had been successfully substituted, in many instances, for each other; but that I had failed in my efforts to engraft a bunch of grapes, by approach, on the leaf-stalk; owing, I conceived, to the operation having been improperly performed. In those experiments, I cut the leaf-stalk into the form of a wedge, and made an incision in the fruit-stalk adapted to receive it; but, under such circumstances, the leaf-stalk (as I had proved by many experiments) has no power to generate new matter; and the wounds of the fruit-stalk heal so slowly that I readily anticipated the ill success of the operation. In the last spring, I pared off similar portions of the leaf-stalk and fruit-stalk; and, bringing the wounded parts into contact, I secured them closely together, by means of a bandage, letting the leaf remain. Under these circumstances a union took place; and the fruit-stalk being then taken off below the point of junction and the leaf-stalk above it, the grapes drew their whole nutriment through the remaining part of the leaf-stalk. They did not, however, acquire their full size; and the seeds were small, and, I think, incapable of vegetating; but this I attribute to the want of nutriment in quantity rather than in quality; for the union of the vessels of the leaf-stalk with those of the fruit-stalk was very imperfect. The grapes, which were the purple Frontignac, possessed their musky flavour in the same degree with others growing on the same plant.

There is another experiment in my last paper, which I will also notice here; because it appears to lead to some important conclusions, and had been tried only in a single instance. I have there stated, that the stem of a young tree became elliptical, by being confined to move only in the segment of a large circle. This experiment was successfully repeated, during the last year, on other trees; but I have nothing to add to the description which I have already given.

V.—OBSERVATIONS ON THE STATE IN WHICH THE TRUE SAP OF TREES IS DEPOSITED DURING WINTER.

[*Read before the* ROYAL SOCIETY, *January* 24, 1805.]

IT is well known that the fluid, generally called the sap in trees, ascends in the spring and summer from their roots, and that in the autumn and winter it is not, in any considerable quantity, found in them; and I have observed in a former paper, that this fluid rises wholly through the alburnum, or sap-wood. But Duhamel and subsequent naturalists have proved, that trees contain another kind of sap, which they have called the true, or peculiar juice, or sap of the plant. Whence this fluid originates does not appear to have been agreed upon by naturalists ; but I have offered some facts to prove that it is generated by the leaf* ; and that it differs from the common aqueous sap owing to changes it has undergone in its circulation through that organ : and I have contended that from this fluid (which Duhamel has called the *suc propre*, and which I will call the true sap) the whole substance, which is annually added to the tree, is derived. I shall endeavour in the present paper to prove that this fluid, in an inspissated state, or some concrete matter deposited by it, exists during the winter in the alburnum, and that from this fluid, or substance, dissolved in the ascending aqueous sap, is derived the matter which enters into the composition of the new leaves in the spring, and thus furnishes those organs, which were not wanted during the winter, but which are essential to the further progress of vegetation.

Few persons at all conversant with timber are ignorant, that the alburnum, or sap-wood of trees, which are felled in the autumn or winter, is much superior in quality to that of other trees of the same species, which are suffered to stand till the spring, or summer : it is at once more firm and tenacious in its texture, and more durable. This superiority in winter-felled wood has been generally attributed to the absence of the sap at that season ; but the appearance and qualities of the wood seem more justly to warrant the conclusion, that some substance has been added to, instead of taken from it, and many circumstances induced me to suspect that this substance is generated, and deposited within it, in the preceding summer and autumn.

Duhamel has remarked, and is evidently puzzled with the circumstance, that trees perspire more in the month of August, when the leaves are full grown, and when the annual shoots have ceased to elongate, than at any earlier period ; and we cannot suppose the powers of vegetation to be thus actively employed, but in the execution of some very important operation. Bulbous and tuberous roots are almost wholly generated after the leaves and stems of the plants to which they belong have attained their full growth : and I have constantly found, in my practice as a farmer, that the produce of my meadows has been immensely increased when the herbage of the preceding year had remained to perform its proper office till the end of the autumn, on ground which had been mowed early in the summer. Whence I have been led to imagine, that the leaves, both of trees and herbaceous plants, are alike employed, during the latter part of the summer, in the preparation of matter calculated to

* See above, Paper No. III.

afford food to the expanding buds and blossoms of the succeeding spring, and to enter into the composition of new organs of assimilation.

If the preceding hypothesis be well founded, we may expect to find that some change will gradually take place in the qualities of the aqueous sap of trees during its ascent in the spring; and that any given portion of winter-felled wood will at the same time possess a greater degree of specific gravity, and yield a larger quantity of extractive matter, than the same quantity of wood which has been felled in the spring or in the early part of the summer. To ascertain these points I made the experiments, an account of which I have now the honour to lay before you.

As early in the last spring as the sap had risen in the sycamore and birch, I made incisions into the trunks of those trees, some close to the ground, and others at the elevation of seven feet, and I readily obtained from each incision as much sap as I wanted. Ascertaining the specific gravity of the sap of each tree, obtained at the different elevations, I found that of the sap of the sycamore with very little variation, in different trees, to be 1.004 when extracted close to the ground, and 1.008 at the height of seven feet. The sap of the birch was somewhat lighter; but the increase of its specific gravity, at greater elevation, was comparatively the same. When extracted near the ground the sap of both kinds was almost free from taste; but when obtained at a greater height, it was sensibly sweet. The shortness of the trunks of the sycamore trees, which were the subjects of my experiments, did not permit me to extract the sap at a greater elevation than seven feet, except in one instance, and in that, at twelve feet from the ground, I obtained a very sweet fluid, whose specific gravity was 1.012.

I conceived it probable, that if the sap in the preceding cases derived any considerable portion of its increased specific gravity from matter previously existing in the alburnum, I should find some diminution of its weight, when it had continued to flow some days from the same incision, because the alburnum in the vicinity of that incision would, under such circumstances, have become in some degree exhausted: and on comparing the specific gravity of the sap which had flowed from a recent and an old incision, I found that from the old to be reduced to 1.002, and that from the recent one to remain 1.004, as in the preceding cases, the incision being made close to the ground. Wherever extracted, whether close to the ground, or at some distance from it, the sap always appeared to contain a large portion of air.

In the experiments to discover the variation in the specific gravity of the alburnum of trees at different seasons, some obstacles to the attainment of any very accurate results presented themselves. The wood of different trees of the same species, and growing in the same soil, or that taken from different parts of the same tree, possesses different degrees of solidity ; and the weight of every part of the alburnum appears to increase with its age, the external layers being the lightest. The solidity of wood varies also with the greater or less rapidity of its growth. These sources of error might apparently have been avoided by cutting off, at different seasons, portions of the same trunk or branch : but the wound thus made might, in some degree, have impeded the due progress of the sap in its ascent, and the part below might have been made heavier by the stagnation of the sap, and that above lighter by privation of its proper quantity of nutriment. The most eligible method therefore which occurred to me, was to select and mark in the winter some of the poles of an oak coppice, where all are of equal age, and where many, of the same size and growing with equal vigour, spring from the same stool. One half of the poles which I marked and numbered were cut on the 31st of December 1803, and the remainder on the 15th of the following May, when the leaves were nearly half grown. Proper marks were put to distinguish the winter-felled from the summer-felled poles, the bark being left on all, and all being placed in the situation to dry.

In the beginning of August I cut off nearly equal portions from a winter and summer-felled pole, which had both grown on the same stool ; and both portions were then put in a situation, where, during the seven succeeding weeks, they were kept very warm by a fire. The summer-felled wood was, when put to dry, the most heavy ; but it evidently contained much more water than the other, and, partly at least from this cause, it contracted much more in drying. In the beginning of October both kinds appeared to be perfectly dry, and I then ascertained the specific gravity of the winter-felled wood to be 0.679, and that of the summer-felled wood to be 0.609 ; after each had been immersed five minutes in water.

This difference of ten per cent. was considerably more than I had anticipated, and it was not till I had suspended and taken off from the balance each portion, at least ten times, that I ceased to believe that some error had occurred in the experiment : and indeed I was not at last satisfied till I had ascertained by means of compasses adapted to the measurement of solids, that the winter-felled pieces of wood were much less than the others which they equalled in weight.

The pieces of wood, which had been the subjects of these experiments, were again put to dry, with other pieces of the same poles, and I yesterday ascertained the specific gravity of both with scarcely any variation in the result. But when I omitted the medulla, and parts adjacent to it, and used the layers of wood which had been more recently formed, I found the specific gravity of the winter-felled wood to be only 0.583, and that of the summer-felled to be 0.533 ; and trying the same experiment with similar pieces of wood, but taken from poles which had grown on a different stool, the specific gravity of the winter-felled wood was 0.588, and that of the summer-felled 0.534.

It is evident that the whole of the preceding difference in the specific gravity of the winter and summer felled wood might have arisen from a greater degree of contraction in the former kind, whilst drying ; I therefore proceeded to ascertain whether any given portion of it, by weight, would afford a greater quantity of extractive matter, when steeped in water. Having therefore reduced to small fractions 1000 grains of each kind, I poured on each portion six ounces of boiling water; and at the end of twenty-four hours, when the temperature of the water had sunk to 60°, I found that the winter-felled wood had communicated a much deeper colour to the water in which it had been infused, and had raised its specific gravity to 1.002. The specific gravity of the water in which the summer-felled wood had, in the same manner, been infused was 1.001. The wood in all the preceding cases was taken from the upper parts of the poles, about eight feet from the ground.

Having observed, in the preceding experiments, that the sap of the sycamore became specifically lighter when it had continued to flow during several days from the same incision, I concluded that the alburnum in the vicinity of such incision had been deprived of a larger portion of its concrete or inspissated sap than in other parts of the same tree : and I therefore suspected that I should find similar effects to have been produced by the young annual shoots and leaves ; and that any given weight of the alburnum in their vicinity would be found to contain less extractive matter than an equal portion taken from the lower parts of the same pole, where no annual shoots or leaves had been produced.

No information could in this case be derived from the difference in the specific gravity of the wood ; because the substance of every tree is most dense and solid in the lower parts of its trunk : and I could on this account judge only from the quantity of extractive matter which equal portions of the two kinds of wood would afford. Having therefore reduced to pieces several equal portions of wood taken from different parts of the same poles, which had been felled in May, I poured on each portion an equal quantity of boiling water, which I suffered to remain twenty hours, as in

I

the preceding experiments : and I then found that in some instances the wood from the lower, and in others that from the upper parts of the poles, had given to the water the deepest colour and greatest degree of specific gravity ; but that all had afforded much extractive matter, though in every instance the quantity yielded was much less than I had, in all cases, found in similar infusions of winter-felled wood.

It appears, therefore, that the reservoir of matter deposited in the alburnum is not wholly exhausted in the succeeding spring : and hence we are able to account for the several successions of leaves and buds which trees are capable of producing when those previously protruded have been destroyed by insects, or other causes, and for the extremely luxuriant shoots which often spring from the trunks of trees, whose branches have been long in a state of decay.

I have also some reason to believe that the matter deposited in the alburnum remains unemployed in some cases during several successive years : it does not appear probable that it can be all employed by trees which, after having been transplanted, produce very few leaves, or by those which produce neither blossoms nor fruit. In making experiments in 1802, to ascertain the manner in which the buds of trees are reproduced, I cut off in the winter all the branches of a very large old pear-tree, at a small distance from the trunk ; and I pared off, at the same time, the whole of the lifeless external bark. The age of this tree, I have good reasons to believe, somewhat exceeded two centuries : its extremities were generally dead ; and it afforded few leaves, and no fruit ; and I had long expected every successive year to terminate its existence. After being deprived of its external bark, and of all its buds, no marks of vegetation appeared in the succeeding spring, or early part of the summer : but in the beginning of July numerous buds penetrated through the bark in every part, many leaves of large size everywhere appeared, and in the autumn every part was covered with very vigorous shoots exceeding, in the aggregate, two feet in length. The number of leaves which, in this case, sprang at once from the trunk and branches appeared to me greatly to exceed the whole of those which the tree had borne in the three preceding seasons ; and I cannot believe that the matter which composed these buds and leaves could have been wholly prepared by the feeble vegetation and scanty foliage of the preceding year.

But whether the substance which is found in the alburnum of winter-felled trees, and which disappears in part in the spring and early part of the summer, be generated in one or in several preceding years, there seem to be strong grounds of probability, that this substance enters into the composition of the leaf : for we have abundant reason to believe that this organ is the principal agent of assimilation ; and scarcely anything can be more contrary to every conclusion we should draw from analogical reasoning and comparison of the vegetable with the animal economy, or in itself more improbable, than that the leaf, or any other organ, should singly prepare and assimilate immediately from the crude aqueous sap that matter which composes itself.

It has been contended * that the buds themselves contain the nutriment necessary for the minute unfolding leaves : but trees possess a power to reproduce their buds, and the matter necessary to form these buds must evidently be derived from some other source ; nor does it appear probable that the young leaves very soon enter on this office, for the experiments of Ingenhouz prove that their action on the air which surrounds them is very essentially different from that of full-grown leaves. It is true that buds in many instances will vegetate, and produce trees, when a very small portion only of alburnum remains attached to them ; but the first efforts of vegetation in such buds are much more feeble than in others to which a larger quantity of alburnum is attached, and therefore we have, in this case, no grounds to suppose that the leaves derive their first nutriment from the crude sap.

It is also generally admitted, from the experiments of Bonnet and Du Hamel, which I have repeated with the same result, that in the cotyledons of the seed is deposited a quantity of nutriment for the bud which every seed contains ; and though no vessels can be traced † which lead immediately from the cotyledons to the bud or plumula, it is not difficult to point out a more circuitous passage, which is perfectly similar to that through which I conceive the sap to be carried from the leaves to the buds in the subsequent growth of the tree ; and I am in possession of many facts to prove that seedling trees, in the first stage of their existence, depend entirely on the nutriment afforded by the cotyledons ; and that they are greatly injured, and in many instances killed, by being put to vegetate in rich mould.

We have much more decisive evidence that bulbous and tuberous rooted plants contain the matter within themselves which subsequently composes their leaves ; for we see them vegetate even in dry rooms on the approach of spring ; and many bulbous rooted plants produce their leaves and flowers with nearly the same vigour by the application of water only, as they do when growing in the best mould. But the water

* Thomson's Chemistry. † Hedwig.

i 2

in this case, provided that it be perfectly pure, probably affords little or no food to the plant, and acts only by dissolving the matter prepared and deposited in the preceding year ; and hence the root becomes exhausted and spoiled : and Hassenfratz found that the leaves and flowers and roots of such plants afforded no more carbon than he had proved to exist in bulbous roots of the same weight, whose leaves and flowers had never expanded.

As the leaves and flowers of the hyacinth, in the preceding case, derived their matter from the bulb, it appears extremely probable that the blossoms of trees receive their nutriment from the alburnum, particularly as the blossoms of many species precede their leaves ; and, as the roots of plants become weakened and apparently exhausted when they have afforded nutriment to a crop of seed, we may suspect that a tree, which has borne much fruit in one season, becomes in a similar way exhausted, and incapable of affording proper nutriment to a crop in the succeeding year.   And I am much inclined to believe that were the wood of a tree in this state accurately weighed, it would be found specifically lighter than that of a similar tree, which had not afforded nutriment to fruit or blossoms in the preceding year or years.

If it be admitted that the substance which enters into the composition of the first leaves in the spring is derived from matter which has undergone some previous preparation within the plant (and I am at a loss to conceive on what grounds this can be denied, in bulbous and tuberous rooted plants at least), it must also be admitted that the leaves which are generated in the summer derive their substance from a similar source ; and this cannot be conceded without a direct admission of the existence of vegetable circulation, which is denied by so many eminent naturalists.   I have not, however, found in their writings a single fact to disprove its existence, nor any great weight in their arguments, except those drawn from two important errors in the admirable works of Hales and Duhamel, which I have noticed in a former memoir.   I shall therefore proceed to point out the channels through which I conceive the circulating fluids to pass.

When a seed is deposited in the ground, or otherwise exposed to a proper degree of heat and moisture and exposure to air, water is absorbed by the cotyledons, and the young radicle or root is emitted· At this period, and in every subsequent stage of the growth of the root, it increases in length by the addition of new parts to its apex, or point, and not by any general distension of its vessels and fibres ; and the experiments of Bonnet and Duhamel leave little grounds of doubt but that the new matter which is added to the point of the root descends from the cotyledons. The first motion therefore of the fluids in plants is downwards, towards the point of the root; and the vessels which appear to carry them are of the same kind with those which are subsequently found in the bark, where I have, on a former occasion, endeavoured to prove that they execute the same office.

In the last spring I examined almost every day the progressive changes which take place in the radicle emitted by the horse-chestnut: I found it, at its first existence and until it was some weeks old, to be incapable of absorbing coloured infusions when its point was taken off, and I was totally unable to discover any alburnous tubes through which the sap absorbed from the ground, in the subsequent growth of the tree, ascends; but when the roots were considerably elongated, alburnous tubes formed; and, as soon as they had acquired some degree of firmness in their consistence, they appeared to enter on their office of carrying up the aqueous sap, and the leaves of the plumula then, and not sooner, expanded.

The leaf contains at least three kinds of tubes :—the first is what in a former paper I have called the central vessel, through which the aqueous sap appears to be carried, and through which coloured infusions readily pass, from the alburnous tubes into the leaf-stalk. These vessels are always accompanied by spiral tubes, which do not appear to carry any liquid; but there is another vessel which appears to take its origin from the leaf, and which descends down the internal bark, and contains the true or prepared sap. When the leaf has attained its proper growth, it seems to perform precisely the office of the cotyledon; but being exposed to the air, and without the same means to acquire, or the substance to retain moisture, it is fed by the alburnous tubes and central vessels.

The true sap now appears to be discharged from the leaf, as it was previously from the cotyledon, into the vessels of the bark, and to be employed in the formation of new alburnous tubes between the base of the leaf and the root. From these alburnous tubes spring other central vessels and spiral tubes, which enter into and possibly give existence to other leaves; and thus by a repetition of the same process the young tree or annual shoot continues to acquire new parts, which apparently are formed from the ascending aqueous sap.

But it has been proved by Duhamel that a fluid similar to that which is found in the true sap-vessels of the bark exists also in the alburnum, and this fluid is extremely obvious in the fig, and other trees, whose true sap is white or coloured. The vessels which contain this fluid in the alburnum are in contact with those which carry up the aqueous sap ; and it does not appear probable that, in a body so porous as wood, fluids so near each other should remain wholly unmixed. I must therefore conclude that when the true sap has been delivered from the cotyledon or leaf into the returning or true sap-vessels of the bark, one portion of it secretes through the external cellular, or more probably glandular substance of the bark, and generates a new epidermis where that is to be formed ; and that the other portion of it secretes through the internal glandular substance of the bark, where one part of it produces the new layer of wood, and the remainder enters the pores of the wood already formed, and subsequently mingles with the ascending aqueous sap ; which thus becomes capable of affording the matter necessary to form new buds and leaves.

It has been proved in the preceding experiments on the ascending sap of the sycamore and birch, that that fluid does not approach the buds and unfolding leaves in the spring, in the state in which it is absorbed from the earth ; and therefore we may conclude that the fluid which enters into and circulates through the leaves of plants, as the blood through the lungs of animals, consists of a mixture of the true sap or blood of the plant with matter more recently absorbed, and less perfectly assimilated.

It appears probable that the true sap undergoes a considerable change on its mixture with the ascending aqueous sap ; for this fluid in the sycamore has been proved to become more sensibly sweet in its progress from the roots in the spring, and the liquid which flows from the wounded bark of the same tree is also sweet ; but I have never been able to detect the slightest degree of sweetness in decoctions of the sycamore wood in winter. I am therefore inclined to believe that the saccharine matter existing in the ascending sap is not immediately, or wholly, derived from the fluid which had circulated through the leaf in the preceding year ; but that it is generated by a process similar to that of the germination of seeds, and that the same process is always going forward during the spring and summer, as long as the tree continues to generate new organs. But towards the conclusion of the summer I conceive that the true sap simply accumulates in the alburnum, and thus adds to the specific gravity of winter-felled wood, and increases the quantity of its extractive matter.

I have some reasons to believe that the true sap descends through the alburnum as well as through the bark, and I have been informed that if the bark be taken from the trunks of trees in the spring, and such trees be suffered to grow till the following winter, the alburnum acquires a great degree of hardness and durability. If subsequent experiments prove that the true sap descends through the alburnum, it will be easy to point out the cause why trees continue to vegetate after all communication between the leaves and roots, through the bark, has been intercepted ; and why some portion of alburnous matter is in all trees * generated below incisions through the bark.

It was my intention this year to have troubled you with some observations on the reproductions of the buds and roots of trees; but as the subject of the paper which I have now the honour to address to you appeared to be of more importance, I have deferred those observations to a future opportunity; and I shall at present only observe, that I conceive myself to be in possession of facts to prove that both buds and roots originate from the alburnous substance of plants, and not, as is, I believe, generally supposed, from the bark.

VI.—ON THE REPRODUCTION OF BUDS.

[*Read before the* ROYAL SOCIETY, *May* 23, 1805.]

EVERY tree, in the ordinary course of its growth, generates in each season those buds which expand in the succeeding spring ; and the buds thus generated contain, in many instances, the whole of the leaves which appear in the following summer. But if these buds be destroyed during the winter or early part of the spring, other buds, in many species of trees, are generated, which in every respect perform the office of those which previously existed, except that they never afford fruit or blossoms. This reproduction of buds has not escaped the notice of naturalists; but it does not appear to have been ascertained by them, from which amongst the various substances of the tree the buds derive their origin.

Duhamel conceived that reproduced buds sprang from pre-organized germs ; but the existence of such germs has not, in any instance, been proved, and it is well known that the roots and trunk, and branches, of many species of trees will, under proper management, afford buds from every part of their surfaces ; and therefore, if this hypothesis be well founded, many millions of such germs must be annually generated in every large tree ; not one of which in the ordinary course of nature will come

---

* I have in a former paper stated that the perpendicular shoots of the vine form an exception. I spoke on the authority of numerous experiments ; but they had been made late in the summer ; and on repeating the same experiments at an earlier period, I found the result in conformity with my experiments on other trees.

into action; and as nature, amidst all its exuberance, does not abound in useless productions, the opinions of this illustrious physiologist are in this case probably erroneous.

Other naturalists have supposed the buds, when reproduced, to spring from the plexus of vessels which constitutes the internal bark; and this opinion is, I believe, much entertained by modern botanists; it nevertheless appears to be unfounded, as the facts I shall proceed to state will evince.

If the fruit-stalks of the sea-cale (Crambe maritima) be cut off near the ground in the spring, the medullary substance within that part of the stalk which remains attached to the root decays; and a cup is thus formed, in which water collects in the succeeding winter. The sides of this cup consist of a woody substance, which in its texture and office, and mode of generation, agrees perfectly with the alburnum of trees; and I conceive it to be as perfect alburnum as the white wood of the oak or elm; and from the interior part of this substance within the cup, I have frequently observed new buds to be generated in the ensuing spring. It is sufficiently obvious that the buds in this case do not spring from the bark; but it is not equally evident that they might not have sprung from some remains of the medulla.

In the autumn of 1802 I discovered that the potato possessed a similar power of reproducing its buds. Some plants of this species had been set rather late in the preceding spring, in very dry ground, where through want of moisture they vegetated very feebly; and the portions of the old roots remained sound and entire till the succeeding autumn. Being then moistened by rain, many small tubers were generated on the surfaces made by the knife in dividing the roots into cuttings; and the buds of these, in many instances, elongated into runners, which gave existence to other tubers, some of which I had the pleasure to send to you.

I have in a former paper remarked, that the potato consists of four distinct substances, the epidermis, the true skin, the bark, and its internal substance, which from its mode of formation, and subsequent office, I have supposed to be alburnous: there is also in the young tuber a transparent line through the centre, which is probably its medulla. The buds and runners sprang from the substance which I conceive to be the alburnum of the root, and neither from the central part of it, nor from the surface in contact with the bark. It must, however, be admitted, that the internal substance of the potato corresponds more nearly with our ideas of a medullary than of an alburnous substance, and therefore this, with the preceding facts, is adduced to prove only that the reproduced buds of these plants are not generated by the cortical substance of the root: and I shall proceed to relate some experiments on the apple, and pear, and plum-tree, which I conceive to prove that the reproduced buds of those plants do not spring from the medulla.

Having raised from seeds a very considerable number of plants of each of these species in 1802, I partly disengaged them from the soil in the autumn, by digging round each plant, which was then raised about two inches above its former level. A part of the mould was then removed, and the plants were cut off about an inch below the points where the seed-leaves formerly grew; and a portion of the root, about an inch long, without any bud upon it, remained exposed to the air and light. In the beginning of April I observed many small elevated points on the bark of these roots, and, removing the whole of the cortical substance, I found that the elevations were occasioned by small protuberances on the surfaces of the alburnum. As the spring advanced, many minute red points appeared to perforate the bark; these soon assumed the character of buds, and produced shoots, in every respect similar to those which would have sprung from the organized buds of the preceding year. Whether the buds thus reproduced derived any portion of their component parts from the bark or not, I shall not venture to decide; but I am much disposed to believe that, like those of the potato, they sprang from the alburnous substance solely.

The space, however, in the annual root, between the medulla and the bark is very small; and therefore it may be contended that the buds in these instances may have originated from the medulla. I therefore thought it necessary to repeat similar experiments on the roots and trunks of old trees, and by these the buds were reproduced precisely in the same manner as the annual roots: and therefore, conceiving myself to have proved in a former memoir*, that the substance which has been called the medullary process does not originate from the medulla, I must conclude that reproduced buds do not spring from that substance.

I have remarked in a paper, laid before the Royal Society in the commencement of the present year, that the alburnous tubes, at their termination upwards, invariably join the central vessels, and that these vessels which appear to derive their origin from the alburnous tubes, convey nutriment, and probably give existence to new buds and leaves. It is also evident, from the facility with which the rising sap is transferred from one side of a wounded tree to the other,

* See above, the Paper No. III.

that the alburnous tubes possess lateral, as well as terminal orifices : and it does not appear improbable that the lateral as well as the terminal orifices of the alburnous tubes may possess the power to generate central vessels ; which vessels evidently feed, if they do not give existence to, the reproduced buds and leaves. And therefore, as the preceding experiments appear to prove that the buds neither spring from the medulla nor the bark, I am much inclined to believe that they are generated by central vessels which spring from the lateral orifices of the alburnous tubes. The practicability of propagating some plants from their leaves may seem to stand in opposition to this hypothesis ; but the central vessel is always a component part of the leaf, and from it the bud and young plant probably originate.

I expected to discover in seeds a similar power to regenerate their buds ; for the cotyledons of these, though dissimilar in organisation, execute the office of the alburnum, and contain a similar reservoir of nutriment, and at once supply the place of the alburnum and the leaf. But no experiments which I have yet been able to make, have been decisive, owing to the difficulty of ascertaining the number of buds previously existing within the seed. Few, if any, seeds, I have reason to believe, contain less than three buds, one only of which, except in cases of accident, germinates, and some seeds appear to contain a much greater number. The seed of the peach appears to be provided with ten or twelve leaves, each of which probably covers the rudiment of a bud, and the seeds, like the buds of the horse-chestnut, contain all the leaves, and apparently all the buds of the succeeding year : and I have never been able to satisfy myself that all the buds were eradicated without having destroyed the base of the plumule, in which the power of reproducing buds probably resides, if such power exists.

Nature appears to have denied to annual and biennial plants (at least to those which have been the subjects of my experiments) the power which it has given to perennial plants to reproduce their buds ; but nevertheless some biennials possess, under peculiar circumstances, a very singular resource, when all their buds have been destroyed. A turnip, bred between the English and Swedish variety, from which I had cut off the greater part of its fruit-stalks, and of which all the buds had been destroyed, remained some weeks in an apparently dormant state ; after which the first seed in each pod germinated, and bursting the seed-vessel, seemed to execute the office of a bud and leaves to the parent plant, during the short remaining term of its existence, when its preternatural foliage perished with it. Whether this property be possessed by other biennial plants in common with the turnip or not, I am not at present in possession of facts to decide, not having made precisely the same experiment on any other plant.

I will take this opportunity to correct an inference that I have drawn in a former paper *, which the facts (though quite correctly stated) do not, on subsequent repetition of the experiment, appear to justify. I have stated, that when a perpendicular shoot of the vine was inverted to a depending position, and a portion of its bark between two circular incisions round the stem removed, much more new wood was generated on the lower lip of the wound, become uppermost by the inverted position of the branch, than on the opposite lip, which would not have happened had the branch continued to grow erect, and I have inferred that this effect was produced by sap which had descended by gravitation from the leaves above. But the branch was, as I have there stated, employed as a layer, and the matter which would have accumulated on the opposite lip of the wound had been employed in the formation of roots, a circumstance which at that time escaped my attention. The effects of gravitation on the motion of the descending sap, and consequent growth of plants, are, I am well satisfied, from a great variety of experiments, very great; but it will be very difficult to discover any method by which the extent of its operation can be accurately ascertained. For the vessels which convey and impel † the true sap, or fluid from which the new wood appears to be generated, pass immediately from the leaf-stalk towards the root; and though the motion of this fluid may be impeded by gravitation, and it be even again returned into the leaf, no portion of it, unless it had been extravasated, could have descended to the part from which the bark was taken off in the experiment I have described. I am not sensible that in the different papers which I have had the honour to address to you, I have drawn any other inference which the facts, on repetition of the experiments, do not appear capable of supporting.

* See above, No. III.   † See the preceding Papers.

VII.—ON THE DIRECTION OF THE RADICLE AND GERMEN DURING THE VEGETATION OF SEEDS.

[*Read before the* ROYAL SOCIETY, *January* 9, 1806.]

IT can scarcely have escaped the notice of the most inattentive observer of vegetation, that in whatever position a seed is placed to germinate, its radicle invariably makes an effort to descend towards the centre of the earth, whilst the elongated germen takes a precisely opposite direction ; and it has been proved by Duhamel* that if a seed, during its germination, be frequently inverted, the points both of the radicle and germen will return to the first direction. Some naturalists have supposed these opposite effects to be produced by gravitation ; and it is not difficult to conceive that the same agent, by operating on bodies so differently organised as the radicle and germen of plants are, may occasion the one to descend and the other to ascend.

The hypothesis of these naturalists does not, however, appear to have been much strengthened by any facts they were able to adduce in support of it, nor much weakened by the arguments of their opponents ; and therefore, as the phenomena observable during the conversion of a seed into a plant are amongst the most interesting that occur in vegetation, I commenced the experiments, an account of which I have now the honour to request you to lay before the Royal Society.

I conceived that if gravitation were the cause of the descent of the radicle, and of the ascent of the germen, it must act either by its immediate influence on the vegetable fibres and vessels during their formation, or on the motion and consequent distribution of the true sap afforded by the cotyledons : and as gravitation could produce these effects only whilst the seed remained at rest, and in the same position relative to the attraction of the earth, I imagined that its operation would become suspended by constant and rapid change of the position of the germinating seed, and that it might be counteracted by the agency of centrifugal force.

Having a strong rill of water passing through my garden, I constructed a small wheel similar to those used for grinding corn, adapting another wheel of a different construction, and formed of very slender pieces of wood, to the same axis. Round the circumference of the latter, which was eleven inches in diameter, numerous seeds of the garden bean, which had been soaked in water to produce their greatest degree of expansion, were bound, at short distances from each other.

* Physique des Arbres.

The radicles of these seeds were made to point in every direction, some towards the centre of the wheel, and others in the opposite direction; others as tangents to its curve, some pointing backwards, and others forwards, relative to its motion; and others pointing in opposite directions in lines parallel with the axis of the wheels. The whole was inclosed in a box, and secured by a lock, and a wire grate was placed to prevent the ingress of any body capable of impeding the motion of the wheels.

The water being then admitted, the wheels performed something more than 150 revolutions in a minute; and the position of the seeds relative to the earth was of course as often perfectly inverted, within the same period of time; by which I conceive that the influence of gravitation must have been wholly suspended.

In a few days the seeds began to germinate, and as the truth of some of the opinions I had communicated to you, and of many others which I had long entertained, depended on the result of the experiment, I watched its progress, with some anxiety, though not with much apprehension; and I had soon the pleasure to see that the radicles, in whatever direction they were protruded from the position of the seed, turned their points outwards from the circumference of the wheel, and in their subsequent growth receded nearly at right angles from its axis. The germens, on the contrary, took the opposite direction, and in a few days their points all met in the centre wheel. Three of these plants were suffered to remain on the wheel, and were secured to its spokes to prevent their being shaken off by its motion. The stems of these plants soon extended beyond the centre of the wheel: but the same cause, which first occasioned them to approach its axis, still operating, their points returned and met again at its centre.

The motion of the wheel being in this experiment vertical, the radicle and germen of every seed occupied, during a minute portion of time in each revolution, precisely the same position they would have assumed had the seeds vegetated at rest; and as gravitation and centrifugal force also acted in lines parallel with the vertical motion and surface of the wheel, I conceived that some slight objections might be urged against the conclusions I felt inclined to draw. I therefore added to the machinery I have described another wheel, which moved horizontally over the vertical wheels; and to this, by means of multiplying wheels of different powers, I was enabled to give many different degrees of velocity. Round the circumference of the horizontal wheel, whose diameter was also eleven inches, seeds of the bean were bound as in the experiment which I have already described, and it was then made to perform 250 revolutions in a minute. By the rapid motion of the water-wheel much water was thrown upwards on the horizontal wheel, part of which supplied the seeds upon it with moisture, and the remainder was dispersed, in a light and constant shower, over the seeds in the vertical wheel, and on others placed to vegetate at rest in different parts of the box.

Every seed on the horizontal wheel, though moving with great rapidity, necessarily retained the same position relative to the attraction of the earth; and therefore the operation of gravitation could not be suspended, though it might be counteracted, in a very considerable degree, by centrifugal force: and the difference, I had anticipated, between the effects of rapid vertical and horizontal motion soon became sufficiently obvious. The radicles pointed downwards about ten degrees below, and the germens as many degrees above, the horizontal line of the wheel's motion; centrifugal force having made both to deviate 80° from the perpendicular direction each would have taken, had it vegetated at rest. Gradually diminishing the rapidity of the motion of the horizontal wheel, the radicles descended more perpendicularly, and the germens grew more upright; and when it did not perform more than eighty revolutions in a minute, the radicle pointed about 45° below, and the germen as much above, the horizontal line, the one always receding from, and the other approaching to, the axis of the wheel.

I would not, however, be understood to assert that the velocity of 250, or of eighty horizontal revolutions in a minute, will always give accurately the degrees of depression and elevation of the radicle and germen which I have mentioned; for the rapidity of the motion of my wheels was sometimes diminished by the collection of fibres of conferva against the wire grate; which obstructed in some degree the passage of the water: and the machinery, having been the workmanship of myself and my gardener, cannot be supposed to have moved with all the regularity it might have done, had it been made by a professional mechanic. But I conceive myself to have fully proved that the radicles of germinating seeds are made to descend, and their germens to ascend, by some external cause, and not by any power inherent in vegetable life: and I see little reason to doubt that gravitation is the principal, if not the only agent employed, in this case, by nature. I shall therefore endeavour to point out the means by which I conceive the same agent may produce effects so diametrically opposite to each other.

The radicle of a germinating seed (as many naturalists have observed) is increased in length only by new parts successively added to its apex or point, and not at all by any general extension of parts already formed : and the new matter which is thus successively added unquestionably descends in a fluid state from the cotyledons*. On this fluid, and on the vegetable fibres and vessels whilst soft and flexible, and whilst the matter which composes them is changing from a fluid to a solid state, gravitation, I conceive, would operate sufficiently to give an inclination downwards to the point of the radicle ; and as the radicle has been proved to be obedient to centrifugal force, it can scarcely be contended that its direction would remain uninfluenced by gravitation.

I have stated that the radicle is increased in length only by parts successively added to its point : the germen, on the contrary, elongates by a general extension of its parts previously organised ; and its vessels and fibres appear to extend themselves in proportion to the quantity of nutriment they receive. If the motion and consequent distribution of the true sap be influenced by gravitation, it follows, that when the germen at its first emission, or subsequently, deviates from a perpendicular direction, the sap must accumulate on its under side : and I have found in a great variety of experiments on the seeds of the horse-chestnut, the bean, and other plants, when vegetating at rest, that the vessels and fibres on the under side of the germen invariably elongate much more rapidly than those on its upper side ; and thence it follows that the point of the germen must always turn upwards. And it has been proved that a similar increase of growth takes place on the external side of the germen when the sap is impelled there by centrifugal force, as it is attracted by gravitation to its under side, when the seed germinates at rest.

This increased elongation of the fibres and vessels of the under side is not confined to the germens, nor even to the annual shoots of trees, but occurs and produces the most extensive effects in the subsequent growth of their trunks and branches. The immediate effect of gravitation is certainly to occasion the further depression of every branch, which extends horizontally from the trunk of the tree ; and, when a young tree inclines to either side, to increase that inclination : but it at the same time attracts the sap to the under side, and thus occasions an increased longitudinal extension of the substance of the new wood on that side †. The depression of the lateral branch is thus prevented ; and it is even enabled to raise itself above its natural level, when the branches above it are removed ; and the young tree, by the same means, becomes more upright,

---

* See the preceding Paper.

† This effect does not appear to be produced in what are called weeping trees ; the cause of which I have endeavoured to point out in a former memoir. (See above, No. IV.)

in direct opposition to the immediate action of gravitation : nature, as usual, executing the most important operations by the most simple means.

I could adduce many more facts in support of the preceding deductions, but those I have stated, I conceive to be sufficiently conclusive. It has however been objected by Duhamel, (and the greatest deference is always due to his opinions,) that gravitation could have little influence on the direction of the germen, were it in the first instance protruded, or were it subsequently inverted, and made to point perpendicularly downwards. To enable myself to answer this objection, I made many experiments on seeds of the horse-chestnut, and of the bean, in the box I have already described ; and as the seeds there were suspended out of the earth, I could regularly watch the progress of every effort made by the radicle and germen to change their positions. The extremity of the radicle of the bean, when made to point perpendicularly upwards, generally formed a considerable curvature within three or four hours, when the weather was warm. The germen was more sluggish; but it rarely or never failed to change its direction in the course of twenty-four hours ; and all my efforts to make it grow downwards, by slightly changing its direction, were invariably abortive.

Another, and apparently a more weighty, objection to the preceding hypothesis, (if applied to the subsequent growth and forms of trees,) arises from the facts that few of their branches rise perpendicularly upwards, and that their roots always spread horizontally ; but this objection I think may be readily answered.

The luxuriant shoots of trees, which abound in sap, in whatever direction they are first protruded, almost uniformly turn upwards, and endeavour to acquire a perpendicular direction ; and to this their points will immediately return, if they are bent downwards during any period of their growth; their curvature upwards being occasioned by an increased extension of the fibres and vessels of their under sides, as in the elongated germens of seeds. The more feeble and slender shoots of the same trees will, on the contrary, grow in almost every direction, probably because their fibres, being more dry, and their vessels less amply supplied with sap, they are less affected by gravitation. Their points, however, generally show an inclination to turn upwards; but the operation of light, in this case, has been proved by Bonnet * to be very considerable.

The radicle tapers rapidly, as it descends into the earth, and its lower part is much compressed by the greater solidity of the mould into which

* Récherches sur l'Usage des Feuilles dans les Plantes.

it penetrates.   The true sap also continues to descend from the cotyledons and leaves, and occasions a continued increase of the growth of
the upper parts of the radicle, and this growth is subsequently augmented by the effects of motion, when the germen has risen above the
ground.   The true sap is therefore necessarily obstructed in its descent;
numerous lateral roots are generated, into which a portion of the
descending sap enters.   The substance of these roots, like that of the
slender horizontal branches, is much less succulent than that of the
radicle first emitted, and they are in consequence less obedient to gravitation: and therefore, meeting less resistance from the superficial soil
than from that beneath it, they extend horizontally in every direction,
growing with most rapidity, and producing the greatest number of ramifications, wherever they find most warmth, and a soil best adapted to
nourish the tree.   As these horizontal, or lateral roots surround the
base of the tree on every side, the true sap descending down its bark,
enters almost exclusively into them, and the first perpendicular root,
having executed its office of securing moisture to the plant, whilst young,
is thus deprived of proper nutriment, and, ceasing almost wholly to grow,
becomes of no importance to the tree.   The tap root of the oak, about
which so much has been written, will possibly be adduced as an exception; but having attentively examined at least 20,000 trees of this species, many of which had grown in some of the deepest and most favourable soils of England, and never having found a single tree possessing a
tap root, I must be allowed to doubt that one ever existed.

As trees possess the power to turn the upper surfaces of their leaves,
and the points of their shoots to the light, and their tendrils in any
direction to attach themselves to contiguous objects, it may be suspected
that their lateral roots are by some means directed to any soil in their
vicinity which is best calculated to nourish the plant, to which they belong;
and it is well known that much the greater part of the roots of an
aquatic plant, which has grown in a dry soil, on the margin of a lake or
river, have been found to point to the water; whilst those of another
species of tree which thrives best in a dry soil, have been ascertained to
take an opposite direction : but the result of some experiments I have
made is not favourable to this hypothesis, and I am rather inclined to
believe that the roots disperse themselves in every direction, and only
become most numerous where they find most employment, and a soil best
adapted to the species of plant.   My experiments have not, however,
been sufficiently varied, or numerous, to decide this question, which I
propose to make the subject of future investigation.

VIII.—ON THE INVERTED ACTION OF THE ALBURNOUS VESSELS OF TREES.

[Read before the ROYAL SOCIETY, *May* 15, 1806.]

I HAVE endeavoured to prove, in several memoirs* laid before the Royal Society, that the fluid by which the various parts (that are annually added to trees, and herbaceous plants whose organization is similar to that of trees,) are generated, has previously circulated through their leaves† either in the same or preceding season, and subsequently descended through their bark ; and after having repeated every experiment that occurred to me, from which I suspected an unfavourable result, I am not in possession of a single fact, which is not perfectly consistent with the theory I have advanced.

There is, however, one circumstance stated by Hales and Duhamel, which appears to militate against my hypothesis ; and as that circumstance probably induced Hales to deny altogether, the existence of circulation in plants, and Duhamel to speak less decisively in favour of it than he possibly might otherwise have done, I am anxious to reconcile the statements of these great naturalists, (which I acknowledge to be perfectly correct,) with the statements and opinions I have on former occasions communicated to you.

Both Hales and Duhamel have proved, that when two circular incisions through the bark, round the stem of a tree, are made at a small distance from each other, and when the bark between these incisions is wholly taken away, that portion of the stem which is below the incisions through the bark continues to live, and in some degree to increase in size, though much more slowly than the parts above the incisions. They have also observed, that a small elevated ridge (bourrelet) is formed round the lower lip of the wound in the bark, which makes some slight advances to meet the bark and wood projected, in much larger quantity from the opposite, or upper lip of the wound.

I have endeavoured in a former memoir‡, to explain the cause why some portion of growth takes place below incisions through the bark,

* See the preceding memoirs, Nos. II. IV. and V.

† During the circulation of the sap through the leaves, a transparent fluid is emitted, in the night, from pores situated on their edges, and on evaporating this liquid obtained from very uxuriant plants of the vine I found a very large residuum to remain, which was similar in external appearance to carbonate of lime. It must however have been a very different substance from the very large portion which the water held in solution. I do not know that this substance has been analysed or observed by any naturalist.

‡ See above, No. III.

by supposing that a small part of the true sap, descending from the leaves, escapes downwards through the porous substance of the alburnum. Several facts stated by Hales, seem favourable to this supposition; and the existence of a power in the alburnum to carry the sap in different directions, is proved in the growth of inverted cuttings of different species of trees*. But I have derived so many advantages, both as gardener and farmer, (particularly in the management of fruit and forest trees,) from the experiments which have been the subject of my former memoirs, that I am confident much public benefit might be derived from an intimate acquaintance with the use and office of the various organs of plants; and thence feel anxious to adduce facts to prove that the conclusions I have drawn, are not inconsistent with the facts stated by my great predecessors.

It has been acknowledged, I believe, by every naturalist who has written on the subject, (and the fact is indeed too obvious to be controverted,) that the matter which enters into the composition of the radicles of germinating seeds existed previously in their cotyledons, and as the radicles increase only in length by parts successively added to their apices, or points most distant from their cotyledons, it follows of necessity, that the first motion of the true sap, at this period, is downwards. And as no alburnous tubes exist in the radicles of germinating seeds during the earlier periods of their growth, the sap in its descent, must either pass through the bark, or the medulla. But the medulla does not apparently contain any vessels calculated to carry the descending sap; while the cortical vessels are during this period much distended and full of moisture; and as the medulla certainly does not carry any fluid in stems or branches of more than one year old, it can scarcely be suspected that it, at any period, conveys the whole current of the descending sap.

As the leaves grow, and enter on their office, cortical vessels, in every respect apparently similar to those which descended from the cotyledons, are found to descend from the bases of the leaves; and there appears no reason with which I am acquainted, to suspect that both do not carry a similar fluid, and that the course of this fluid is, in the first instance, always towards the roots.

The ascending sap, on the contrary, rises wholly through the alburnum and central vessels; for the destruction of a portion of the bark, in a circle round the tree, does not immediately, in the slightest degree check the growth of its leaves and branches; but the alburnous vessels appear, from the experiments I have stated in a former paper†, and from those

* See above, No. IV.　　　　　　　　† See above, No. IV.

I shall now proceed to relate, to be also capable of an inverted action, when that becomes necessary to preserve the existence of the plant.

As soon as the leaves of the oak were nearly full grown in the last spring, I selected in several instances, two poles of the same age, and springing from the same roots in a coppice, which had been felled about six years preceding; and making two circular incisions at the distance of three inches from each other, through the bark of one of the poles on each stool, I destroyed the bark between the incisions, and thus cut off the communication between the leaves, and the lower parts of the stem and roots, through the bark. Much growth, as usual, took place above the space from which the bark had been taken off, and very little below it.

Examining the state of the experiment in the succeeding winter, I found it had not succeeded according to my hopes; for a portion of the alburnum, in almost every instance was lifeless, and almost dry, to a considerable distance below the space from which the bark had been removed. In one instance the whole of it was, however, perfectly alive; and in this I found the specific gravity of the wood above the decorticated space to be 1.114, and below it, 1.111; and the wood of the unmutilated pole, at the same distance from the ground, to be 1.112, each being weighed as soon as it was detached from the root.

Had the true sap in this instance wholly stagnated above the decorticated space, the specific gravity of the wood there ought to have been, according to the result of former experiments*, comparatively much greater; but I do not wish to draw any conclusion from a single experiment; and indeed, I see very considerable difficulty in obtaining any very satisfactory, or decisive facts from any experiments on plants, in this case, in which the same roots and stems collect and convey the sap during the spring and summer; and retain within themselves that which is, during the autumn and winter, reserved to form new organs of assimilation in the succeeding spring. In the tuberous-rooted plants, the roots and stems which collect and convey the sap in one season, and those in which it is deposited and reserved for the succeeding season, are perfectly distinct organs; and from one of these, the potato, I obtained more interesting and decisive results.

My principal object was to prove, that a fluid descends from the leaves and stem, to form the tuberous roots of this plant; and that this fluid will in part escape down the alburnous substance of the stem, when the continuity of the cortical vessel is interrupted. But I had also another object in view.

* See above, No. V.

Every gardener knows that early varieties of the potato never afford either blossom or seeds ; and I attributed this peculiarity to privation of nutriment, owing to the tubers being formed preternaturally early, and thence drawing off that portion of the true sap which, in the ordinary course of nature, is employed in the formation and nutrition of blossoms and seeds.

I therefore planted, in the last spring, some cuttings of a very early variety of the potato, which had never been known to blossom, in garden pots, having heaped the mould as high as I could above the level of the pot, and planted the portion of the root nearly at the top of it. When the plants had grown a few inches high, they were secured to strong sticks, which had been fixed erect in the pots for that purpose, and the mould was then washed away from the base of their stems by a strong current of water.

Each plant was now suspended in air, and had no communication with the soil in the pots, except by its fibrous roots, and as these are perfectly distinct organs from the runners which generate and feed the tuberous roots, I could readily prevent the formation of them. Efforts were soon made by every plant to generate runners, and tuberous roots ; but these were destroyed as soon as they became perceptible. An increased luxuriance of growth now became visible in every plant, numerous blossoms were emitted, and every blossom afforded fruit.

Conceiving, however, that a small portion only of the true sap would be expended in the production of blossoms and seeds, I was anxious to discover what use nature would make of that which remained ; and I therefore took effectual means to prevent the formation of tubers on any part of the plants, except the extremities of the lateral branches, those being the points most distant from the earth, in which the tubers are naturally deposited. After an ineffectual struggle of a few weeks, the plants became perfectly obedient to my wishes, and formed their tubers precisely in the places I had assigned them. Many of the joints of the plants during the experiment became enlarged and turgid ; and I am much inclined to believe, that if I had totally prevented the formation of regular tubers, these joints would have acquired an organization capable of retaining life, and of affording plants in the succeeding spring.

I had another variety of the potato, which grew with great luxuriance, and afforded many lateral branches ; and just at that period, when I had ascertained the first commencing formation of the tubers beneath the soil, I nearly detached many of these lateral branches from the principal stems, letting them remain suspended by such a portion only of alburnous and cortical fibres and vessels as were sufficient to preserve life. In this position I conceived that if their leaves and stems contained any unemployed true sap, it could not readily find its way to the tuberous roots, its passage being obstructed by the rupture of the vessels, and by gravitation; and I had soon the pleasure to see, that instead of returning down the principal stem into the ground, it remained and formed small tubers at the base of the leaves of the depending branches.

The preceding facts are, I think, sufficient to prove that the fluid, from which the tuberous root of the potatoe, when growing beneath the soil, derives its component matter, exists previously either in the stems or leaves; and that it subsequently descends into the earth : and as the cortical vessels during every period of the growth of the tuber are filled with the true sap of the plant, and as these vessels extend into the runners, which carry nutriment to the tuber, and in other instances evidently convey the true sap downwards, there appears little reason to doubt that through these vessels the tuber is naturally fed.

To ascertain, therefore, whether the tubers would continue to be fed when the passage of the true sap down the cortical vessels was interrupted, I removed a portion of bark of the width of five lines, and extending round the stems of several plants of the potato, close to the surface of the ground, soon after that period when the tubers were first formed. The plants continued some time in health, and during that period the tubers continued to grow, deriving their nutriment, as I conclude, from the leaves by an inverted action of the alburnous vessels. The tubers, however, by no means attained their natural size, partly owing to the declining health of the plant, and partly to the stagnation of a portion of the true sap above the decorticated space.

The fluid contained in the leaf has not, however, been proved, in any of the preceding experiments, to pass downwards through the decorticated space, and to be subsequently discharged into the bark below it ; but I have proved with amputated branches of different species of trees that the water which their leaves absorb, when immersed in that fluid, will be carried downwards by the alburnum, and conveyed into a portion of bark below the decorticated space ; and that the insulated bark will be preserved alive and moist during several days * ; and if the moisture absorbed by a leaf can be thus transferred, it appears extremely probable that the true sap will pass through the same channel. This power in alburnum to carry fluids in different directions probably answers very

---

* This experiment does not succeed till the leaf has attained its full growth and maturity and the alburnum of the annual shoot its perfect organisation.

important purposes in hot climates, where the dews are abundant, and the soil very dry; for the moisture the dews afford may thus be conveyed to the extremities of the roots : and Hales has proved that the leaves absorb most when placed in humid air; and that the sap descends, either through the bark or alburnum during the night.

If the inverted action of the alburnous vessels in the decorticated space be admitted, it is not difficult to explain the cause why some degree of growth takes place below such decorticated spaces on the stems of trees; and why a small portion of bark and wood is generated on the lower lip of the wound. A considerable portion of the descending true sap certainly stagnates above the wound, and of that which escapes into the bark below it, the greater part is probably carried towards, and into, the roots; where it preserves life, and occasions some degree of growth to take place. But a small portion of that fluid will be carried upwards by capillary attraction, between the bark and the alburnum, exclusive of the immediate action of the latter substance, and the whole of this will stagnate on the lower lip of the wound; where I conceive it generates the small portion of wood and bark, which Hales and Duhamel have described.

I should scarcely have thought an account of the preceding experiments worth sending to you, but that many of the conclusions I have drawn in former memoirs appear, at first view, almost incompatible with the facts stated by Hales and Duhamel, and that I had one fact to communicate relative to the effects produced by the stagnation of the descending sap of resinous trees, which appeared to lead to important consequences. I have in my possession a piece of a fir-tree, from which a portion of bark, extending round its whole stem, had been taken off several years before the tree was felled; and of this portion of wood, one grew above, and the other below, the decorticated space. Conceiving that the wood above the decorticated space ought to be much heavier than that below it, owing to the stagnation of the descending sap, I ascertained the specific gravity of both kinds, taking a wedge of each as nearly of the same form, as I could obtain, and I found the difference greatly more than I had anticipated, the specific gravity of the wood above the decorticated space being 0·590, and of that below only 0·491 : and having steeped pieces of each, which weighed a hundred grains, during twelve hours in water, I found the latter had absorbed 69 grains, and the former only 51.

The increased solidity of the wood above the decorticated space, in this instance, must, I conceive, have arisen from the stagnation of the true sap in its descent from the leaves; and therefore in felling firs, or other resinous trees, considerable advantages may be expected from stripping off a portion of their bark all round their trunks, close to the surface of the ground, about the end of May, or beginning of June, in the summer preceding the autumn in which they are to be felled. For much of the resinous matter contained in the roots of these is probably carried up by the ascending sap in the spring, and the return of a large portion of this matter to the roots would probably be prevented *; the timber, I have however very little doubt, would be much improved by standing a second year, and being then felled in the autumn; but some loss would be sustained owing to the slow growth of the trees in the second summer. The alburnum of other trees might probably be rendered more solid and durable by the same process; but the descending sap of these, being of a more fluid consistence than that of the resinous tribe, would escape through the decorticated space into the roots in much larger quantity.

It may be suspected that the increased solidity of the wood in the fir-tree I have described was confined to the part adjacent to the decorticated space; but it has been long known to gardeners, that taking off a portion of bark round the branch of a fruit-tree occasions the production of much blossom on every part of the branch in the succeeding season. The blossom in this case probably owes its existence to a stagnation of the true sap extending to the extremities of the branch above the decorticated space; and it may therefore be expected that the alburnous matter of the trunk and branches of a resinous tree will be rendered more solid by a similar operation.

IX.—ON THE FORMATION OF THE BARK OF TREES.

[*Read before the* ROYAL SOCIETY, *February* 19, 1807.]

An extraordinary diversity of opinion appears to have prevailed among naturalists, respecting the production and subsequent state of the bark of trees.

According to the theory of Malpighi, the cortical substance, which is

---

* The roots of trees, though of much less diameter than their trunks and branches, probably contain much more alburnum and bark, because they are wholly without heart wood, and extend to a much greater length than the branches; and thence it may be suspected that when fir-trees are felled, their roots contain at least as much resinous matter, in a fluid moveable state, as their trunks and branches; though not so much as is contained, in a concrete state, in the heart wood of those.

annually generated, derives its origin from the older bark; and the interior part of this new substance is annually transmuted into alburnum, or sap wood; whilst the exterior part, becoming dry and lifeless, forms the exterior covering, or cortex.

The opinions of Grew do not appear to differ much from those of Malpighi; but he conceives the interior bark to consist of two distinct substances, one of which becomes alburnum, whilst the other remains in the state of bark: he, however, supposes the insertments in the wood, the " utriculi" of Malpighi, and the " tissu cellulaire" of Duhamel, to have originally existed in the bark.

Hales on the contrary contends, that the bark derives its existence from the alburnum, and that it does not undergo any subsequent transformation.

The discoveries of Duhamel have thrown much light on the subject; but his experiments do not afford any conclusive result, and some of them may be adduced in support of either of the preceding hypotheses: and a modern writer (Mirbel*) has endeavoured to combine and reconcile, in some degree, the apparently discordant theories of Malpighi and Hales. He contends, with Hales, that the alburnum gives existence to the new layer of bark; but that this bark subsequently changes into alburnum, though not precisely in the manner described by Malpighi.

So much difference of opinion, amongst men so capable of observing, sufficiently evinces the difficulty of the subject they endeavoured to investigate: and in a course of experiments, which has occupied more than twenty years, I have scarcely felt myself prepared, till the present time, even to give an opinion respecting the manner, in which the cortical substance is generated in the ordinary course of its growth; or reproduced, when that, which previously existed, has been taken off.

Duhamel has shown, that the bark of some species of trees is readily reproduced, when the decorticated surface of the alburnum is secluded from the air; and I have repeated similar experiments on the apple, the sycamore, and other trees, with the same result; I have also often observed a similar reproduction of bark on the surface of the alburnum of the *Wych elm* (Ulmus montana) *in shady situations*, when no covering whatever was applied. A glareous fluid, as Duhamel has stated, exudes from the surface of the alburnum: this fluid appears to change into a pulpous unorganised mass which subsequently becomes organised and cellular; and the matter, which enters into the composition of this cellular substance, is evidently derived from the alburnum.

* Traité d'Anatomie et de Physiologie végétales.

These facts are therefore extremely favourable to the theory of Hales ; but other facts may be adduced which are scarcely consistent with that theory.

The internal surface of pieces of bark, when detached from contact with the alburnum, provided they remain united to the tree at their upper ends, much more readily generate a new bark, than the alburnum does under similar circumstances : a similar fluid exudes from the surfaces of both, and the same phenomena are observable in both cases. The cellular substance, however, which is thus generated, though it presents every external appearance of a perfect bark, is internally very imperfectly organised; and the vessels which contain the true sap in the bark, are still wanting ; and I have found, that these may be made, by appropriate management, to traverse the new cellular substance in almost any direction. When I cut off all communication above, and on one side, between the old bark and that substance, I observed that the vessels proceeded across it, from the old bark on the other side, taking always in a greater or less degree an inclination downwards ; and when the cellular substance remained united to the bark at its upper end only, the vessels descended nearly perpendicularly down it ; but they did not readily *ascend* into it, *when it was connected with the bark at its lower extremity only ;* the result of similar experiments, when made on different species of trees, was, however, subject to some variations.

Pieces of bark of the walnut-tree, which were two inches broad, and four long, having been detached from contact with the alburnum, except at their upper ends, and covered with a plaster composed of bees-wax and turpentine, in some instances, and with clay only in others, readily generated the cellular substance of a new bark; and between that and the old detached bark, very nearly as much alburnum was deposited as in other parts of the tree, where the bark retained its natural position ; which, I think, affords very decisive evidence of the descent of the sap through the bark. Similar pieces of bark, under the same mode of treatment, but united to the tree at their lower ends only, did not long remain alive, except at their lower extremities ; and there a very little alburnum only was generated. Other pieces of bark of the same dimensions, which were laterally united to the tree, continued alive almost to their extremities ; and a considerable portion of alburnum was generated, particularly near their lower edges; the sap appearing in its passage across the bark to have been given a considerable inclination downwards : probably owing to an arrangement in the organisation of the bark, that I have noticed in a former memoir*, which renders it better calculated to transmit the sap towards the roots than in any other direction.

I have in very few instances been able to make the walnut-tree reproduce its bark from the alburnum, though under the same management I rarely failed to succeed with the sycamore and apple-tree. Pieces of the bark of the apple-tree will also live, and generate a small portion of alburnum, though only attached to the tree at their lower extremities; probably owing to a small part of the true sap being carried upwards by capillary attraction, when the proper action of the cortical vessels is necessarily suspended.

The preceding experiments, and the authority of Duhamel, having perfectly satisfied me, that both the alburnum and bark of trees are capable of generating a new bark, or at least of transmitting a fluid capable of generating a cellular substance, to which the bark in its more perfectly organised state owes its existence, my attention was directed to discover the sources from which this fluid is derived. Both the bark and the alburnum of trees are composed principally of two substances; one of which consists of long tubes, and the other is cellular; and the cellular substance of the bark is in contact with the similar substance in the alburnum, and through these I have long suspected the true sap to pass from the vessels of the bark to those of the alburnum†. The intricate mixture of the cellular and vascular substances long baffled my endeavours to discover from which of them, in the preceding cases, the sap, and consequently the new bark, proceeded; but I was ultimately successful.

The cellular substance, both in the alburnum and bark of old pollard oaks, often exists in masses of near a line in width, and this organisation was peculiarly favourable to my purpose. I therefore repeated on the trunks of trees of this kind experiments similar to those above-mentioned which were made on the walnut-tree.

Apparently owing to the small quantity of sap, which the old pollard trees contained, their bark was very imperfectly reproduced; but I observed a fluid to ooze from the cellular substance, both of the bark and alburnum; and on the surface of these substances alone, in many instances, the new bark was reproduced in small detached pieces.

I have endeavoured to prove in former communications‡, that the true sap of trees acquires those properties which distinguish it from the fluid recently absorbed, by circulating through the leaf; and that it descends

* See above, No. IV.   † See above, No. V. p. 118.
‡ See above, Nos. II. V. and VII.

down the bark, where part of it is employed in generating the new substances annually added to the tree; and that the remainder, not thus
expended, passes into the alburnum, and there joins the ascending current
of sap.   The cellular substance, both of the bark and alburnum, has been
proved, in the preceding experiments, to be capable of affording the sap
a passage through it; and therefore it appears not very improbable, that
it executes an office similar to that of the anastomosing vessels of the
animal economy, when the cellular surfaces of the bark and alburnum are
in contact with each other; and, when detached, it may be inferred, that
the passing fluid will exude from both surfaces: because almost all the
vessels of trees appear to be capable of an inverted action in giving
motion to the fluids which they carry.

As the power of generating a new bark appeared in the preceding cases
to exist alike in the sap of the bark and of the alburnum, I was anxious
to discover how far the fluid, which ascends through the central vessels of
the succulent annual shoot, is endued with similar powers. Having therefore made two circular incisions through the bark, round the stems of
several annual shoots of the vine, as early in the summer as the alburnum
within them had acquired sufficient maturity to perform its office of carrying up the sap, I took off the bark between these incisions; and I abraded
the surface of the alburnum to prevent a reproduction of it. The alburnum
in the decorticated spaces soon became externally dry and lifeless; and
several incisions were then made longitudinally through it.   The incisions
commenced a little above, and extended below the decorticated spaces, so
that, if the sap of the central vessels generated a cellular substance (as I
concluded it would), that substance might come into contact and form a
union with the substance of the same kind emitted by the bark above and
below.

The experiment succeeded perfectly, and the cellular substances generated by the central vessels, and the bark, soon united, and a perfect
vascular bark was subsequently formed beneath the alburnum, and
appeared perfectly to execute the office of that which had been taken off;
the medulla appeared to be wholly inactive.

I have already observed, that the vessels, which were generated in the
cellular substance on the surface of the alburnum of the sycamore and the
apple-tree, traversed that substance in almost every direction; and the
same thing appears to occur beneath the old bark, when united to the
alburnum.   For having attentively examined, through every part of the
spring and summer, the formation of the internal bark, and alburnous
layer beneath it, round the basis of regenerated buds, which I had made

to spring from smooth spaces on the roots and stems of trees, I found every appearance perfectly consistent with the preceding observations. A single shoot only was suffered to spring from each root and stem, and from the base of this, in every instance, the cortical vessels dispersed themselves in different directions. Some descended perpendicularly downwards, whilst others diverged on each side, round the alburnum, with more or less inclination downwards, and met on the opposite side of it. The same pulpous and cellular substance appeared to cover the surfaces of the bark and alburnum, when in contact with each other, as when detached; and through this substance the ramifications of the vessels of the new bark extended themselves, appearing to receive their direction from the fluid sap which descended from the bark of the young shoots, and not to be, in any degree, influenced in their course by the direction taken by the cortical and alburnous vessels of the preceding year.

Whenever the vessels of the bark, which proceeded from different points, met each other, an interwoven texture was produced, and the alburnum beneath acquired a similar organisation : and the same thing occurs, and is productive of very important effects, in the ordinary course of the growth of trees. The bark of the principal stem, and of every lateral branch, contains very numerous vessels, which are charged with the descending true sap ; and at the juncture of the lateral branch with the stem, these vessels meet each other. A kind of pedestal of alburnum, the texture of which is much interwoven, is in consequence formed round the base of the lateral branch, which thus becomes firmly united to the tree. This pedestal, though apparently a part of the branch, derives a large portion of the matter annually added to it from the cortical vessels of the principal stem ; and thence, in the event of the death of the lateral branch, it always continues to live. But it not unfrequently happens, that a lateral branch forms a very acute angle with the principal stem, and, in this case, the bark between them becomes compressed and inactive; no pedestal is in consequence formed, and the attachment of such a branch to the stem becomes extremely feeble and insecure *.

---

* The advantages which may be obtained by pruning timber trees judiciously, appear to be very little known. I have endeavoured to ascertain the practicability of giving to trees such forms as will render their timber more advantageously convertible to naval or other purposes. The success of the experiments on small trees has been complete, and the results perfectly consistent, in every case, with the theory I have endeavoured to support in former memoirs ; and I am confident, that by appropriate management, the trunks and branches of growing trees may be moulded into the various forms best adapted to the use of the ship-builder, and that the growth of the trees may at the same time be rendered considerably more rapid, without any expense or temporary loss to the proprietor.

Instead of the reproduced buds of the preceding experiment, buds were inserted in the foregoing summer, or attached by grafting in the spring; and, when these succeeded, though they were in many instances taken from trees of different species, and even of different genera, no sensible difference existed in the vessels, which appeared to diverge into the bark of the stock, from these buds and from those reproduced in the preceding experiments.

It appears, therefore, probable, that a pulpous organisable mass first derives its matter either from the bark or the alburnum, and that this matter subsequently forms the new layer of bark; for, if the vessels had proceeded, as radicles *, from the inserted buds, or grafts, such vessels would have been, in some degree, different from the natural vessels of the bark of the stocks; and it does not appear probable, even without referring to the preceding facts, that vessels should be extended, in a few days, by parts successively added to their extremities, from the leaves to the extremities of the roots; which are, in many instances, more than 200 feet distant from each other. I am, therefore, inclined to believe, that, as the preceding facts seem to indicate, the matter which composes the new bark acquires an organisation calculated to transmit the true sap towards the roots, as that fluid progressively descends from the leaves in the spring; but whether the matter which enters into the composition of the new bark, be derived from the bark or alburnum, in the ordinary course of the growth of the tree, it will be extremely difficult to ascertain.

It is, however, no difficult task to prove, that the bark does not, in all cases, spring from the alburnum; for many cases may be adduced in which it is always generated previously to the existence of the alburnum beneath it: but none, I believe, in which the external surface of the alburnum exists previously to the bark in contact with it, except when the cortical substance has been taken off, as in the preceding experiments. In the radicle of germinating seeds, the cortical vessels elongate, and new portions of bark are successively added to their points, many days before any alburnous substance is generated in them; and in the succulent annual shoot the formation of the bark long precedes that of the alburnum. In the radicle the sap appears also evidently to descend†, through the cortical vessels‡, and in the succulent annual shoot it as

---

* Darwin's Phytologia.   † See above, Nos. V. and VII.

‡ I wish it to be understood, that I exclude in these remarks, and in those contained in my former memoirs, all trees of the palm kind, with the organisation of which I am almost wholly unacquainted.

evidently passes through the central vessels*, which surround the medulla. In both cases a cellular substance, similar to that which was generated in the preceding experiments, is first formed, and this cellular substance in the same manner subsequently becomes vascular; whence it appears, that the true sap, or blood of the plant, produces similar effects, and passes through similar stages of organisation, when it flows from different sources, and that the power of generating a new bark, properly speaking, belongs neither to the bark nor alburnum, but to a fluid which pervades alike the vessels of both.

I shall, therefore, not attempt to decide on the merits of the theory of Malpighi, or of Hales, respecting the reproduction of the interior bark; but I cannot by any means admit the hypothesis of Malpighi and other naturalists, relative to the transmutation of bark into alburnum; and I propose, in my next communication, to state my reasons for rejecting that hypothesis.

X.—ON THE INCONVERTIBILITY OF BARK INTO ALBURNUM.

[Read before the ROYAL SOCIETY, Feb. 4, 1808.]

In a letter which I had the honour to address to you in the end of the last year†, I endeavoured to prove that the matter which composes the bark of trees previously exists in the cells both of their bark and alburnum, in a fluid state, and that this fluid, even when extravasated, is capable of changing into a pulpous and cellular, and ultimately a vascular substance; the direction taken by the vessels being apparently dependent on the course which the descending fluid sap is made to take‡. The object of the present memoir is to prove, that the bark thus formed always remains in the state of bark, and that no part of it is ever transmuted into alburnum, as many very eminent naturalists have believed.

Having procured, by grafting, several trees of a variety of the apple and crab tree, the woods of which were distinguishable from each other by

---

* See above, No. V. Mirbel has called the tubes, which I call the central vessels, the " tissu tubulaire" of the medulla.

† See the preceding paper.

‡ I had observed this circumstance in many successive seasons; but I was not by any means prepared to believe that such an arrangement could take place in the coagulum afforded by an extravasated fluid; and I am indebted to Mr. Carlisle for having pointed out to me many circumstances in the motion and powers of the blood of animals, which induced me to give credit to the accuracy of my observations; and to that gentleman, and to Mr. Home, I have also subsequently to acknowledge many obligations.

their colours, I took off, early in the spring, portions of bark of equal length, from branches of equal size, and I transposed these pieces of bark, inclosing a part of the stem of the apple tree with a covering of the bark of the crab tree, which extended quite round it, and applying the bark of the apple tree to the stem of the crab tree in the same manner. Bandages were then applied to keep the transposed bark and the alburnum in contact with each other; and the air was excluded by a plaster composed of bees-wax and turpentine, and with a covering of tempered clay.

The interior surface of the bark of the crab tree presented numerous sinuosities, which corresponded with similar inequalities on the surface of the alburnum, occasioned by the former existence of many lateral branches. The interior surface of the bark of the apple tree, as well as the external surface of the alburnum, was, on the contrary, perfectly smooth and even. A vital union soon took place between the transposed pieces of bark, and the alburnum and bark of the trees to which they were applied; and in the autumn it appeared evident, that a layer of alburnum had been, in every instance, formed beneath the transposed pieces of bark, which were then taken off.

Examining the organisation of the alburnum, which had been generated beneath the transposed pieces of bark of the crab tree, and which had formed a perfect union with the alburnum of the apple tree, I could not discover any traces of the sinuosities I had noticed; nor was the uneven surface of the alburnum of the crab tree more changed by the smooth transposed bark of the apple tree. The newly generated alburnum, beneath the transposed bark, appeared perfectly similar to that of other parts of the stock, and the direction of the fibres and vessels did not in any degree correspond with those of the transposed bark *.

Repeating this experiment, I scraped off the external surface of the alburnum in several spaces, about three lines in diameter, and in these spaces no union took place between the transposed bark and the alburnum of the stock, nor was there any alburnum deposited in the abraded spaces; but the newly generated cortical and alburnous layers took a circular, and rather elliptical, course round those spaces, and appeared to have been generated by a descending fluid, which had divided into two

---

* Duhamel having taken off, and immediately replaced, similar pieces of the bark of young elms, subsequently found that the alburnum, which was generated beneath such pieces of bark, had not formed any union with the alburnum of the tree beneath it. But this great naturalist did not employ ligatures of sufficient power to bring the bark and alburnum into close contact, or the result would have been different.

currents when it came into contact with the spaces from which the surface had been scraped off, and to have united again immediately beneath them.

In each of these experiments, a new cortical and alburnous layer was evidently generated, and apparently by the same means that similar substances were generated beneath a plaister composed of bees-wax and turpentine, in former experiments *; and the only obvious difference in the result appears to be, that the transposed and newly-generated bark formed a vital union with each other : and it is sufficiently evident, that if bark of any kind was converted into alburnum, it must have been that newly generated.   For it can scarcely be supposed, that the bark of a crab-tree was transmuted into the alburnum of an apple-tree, or that the sinuosities of the bark of the crab-tree could have been obliterated, had such transmutation taken place.   There is not, however, anything in the preceding cases calculated to prove that the newly-generated bark was not converted into alburnum ; and the elaborate experiments of Duhamel sufficiently evince the difficulty of producing any decisive evidence in this case: nevertheless I trust that I shall be able to adduce such facts as, in the aggregate, will be found nearly conclusive.

Examining almost every day, during the spring and summer, the progressive formation of alburnum in the young shoots of an oak coppice which had been felled two years preceding, I was wholly unable to discover anything like the transmutation of bark into alburnum.   The commencement of the alburnous layers in the oak (*Quercus robur*) is distinguished by a circular row of very. large tubes.   These tubes are of course generated in the spring ; and during their formation, I found the substance through which they passed to be soft and apparently gelatinous, and much less tenacious and consistent than the substance of the bark itself ; and, therefore, if the matter which gave existence to the alburnum previously composed the bark, it must have been, during its change of character, nearly in a state of solution ; but it is the transmutation of one organised substance into the other, and not the identity only of the matter of both, for which the disciples of Malpighi contend ; and if the fibres and vessels of the bark really became those of the alburnum, a very great degree of similarity ought to be found in the organisation of those substances.   No such similarity, however, exists ; and not anything at all corresponding with the circular row of large tubes in the alburnum of the oak is discoverable in the bark of that tree.   These tubes are also generated within the interior surface of the bark, which is well defined ; and during their

* See the preceding Paper.

L

formation the vessels of the bark are distinctly visible, as different organs; and had the one been transmuted into the other, their progressive changes could not, I think, possibly have escaped my observation : nor does the organisation of the bark in other instances in any degree indicate the character of the wood that is generated beneath it : the bark of the wych elm (*Ulmus montana*) is extremely tough and fibrous ; and it is often taken from branches of six or eight years old, to be used instead of cords; that of the ash (*Fraxinus excelsior*), on the contrary, when taken from branches of the same age, breaks almost as readily in any one direction as in another, and scarcely presents a fibrous texture ; yet the alburnum of these trees is not very dissimilar, and the one is often substituted for the other in the construction of agricultural instruments.

Mirbel has endeavoured to account for the dissimilar organisation of the bark, and of the wood into which he conceives it to be converted, by supposing that the cellular substance of the bark is always springing from the alburnum, whilst the tree is growing, and that it carries with it part of the tubular substance (*tissu tubulaire*) of the liber, or interior bark. These parts of the interior bark, which are thus removed from contact with the alburnum, he conceives to constitute the external bark or cortex, whilst the interior part of the liber progressively changes into alburnum.

But if this theory (which I believe I have accurately stated, though I am not quite certain that I fully comprehend its author*) were well founded, the texture of the alburnum must surely be much more intricate and interwoven than it is, and its tubes would lie less accurately parallel with each other than they do : and were the fibrous substance of the bark progressively changing into alburnum, the bark must of necessity be firmly attached to the alburnum during the spring and summer by the continuity, and indeed identity, of the vessels and fibres of both these substances. This, however, is not in any degree the case, and the bark is in those seasons very easily separated from the alburnum; to which it appears to be attached by a substance that is apparently rather gelatinous than fibrous or vascular : and the obvious fact, that the adhesion of the cortical vessels and fibres to each other is much more strong than the adhesion of the bark to the alburnum, affords another circumstance almost as inconsistent with the theory of Malpighi, as with that of Mirbel.

Many of the experiments of Duhamel are, however, apparently favourable to the theory of Malpighi, respecting the conversion of bark into alburnum ; and Mirbel has cited two, which he appears to think conclu-

* Traité d'Anatomie et de Physiologie Végétale, Chap. iii. Article 5.

sive*. In the first of these, Duhamel shows that pieces of silver wire, inserted in the bark of trees, were subsequently found in their alburnum; but Duhamel himself has shown, with his usual acuteness and candour, that the evidence afforded by this experiment is extremely defective; and he declares himself to be uncertain that the pieces of wire did not, at their first insertion, pass between the bark and the alburnum; in which case they would necessarily have been covered by every successive layer of alburnum, without any transmutation of bark into that substance†.

In the second experiment cited by Mirbel, Duhamel has shown that when a bud of the peach-tree, with a piece of bark attached to it, is inserted in a plum stock, a layer of wood perfectly similar to that of the peach-tree will be found, in the succeeding winter, beneath the inserted bark. The statement of Duhamel is perfectly correct; but the experiment does not by any means prove the conversion of bark into wood; for if it be difficult to conceive (as he remarks) that an inserted piece of bark can deposit a layer of alburnum, it is at least as difficult to conceive how the same piece of bark can be converted into a layer of alburnum of more than twice its own thickness (and the thickness of the alburnum deposited frequently exceeds that of the bark in this proportion), without any perceptible diminution of its own proper substance. The probable operation of the inserted bud, which is a well-organised plant, at the period when it becomes capable of being transposed with success, appears also, in this case, to have been overlooked; for I found that when I destroyed the buds in the succeeding winter, and left the bark which belonged to them uninjured, this bark no longer possessed any power to generate alburnum. It nevertheless continued to live, though perfectly inactive, till it became covered by the successive alburnous layers of the stock; and it was found many years afterwards inclosed in the wood. It was, however, still bark, though dry and lifeless, and did not appear to have made any progress towards conversion into wood.

In the course of very numerous experiments which were made to ascertain the manner in which vessels are formed in the reproduced bark‡, many circumstances came under my observation which I could adduce in support of my opinion, that bark is never transmuted into alburnum; but I do not think it necessary to trouble you with an account of them; for though much deference is certainly due to the opinions of those naturalists who have adopted the opposite theory, and to the doubts of Duhamel, I am not acquainted with a single experiment which warrants

* Chap. iii. Article 5.   † Physique des Arbres, Liv. IV. chap. iii.
‡ See the last Paper.

the conclusions they have drawn; and I think that were bark really transmuted into alburnum, its progressive changes could only have escaped the eyes of prejudiced or inattentive observers. In the course of the ensuing spring, I hope to address to you some observations respecting the manner in which the alburnum is generated.

XI.—ON THE ORIGIN AND OFFICE OF THE ALBURNUM OF TREES.

[*Read before the* ROYAL SOCIETY, *June* 30, 1808.]

IN my last communication I endeavoured to prove that the bark of trees is not subsequently transmuted into alburnum; and if the statements that I have there given be correct, they are, I conceive, decisive on the point for which I contended: and if the bark be not converted into alburnum, the experiments of Duhamel and subsequent naturalists, and those of which I have given an account in former memoirs, afford sufficient evidence that the bark deposits the alburnous matter. If the succulent shoot of a horse chestnut, or other tree, be examined at successive periods in the spring, it will be seen that the alburnum is deposited, and its tubes arranged, in ridges beneath the cortical vessels; and the number of these ridges, at the base of each leaf, will be found to correspond accurately with the number of apertures through which the vessels pass from the leaf-stalks into the interior bark, the alburnous matter being apparently deposited (as I have endeavoured to prove in former memoirs) by a fluid which descends from the leaves, and subsequently secretes through the bark*. I shall therefore venture to conclude that it is thus deposited, and shall proceed to inquire into the origin and office of the alburnous tubes.

The position and direction of these tubes have induced almost all naturalists to consider them as the passages through which the sap ascends; and at their first formation, when the substance which surrounds them is still soft and succulent, they are always filled with the fluid, which has apparently secreted from the bark. They appear to be formed in the soft cellular mass, which becomes the future alburnum, as receptacles of this fluid, to which they may either afford a passage upwards, or simply retain it as reservoirs, till absorbed, and carried off, by the surrounding cellular substance. The former supposition is, at first view, the most probable; but the latter is much more consistent with the circumstances that I shall proceed to state.

* See above, No. II.

Many different hypotheses have been offered by naturalists to account for the force with which the sap ascends in the spring; of these hypotheses two only appear in any degree adequate to the effects produced. Saussure, jun. supposes that the tubes contract as soon as they have received the sap in the root, and that this contraction, commencing in the root, proceeds upwards, impelling the sap before it : and I have suggested that the expansion and contraction of the compressed cellular, or laminated substance (the *tissu cellulaire* of Duhamel and Mirbel) which expands and contracts with change of temperature* after the tree has ceased to live, might produce similar effects by occasioning nearly a similar motion and compression of the tubes, the coats of which are, I believe, universally admitted not to be membranous. But both these hypotheses are inconsistent with the facts that I have now the pleasure to communicate to you.

Selecting parts of the stems of young trees from which annual branches had sprung in the preceding year, I ascertained, by injecting coloured infusions into the stems through the annual shoots, that the tubes which descended from the latter, were, at their bases, confined to that side of the stem from which they sprang, and to the external annual layer of wood. Deep incisions were then made into the stems of other trees immediately beneath the bases of similar annual shoots, by which I am quite confident that all communication through the alburnous tubes, with the stem, was wholly cut off; yet the sap passed into the annual shoots in the succeeding spring, all of which lived, and some grew with considerable vigour. I, at the same time, selected many lateral branches, about three lines in diameter, in a nursery of apple trees, which I could easily secure to the stems of the adjoining trees to prevent their being broken. I then made an incision, more than two lines deep in each, on one side, and at the distance of six or seven lines another incision, equally deep, on the opposite side; and as I am quite certain, from the texture of these branches, that the alburnous tubes passed straight through them, I am equally certain that every alburnous tube was at least once intersected. Yet the sap passed into these branches, and their buds unfolded in the succeeding spring, the incisions having been made in the winter. But I have repeated the same experiment after the leaves have been full-grown in the summer, and still the branches have continued to live.

All naturalists have agreed in stating that trees perspire most in the summer, when their leaves have attained their full growth, and of

* See above, p. 92.

course that much sap must ascend at this period; yet at this period the tubes of the alburnum appear dry, and to contain air only; which induced Grew to suppose that the sap rose in the state of vapour; a supposition by no means admissible. Yet it is, I conceive, evident that the sap cannot rise, as a liquid, through dry tubes, nor in any state through intersected tubes; and therefore it appears probable that it does not rise at all through the tubes of the alburnum, and that those tubes are intended to execute a different office.

If the sap do not rise through the tubes of the alburnum, it must rise through the cellular substance; yet the passage of any fluid through this has been denied by almost every naturalist, probably because coloured infusions have not been observed to penetrate it, and because many naturalists have considered it as mere compressed medulla. Mirbel, however, contends that the fluid which generates the new bark exudes from it; and although a fluid capable of producing the same effects exudes from the bark when detached from the alburnum, I am much disposed to coincide with him in opinion, having observed a new bark to be generated on the surface of the cellular substance of pollard oaks, in detached spaces*. And if the sap in sufficient quantity to generate a new bark can pass through the cellular substance of an oak, it appears possible at least that the whole of the sap may ascend through it. Coloured infusions do not, I think, in any degree, pass through the bark of trees, yet it is evident that the sap passes readily through it; and therefore, should it be proved that such infusions do not penetrate the cellular substance of the alburnum, the evidence which this circumstance would afford would be very defective.

Amongst other experiments that I made to ascertain whether the cellular substance of the alburnum would imbibe coloured infusions, I took off branches of two years old with the annual shoots and leaves attached to them, in the summer, from trees of different species; and I effectually closed the alburnous tubes with a composition formed of calcined oyster shells and cheese† ; and this was covered with a mixture of bees-wax and turpentine, so as to effectually exclude all moisture. A part of the bark was taken off each branch, in a circle round it, a few lines distant from its lower end, where the tubes had been closed; and each branch was then placed in a decoction of logwood, in a vessel

---

* See above, p. 139.

† I have found this composition, and this only, to be capable of instantaneously stopping the effusion of sap from the vine, or other tree, in the bleeding season.

deep enough to cover the decorticated spaces. At the end of twenty hours, or somewhat longer periods, these branches were examined, and the coloured infusion was found to have insinuated itself between the alburnous tubes, in many instances apparently through the cellular substance. This was most obvious in the walnut-tree, the young wood of which is very white. The principal object I had in view in making this experiment, was to detect the passages through which I conceived the sap to pass from the bark into the alburnum*.

From the preceding circumstances, I am disposed to infer that the sap secretes through the cellular substance of the alburnum; and through this I conceived that it must ascend when the tubes were intersected in the preceding experiments, and in those seasons of the year when the alburnous tubes are empty, though the sap must be rising with great rapidity: and I shall endeavour to show that the presence of the sap in the alburnous tubes, during that part of the year in which trees, when wounded, bleed abundantly, does not afford any decisive evidence of the ascent of the sap through those tubes.

In the last spring, when the buds of the sycamore first began to prepare for unfolding, I found that the sap abounded in the points at the annual branches; and at the same time it flowed abundantly from incisions made into the alburnum near the root. But when similar incisions were made at the distance of eight or ten feet from the ground, not the least moisture flowed; and the tubes of the alburnum appeared to contain air only. I also observed that the sap flowed as abundantly from the upper as from the under side of the lower incisions, if not more abundantly, and so it continued to flow to the end of the bleeding season.

The sap must therefore have been, by some means, thrown into the tubes above the incisions, for the quantity discharged from them exceeded more than a hundred times that which the tubes could have contained at the time the incisions were made, even had every tube been filled to the extremity of the most distant branch. And, as it has been shown that the sap can pass up when all the alburnous tubes are intersected, there appears, I think, sufficient evidence that it must in this case have been raised by some other agent than those tubes.

Through the cellular substance I therefore venture to conclude that the sap ascends; and it is not, I think, difficult to conceive that this substance may give the impulse with which the sap is known to ascend in the spring. I have shown that the bark more readily transmits the

* See above, p. 139.

descending sap towards the roots than towards the points of the branches* ; and if the cellular substance of the alburnum expand and contract, and be so organised as to permit the sap to escape more easily upwards from one cell to another than in any other direction, it will be readily impelled to the extremities of the branches : and I have shown that the statement, so often repeated in the writings of naturalists, of a power in the alburnum to transmit the sap with equal facility in opposite directions, and as well through inverted cuttings as others, is totally erroneous†.

If the sap be raised in the manner I have suggested, much of it will probably accumulate in the alburnum in the spring ; because the powers of vegetable life are, at that period, more active than at any other season, and the leaves are not then prepared to throw off any part of it by trans-piration. And the cellular substance, being then filled, may discharge a part of its contents into the alburnous tubes, which again become reser-voirs, and are filled to a greater or less height, in proportion to the vigour of the tree, and the state of the soil and season : and if the tubes which are thus filled be divided, the sap will flow out of them, and the tree will be said to bleed. But as soon as the leaves are unfolded, and begin to execute their office, the sap will be drawn from its reservoirs, and the tree will cease to bleed, if wounded.

The alburnous tubes appear to answer another purpose in trees, and to be analogous, in some degree, in their effects, to the cavities in the bones of animals ; by which any degree of strength that is necessary, is given with less expenditure of materials, or the incumbrance of unnecessary weight ; and the wood of many different species of trees is thus made, at the same time, very light, and very strong, the rigid vegetable fibres being placed at greater distances from each other by the intervention of alburnous tubes, and consequently acting with greater mechanical advan-tage, than they would if placed immediately in contact with each other.

I have shown in a former communication, that the specific gravity of the sap increases during its ascent in the spring, and that saccharine matter is generated, which did not previously exist in the alburnum, nor in the sap, as it rose from the root : and I conceive it not to be impro-bable, that the air contained in the alburnous tubes may be instrumental in the generation of this saccharine matter. For I discovered in the last autumn, that much air is absorbed, or at least disappears, during the process of grinding apples for the purpose of making cider, and that during this absorption of air, the juice of acid apples becomes very sweet.

See above, No. IV.   † Ibid.

and acquires many degrees of increased specific gravity; and a similar absorption of air, with corresponding effects, is well known to take place in the process of malting.

I shall conclude with observing, that in retracting the opinion I formerly entertained respecting the ascent of the sap in the alburnous tubes, I do not mean to retract any opinion that I have given in former communications respecting the subsequent motion of the sap through the central vessels, the leaves, and bark; or the subsequent junction of the descending with the ascending current in the alburnum : every experiment that I have made has, on the contrary, tended to confirm my former conclusions.

XII.—ON THE ORIGIN AND FORMATION OF ROOTS.

[*Read before the* ROYAL SOCIETY, *February*, 23, 1809.]

IN a former communication I have given an account of some experiments, which induced me to conclude that the buds of trees invariably spring from their alburnum, to which they are always connected by central vessels of greater or less length ; and in the course of much subsequent experience, I have not found any reason to change the opinion that I have there given *. The object of the present communication is to show, that the roots of trees are always generated by the vessels which pass from the cotyledons of the seed, and from the leaves, through the leaf-stalks and the bark, and that they never, under any circumstances, spring immediately from the alburnum.

The organ which naturalists have called the radicle in the seed, is generally supposed to be analogous to the root of the plant, and to become a perfect root during germination; and I do not know that this opinion has ever been controverted, though I believe that, when closely investigated, it will prove to be founded in error.

A root, in all cases with which I am acquainted, elongates only by new parts which are successively added to its apex or point, and never, like the stem or branch, by the extension of parts previously organised ; and I have endeavoured to show, in a former memoir, that owing to this difference in the mode of the growth of the root and lengthened plumule of germinating seeds, the one must ever be obedient to gravitation, and point towards the centre of the earth, whilst the other must take the opposite direction†. But the radicle of germinating seeds elongates by

---

* See above, No. VI. † Ibid. No. VII.

the extension of parts previously organised, and in a great number of cases, which must be familiar to every person's observation, raises the cotyledons out of the mould in which the seed is placed to vegetate. The mode of growth of the radicle is therefore similar to that of the substance which occupies the spaces between the buds near the point of the succulent annual shoot, and totally different from that of the proper root of the plant, which I conceive to come first into existence during the germination of the seed, and to spring from the point of what is called the radicle. At this period, neither the radicle nor cotyledons contain any alburnum, and therefore the first root cannot originate from that substance; but the cortical vessels are then filled up with sap, and apparently in full action, and through these the sap appears to descend which gives existence to the true root.

When first emitted, the root consists only of a cellular substance, similar to that of the bark of other parts of the future tree; and within this the cortical vessels are subsequently generated in a circle, inclosing within it a small portion of the cellular substance, which forms the pith or medulla of the root. The cortical vessels soon enter on their office of generating alburnous matter; and a transverse section of the root then shows the alburnum arranged in the form of wedges round the medulla, as it is subsequently deposited on the central vessels of the succulent annual shoot, and on the surface of the alburnum of the stems and branches of older trees *.

If a leaf-stalk be deeply wounded, a cellular substance, similar to that of the bark and young root, is protruded from the upper lip of the wound, but never from the lower; and the leaf-stalks of many plants possess the power of emitting roots, which power cannot have resided in alburnum, for the leaf-stalk does not contain any; but vessels, similar to those of the bark and radicle, abound in it, and apparently convey the returning sap; and from these vessels, or perhaps more properly from the fluid they convey, the roots emitted by the leaf-stalk derive their existence †.

If a portion of the bark of a vine, or other tree, which readily emits roots, be taken off in a circle extending round its stem, so as to intercept entirely the passage of any fluid through the bark, and any body which contains much moisture be applied, numerous roots will soon be emitted into it immediately above the decorticated space, but never immediately beneath it: and when the alburnum in the decorticated spaces has become lifeless to a considerable depth, buds are usually protruded beneath, but never immediately above it, apparently owing to the

---

* See above, No. II. Plate 4.   † Ibid. No. II.

obstruction of the ascending sap. The roots which are emitted in the preceding case do not appear in any degree to differ from those which descend from the radicles of generating seeds, and both apparently derive their matter from the fluid which descends through the cortical vessels.

There are several varieties of the apple-tree, the trunks and branches of which are almost covered with rough excrescences, formed by congeries of points which would have become roots under favourable circumstances; and such varieties are always very readily propagated by cuttings. Having thus obtained a considerable number of plants of one of these varieties, the excrescences began to form upon their stems when two years old, and mould being then applied to them in the spring, numerous roots were emitted into it early in the summer. The mould was at the same time raised around, and applied to, the stems of other trees of the same age and variety, and in every respect similar, except that the tops of the latter were cut off a short distance above the lowest excrescence, so that there was no buds or leaves from which sap could descend to generate or feed new roots; and under these circumstances no roots, but numerous buds were emitted, and these buds all sprang from the spaces and points, which under different circumstances had afforded roots. The tops of the trees last mentioned, having been divided into pieces of ten inches long, were planted as cuttings, and roots were by these emitted from the lowest excrescences beneath the soil, and buds from the uppermost of those above it.

I had anticipated the result of each of the preceding experiments; not that I supposed, or now suppose, that roots can be changed into buds, or buds into roots; but I had before proved that the organisation of the alburnum is better calculated to carry the sap it contains, from the root upward, than in any other direction, and I concluded that the sap when arrived at the top of the cutting through the alburnum would be there employed, as I had observed in many similar cases, in generating buds, and that these buds would be protruded where the bark was young and thin, and consequently afforded little resistance *. I had also proved the bark to be better calculated to carry the sap towards the roots than in the opposite direction, and I thence inferred that as soon as any buds, emitted by the cuttings, afforded leaves, the sap would be conveyed from these to the lower extremity of the cuttings by the cortical vessels, and be there employed in the formation of roots †.

Both the alburnum and bark of trees evidently contain their true sap;

* See above, No. VI.

but whether the fluid which ascends in such cases as the preceding through the alburnum, to generate buds, be essentially different from that which descends down the bark to generate roots, it is perhaps impossible to decide. As nature, however, appears in the vegetable world to operate by the simplest means; and as the vegetable sap, like the animal blood, is probably filled with particles which are endued with life; were I to offer a conjecture, I am much more disposed to believe that the same fluid, even by merely acquiring different motions, may generate different organs, than that two distinct fluids are employed to form the root, and the bud and leaf.

When alburnum is formed in the root, that organ possesses, in common with the stem and branches, the power of producing buds, and of emitting fibrous roots; and when it is detached from the tree, the buds always spring near its upper end, and the roots near the opposite extremity, as in the cuttings above mentioned. The alburnum of the root is also similar to that of other parts of the tree, except that it is more porous, probably owing to the presence of abundant moisture during the period in which it is deposited*. And possibly the same cause may retain the wood of the root permanently in the state of alburnum; for I have shown, in a former memoir, that if the mould be taken away, so that the parts of the larger roots, which adjoin the trunk, be exposed to the air, such parts are subsequently found to contain much heart wood†.

I would wish the preceding observations to be considered as extending to trees only, and exclusive of the palm tribe: but I believe they are nevertheless generally applicable to perennial herbaceous plants, and that the buds and fibrous roots of these originate from substances which correspond with the alburnum and bark of trees. It is obvious, that the roots which bulbs emit in the spring, are generated by the sap which descends from the bulb, when that retains its natural position; and such tuberous-rooted plants as the potato offer rather a seeming than a real obstacle to the hypothesis I am endeavouring to establish. The buds of these are generally formed beneath the soil; but I have shown, in a former memoir, that the buds on every part of the stem may be made to generate tubers, which are similar to those usually formed beneath the soil; and I have subsequently seen, in many instances, such emitted by a re-produced bud, without the calix of a blossom, which had failed to produce fruit; but I have never, under any circumstances, been able to obtain tubers from the fibrous roots of the plant.

* See above, No. VI. for 1805.   † Ibid.

The tuber therefore appears to differ little from a branch, which has dilated instead of extending itself, except that it becomes capable of retaining life during a longer period ; and when I have laboured through a whole summer to counteract the natural habits of the plant, a profusion of blossoms has in many instances sprung from the buds of a tuber.

The runners also, which, according to the natural habit of the plant, give existence to the tubers beneath the soil, are very similar in organisation to the stem of the plant, and readily emit leaves and become converted into perfect stems in a few days, if the current of ascending sap be diverted into them ; and the mode in which the tuber is formed above, and beneath the soil, is precisely the same. And when the sap, which has been deposited at rest during the autumn and winter, is again called into action to feed the buds, which elongate into parts of the stems of the future plants in the spring, fibrous roots are emitted from the basis of these stems, whilst buds are generated at the opposite extremities, as in the cases I have mentioned respecting trees.

Many naturalists* have supposed the fibrous roots of all plants to be of annual duration only; and those of bulbous and tuberous-rooted plants certainly are so : as in these nature has provided a distinct reservoir for the sap which is to form the first leaves and fibrous roots of the succeeding season; but the organisation of trees is very different, and the alburnum and bark of the roots and stems of these are the reservoirs of their sap during the winter†. When, however, the fibrous roots of trees are crowded together in a garden-pot, they are often found lifeless in the succeeding spring; but I have not observed the same mortality to occur, in any degree, in the roots of trees when growing, under favourable circumstances, in their natural situation.

XIII.—ON THE CAUSES WHICH INFLUENCE THE DIRECTION OF THE GROWTH OF ROOTS.

[*Read before the* ROYAL SOCIETY, *March* 7, 1811.]

I HAVE shown, in a former communication, the effects of centrifugal force upon germinating seeds ; from which I have inferred that the radicles are made to descend towards the earth, and the germs, or elongated plumules, to take the opposite direction, by the influence of gravitation ; and I believe the facts I have stated to be sufficient to support the inferences

* M. Mirbel's Traité d'Anatomie, &c. &c. Dr. Smith's Introduction to Botany.
† See above, No. V.

I have drawn*.   But the fibrous roots of plants, being much less succulent, though not uninfluenced in the directions they take by gravitation, are, to a great extent, obedient to other laws, and are generally found to extend themselves most rapidly, and to the greatest length, in whatever direction the soil is most favourable : whence many naturalists have been disposed to believe that these are guided by some degrees of feeling and perception, analogous to those of animal life.

I shall proceed to state some of the facts upon which this hypothesis has been founded, and others which have occurred in the course of my own experience, and which are favourable to it ; after which I shall endeavour to trace the effects observed to the operation of different causes.

When a tree which requires much moisture has sprung up, or been planted, in a dry soil in the vicinity of water, it has been observed that much the largest portion of its roots has been directed towards the water ; and that when a tree of a different species, and which requires a dry soil, has been placed in a similar situation, it has appeared, in the direction given to its roots, to have avoided the water and moist soil.

A tree growing upon a wall, at some distance from the ground, and consequently ill supplied with food and water, has also been observed to adapt its habits to its situation, and to make very singular and well-directed efforts to reach the soil beneath, by means of its roots†.   During the period in which it is making such efforts, little addition is made to its branches, and almost the whole powers of the plant appear to be directed to the growth of one or more of its principal roots.   To these much is in consequence annually added, and they proceed perpendicularly towards the earth, unless made to deviate by some opposing body : and as soon as the roots have attached themselves to the soil, the branches grow with vigour and rapidity, and the plant assumes the ordinary habits of its species.

Duhamel caused two trenches to be made so as to intersect each other at right angles, and a tree to be planted at the point of intersection ; and taking up this tree some years afterwards, he found that the roots had almost wholly confined themselves to the trenches, in which the soil of the former surface must have been buried.

A trench which was twenty feet long, six wide, and about two deep, was prepared in my garden, in the bottom of which trench was placed a layer, about six inches deep, of very rich mould, incorporated with much fresh vegetable matter.   This was covered, eighteen inches deep, with

* See above, No. VII.   † Smith's Introduction to Botany.

light and poor loam, and upon the bed thus formed, seeds of the common carrot (*Daucus carota*) and parsnep (*Pastinaca sativa*) were sowed. The plants grew feebly till near the end of the summer, when they assumed a very luxuriant growth, grew rapidly till late in the autumn, and till their leaves were injured by frost. The roots were then examined, and were found of an extraordinary length, and in form almost perfectly cylindrical, having scarcely emitted any lateral fibrous roots into the poor soil, whilst the rich mould beneath was filled with them.

In another experiment of the same season, the preceding process was reversed, the rich soil being placed upon the surface, and the poor beneath. The plants here grew very luxuriantly, and acquired a considerable size early in the summer; and when the roots were taken up in the autumn, they were found to have assumed very different forms. The greater part had divided into two or more unequal ramifications, very near the surface of the ground; and those which were not thus divided tapered rapidly to a point at the surface of the poor soil, into which few of their fibrous roots had entered.

In other experiments seeds of almost all the common esculent plants of a garden were so placed that the young plants had an opportunity of selecting either rich or poor soil; which was disposed, in almost every possible way, within their reach: and I always found abundant fibrous roots in the rich soil, and comparatively few in the poor.

The following experiment afforded the most remarkable result, and one of the least favourable to the hypothesis which I have advanced in a former paper*, and to the conclusion which I shall now endeavour to support; and therefore I think it necessary to describe it very minutely. Some seeds of the common bean (*Vicia faba*), the plant with which many former experiments were made, were placed upon the surface of the mould in garden pots, in rows which were about four inches distant from each other. A grate, formed of slender bars of wood, was then adapted to the surface of each pot, so as to prevent both the mould and the seeds falling out, in whatever position the pots might be placed; and the bars were so disposed as not at all to interfere with the radicles of the seeds, when protruding. The pots were then directly inverted, and the seeds were consequently placed beneath the mould; but each seed was so far depressed into the mould as to be about half covered: by which means each radicle, when first emitted, was in contact with the mould above, and the air below. Water was then introduced through

* See above, No. VII.

the bottom of the inverted pot, in sufficient quantity to keep the mould moderately moist ; and the pots being suspended from the roof of a forcing-house, the seeds soon vegetated.

In former experiments*, wherever the seeds were placed to vegetate at rest, the radicles descended perpendicularly downwards, in whatever direction they were first protruded ; but under the preceding circumstances they extended horizontally along the surface of the mould, and in contact with it ; and in a few days emitted many fibrous roots upwards into it : just as they would have done, if guided by the instinctive faculties and passions of animal life ; and as I concluded before I made the experiment that they would do, under the guidance of much more simple laws, whose mode of operating I shall endeavour to explain.

Whatever be the machinery by which the sap of trees is raised to the extremities of their branches, it is obvious that this machinery is first put into action by the stems and branches, and not by the roots : for the graft or bud, whenever it has become fully united to the stock, wholly regulates the season and temperature, in which the sap is to be put in motion, in perfect independence of the habits of the stock ; whether those be late or early. If all the branches of a tree, exclusive of one, be much shaded by contiguous trees†, or other objects, the branch which is exposed to the light attracts to itself a large portion of the ascending sap, which it employs in the formation of leaves and vigorous annual shoots, whilst the shaded branches become languid and unhealthy. The motion of the ascending current of sap appears therefore to be regulated by the ability to employ it in the trunk and branches of the tree ; and this current passes up through the alburnum, from which substance the buds and leaves spring. Bat the sap which gives existence to, and feeds the root, descends through the bark‡ : and if the operation of light give ability to the exposed branch to attract and employ the ascending or alburnous current of sap, it appears not improbable that the operation of proper food and moisture in the soil, upon the bark of the root, may give ability to that organ to attract and employ the descending, or cortical current of sap ; and if this be the case, an easy explanation of all the preceding phenomena immediately presents itself.

A tree growing upon a wall, and unconnected with the earth, will almost of necessity grow slowly, and as it must be scantily supplied with moisture during the summer, it will rarely produce any other leaves than those which the buds contained, which were formed in the preceding year. Some of the roots of a tree, thus circumstanced, will be less well supplied

---

* See above, No. VII.   † Ibid. No. VI. and XII.   ‡ See the last Paper.

with moisture than others, and these will be first affected by drought : their points will in consequence become rigid and inexpansible, and they will thence generally cease to elongate at an early period of the summer. The descending current of sap will be then employed in promoting the growth and elongation of those roots only, which are more favourably situated, and which, comparatively with other parts of the tree, will grow rapidly. Gravitation will direct these roots perpendicularly downwards, and the tree will appear to have adopted the wisest and best plan of connecting itself with the ground : and it will really have employed the readiest means of doing so, as effectively as it could have done, if it had possessed all the feelings and instinctive passions and powers of animal life. The subsequent vigorous growth of such a tree is the natural consequence of an improved and more extensive pasture.

When the seeds of the carrot and parsnip, in the experiments I have stated, were placed in a poor superficial soil, but which permitted the roots of the plants to pass readily through it, these were conducted downwards by gravitation ; whilst the plants grew feebly, because they received but little nutriment. The roots were in a situation analogous to that of the stems of trees in a crowded forest ; and when the leading fibres of the roots came into contact with the rich mould, they acquired a situation correspondent to that of the leading branches of such trees, which are alone exposed to the light. The form of the roots of the plants was consequently long, slender, and cylindrical, like the stems of such trees. The roots of the one required the actual contact of proper soil and nutriment ; and the branches of the other required the actual contact of light to promote their growth.

When, on the contrary, the seeds of the preceding species of plants were placed in a rich superficial soil, their situation was analogous to that of a tree fully exposed, on every side, to the light, whose branches would be extended, in every direction, immediately above the surface of the ground : and as the fibrous roots of the plants came into contact with the subsoil, which was not well calculated to promote their growth, their situation became analogous to that of shaded branches ; and they consequently ceased to extend downwards. The fibrous roots of a tree, under similar circumstances, would have extended along the lower surface of the favourable soil ; but after these roots had much increased in bulk, they would be found partly compressed into the subsoil, however poor and unfavourable, provided it contained no ingredients actually noxious. in obedience to similar laws, the roots of an aquatic tree will not extend freely in dry soil, nor those of a tree which requires but little moisture

M

in a wet soil; and on this account the roots of the one will appear to
have sought, and those of the other to have avoided, the contiguous
water; though both, in the first period of their growth, pointed their
roots alike in every direction.

When the seeds of the bean, in the experiment I have described, were
placed to vegetate beneath the mould of an inverted pot, a sufficient
quantity of moisture was afforded by the mould to occasion the protrusion
of the radicles : but as soon as the under points of these had penetrated
through the seed-coats, their surfaces were necessarily exposed to dry air,
and were consequently rendered rigid and inexpansible ; whilst their
upper surfaces, being in contact with moist mould, remained soft and
expansible.   If both the upper and lower surfaces of the radicles, at their
points, had been equally well supplied with moisture, gravitation would
have attracted the sap to the lower sides, where new matter would have
been added; and the radicles would have extended perpendicularly
downwards, as in former experiments : but the influence of gravitation
was, to a great extent, counteracted by the effects of drought upon the
lower sides of the radicles, nearly as it was counteracted by centrifugal
force, when made to act horizontally *

As soon as the radicles had acquired sufficient age and maturity, efforts
were made by them to emit fibrous roots ; when want of proper moisture
on the lower sides prevented their being protruded, in any other direction,
except upwards.   In that direction therefore they were alone emitted,
(as I was confident that they would before I began the experiment) and
having found proper food and moisture in the pots, they extended
themselves upwards through more than half the mould, which these
contained.

This experiment was repeated, and water was so constantly and abundantly given, that every part of the radicles was kept equally wet; and
they then became perfectly obedient to gravitation, without being at all
influenced by the mould above them.

In other experiments pieces of alum and of the sulphates of iron and
copper were placed at small distances perpendicularly beneath the radicles
of germinating seeds, of different species, to afford an opportunity of
observing whether any efforts would be made by them to avoid poisons ;
but they did not appear to be at all influenced, except by actual contact
of the injurious substances.   The growth of their fibrous lateral roots
was, however, obviously accelerated, when their points approached any
considerable quantity of decomposing vegetable or animal matter : and

* Above, p. 126

when the growth of the roots was retarded by want of moisture, the contiguity of water, in the adjoining mould, though not apparently in actual
contact with them, operated beneficially : but I had reason to suspect
that the growth of roots was, under these circumstances, promoted by
actual contact with the detached and fugitive particles of the decomposing
body, and of the evaporating water.

The growth and forms assumed by the roots of trees, of every species,
are to a great extent, dependent upon the quantity of motion, which their
stems and branches receive from winds ; for the effects of motion upon
the growth of the root, and of the trunk and branches, which I have
described in a former memoir, are perfectly similar*. Whatever part of
a root is moved and bent by winds, or other causes, an increased deposition of alburnous matter upon that part soon takes place, and consequently the roots which immediately adjoin the trunk of an insulated tree,
in an exposed situation, become strong and rigid ; whilst they diminish
rapidly in bulk, as they recede from the trunk, and descend into the
ground. By this sudden diminution of the bulk of the roots, the passage
of the descending sap, through their bark, is obstructed ; and it in consequence generates, and passes into many lateral roots ; and these, if the
tree be still much agitated by winds, assume a similar form, and consequently divide into many others. A kind of net-work composed of thick
and strong roots is thus formed, and the tree is secured from the dangers
to which its situation would otherwise expose it.

In a sheltered valley, on the contrary, where a tree is surrounded and
protected by others, and is rarely agitated by winds, the roots grow long
and slender, like the stem and branches, and comparatively much less of
the circulating fluid is expended in the deposition of alburnum beneath
the ground ; and hence it not unfrequently happens, that a tree, in the
most sheltered part of a valley, is uprooted; whilst the exposed and
insulated tree, upon the adjoining mountain, remains uninjured by the
fury of the storm.

In all the preceding arrangements, the wisdom of nature, and the
admirable simplicity of the means it employs, are conspicuously displayed;
but I am wholly unable to trace the existence of anything like sensation or
intellect in the plants : and I therefore venture to conclude, that their roots
are influenced by the immediate operation and contact of surrounding
bodies, and not by any degrees of sensation and passion analogous to those
of animal life ; and I reject the latter hypothesis, not only because it is
founded upon assumptions which cannot be granted, but because it is

* Above, p. 99.

M 2

insufficient to explain the preceding phenomena, unless seedling plants be admitted to possess more extensive intellectual powers, than are given to the offspring of the most acute animal. A young wild-duck or partridge, when it first sees the insect upon which nature intends it to feed, instinctively pursues and catches it ; but nature has given to the young bird an appropriate organization. The plant, on the contrary, if it could feel and perceive the objects of its wants, and will the possession of them, has still to contrive and form the organ by which these are to be approached. The writers who have contended for the existence of sensation in plants, appear to have been sensible of the preceding and other obstacles, and have all betrayed the weakness of their hypothesis, in adducing a few facts only which are favourable to it, and waiving wholly the investigation of all others.

In the description of the preceding experiments, I fear that I have been tediously minute; but as I have selected a few facts only from a great number, which I could have adduced, I was anxious to give as accurate and distinct a view of those I stated, as possible.

XIV.—ON THE MOTIONS OF THE TENDRILS OF PLANTS.

[*Read before the* ROYAL SOCIETY, *May 4th*, 1812.]

THE motions of the tendrils of plants, and the efforts they apparently make to approach and attach themselves to contiguous objects, have been supposed by many naturalists to originate in some degrees of sensation and perception : and though other naturalists have rejected this hypothesis, few, or no experiments have been made by them to ascertain with what propriety the various motions of tendrils, of different kinds, can be attributed to peculiarity of organisation, and the operation of external causes. I was consequently induced, during the last summer, to employ a considerable portion of time to watch the motions of the tendrils of different species of plants ; and I have now the pleasure to address to you an account of the observations I was enabled to make.

The plants selected were, the Virginia creeper (the Ampelopsis quinquefolia of Michaux,) the ivy, and the common vine and pea.

A plant of the ampelopsis, which grew in a garden pot, was removed to a forcing-house in the end of May, and a single shoot from it was made to grow perpendicularly upwards, by being supported in that position by a very slender bar of wood, to which it was bound. The plant was placed in the middle of the house, and was fully exposed to the sun ; and every object around it was removed far beyond the reach of its tendrils. Thus circumstanced, its tendrils, as soon as they were nearly full grown, all pointed towards the north, or back wall, which was distant about eight feet : but not meeting with any thing in that direction, to which they could attach themselves, they declined gradually towards the ground, and ultimately attached themselves to the stems beneath, and the slender bar of wood.

A plant of the same species was placed at the east end of the house, near the glass, and was in some measure skreened from the perpendicular light ; when its tendrils pointed towards the west, or centre of the house, as those under the preceding circumstances had pointed towards the north and back wall. This plant was removed to the west end of the house, and exposed to the evening sun, being skreened, as in the preceding case, from the perpendicular light ; and its tendrils, within a few hours, changed their direction, and again pointed to the centre of the house, which was partially covered with vines. This plant was then removed to the centre of the house, and fully exposed to the perpendicular light, and to the sun ; and a piece of dark-coloured paper was placed upon one side of it just within the reach of its tendrils ; and to this substance they soon appeared to be strongly attracted. The paper was then placed upon the opposite side, under similar circumstances, and there it was soon followed by the tendrils. It was then removed, and a piece of plate glass was substituted ; but to this substance the tendrils did not indicate any disposition to approach. The position of the glass was then changed, and care was taken to adjust its surface to the varying position of the sun, so that the light reflected might continue to strike the tendrils ; which then receded from the glass, and appeared to be strongly repulsed by it.

The tendrils of the ampelopsis very closely resemble those of the vine, in their internal organisation, and in originating from the alburnous substance of the plant ; and in being, under certain circumstances, convertible into fruit-stalks. The claws, or claspers of the ivy, to experiments upon which I shall now proceed, appear to be cortical protrusions only ; but to be capable (I have reason to believe) of becoming perfect roots, under favourable circumstances. Experiments, in every respect very nearly similar to the preceding, were made upon this plant ; but I found it necessary to place the different substances, to which I proposed that the claws should attempt to attach themselves, almost in contact with the stems of the plants. I observed that the claws of this plant evaded the light, just as the tendrils of the ampelopsis had done; and that they sprang only from such parts of the stems as were fully, or partially, shaded.

A seedling plant of the peach tree, and one of the ampelopsis and ivy were placed nearly in the centre of the house, and under similar circumstances; except that supports, formed of very slender bars of wood, about four inches high, were applied to the ampelopsis and ivy. The peach tree continued to grow nearly perpendicularly, with a slight inclination towards the front and south side of the house, whilst the stems of the ampelopsis and ivy, as soon as they exceeded the height of their supports, inclined many points from the perpendicular line, in the opposite direction.

It appears therefore that not only the tendrils and claws of these creeping dependent plants, but that their stems also, are made to recede from light, and to press against the opake bodies, which nature intended to support and protect them.

M. De Candolle, I believe, first observed that the succulent shoots of trees and herbaceous plants, which do not depend upon others for support, are bent towards the point from which they receive light, by the contraction of the cellular substance of their bark upon that side, and I believe his opinion to be perfectly well founded. The operation of light upon the tendrils and stems of the ampelopsis and ivy appears to produce diametrically opposite effects, and to occasion an extension of the cellular bark, wherever that is exposed to its influence; and this circumstance affords, I think, a satisfactory explanation why these plants appear to seek and approach contiguous opake objects, just as they would do, if they were conscious of their own feebleness, and of power in the objects, to which they approach, to afford them support and protection.

The tendril of the vine, as I have already stated, is internally similar to that of the ampelopsis, though its external form, and mode of attaching itself by twining round any slender body, are very different. Some young plants of this species, which had been raised in pots in the preceding year, and had been headed down to a single bud, were placed in a forcing-house, with the plants I have already mentioned; and the shoots from these were bound to slender bars of wood, and trained perpendicularly upwards. Their tendrils, like those of the ampelopsis, when first emitted, pointed upwards; but they gradually formed an increasing angle with the stems, and ultimately pointed perpendicularly downwards; no object having presented itself to which they could attach themselves.

Other plants of the vine, under similar circumstances, were trained horizontally; when their tendrils gradually descended beneath their stems, with which they ultimately stood very nearly at right angles.

A third set of plants were trained almost perpendicularly downwards; but with an inclination of a few degrees towards the north; and the tendrils of these permanently retained very nearly their first position, relatively to their stems; whence it appears that these organs, like the tendrils of the ampelopsis, and the claws of the ivy, are to a great extent under the control of light.

A few other plants of the same species were trained in each of the preceding methods; but proper objects were placed, in different situations, near them, with which their tendrils might come into contact; and I was by these means afforded an opportunity of observing, with accuracy, the difference between the motions of these and those of the ampelopsis, under similar circumstances. The latter almost immediately receded from light, by whatever means that was made to operate upon them; and they did not subsequently show any disposition to approach the points, from which they once receded. The tendrils of the vine, on the contrary, varied their positions in every period of the day, and after returned again during the night to the situations they had occupied in the preceding morning; and they did not so immediately, or so regularly, bend towards the shade of contiguous objects. But as the tendrils of this plant, like those of the ampelopsis, spring alternately from each side of the stem, and as one point only in three is without a tendril, and as each tendril separates into two divisions, they do not often fail to come into contact with any object within their reach; and the effects of contact upon the tendril are almost immediately visible. It is made to bend towards the body it touches, and if that body be slender, to attach itself firmly by twining round it, in obedience to causes which I shall endeavour to point out.

The tendril of the vine, in its internal organization, is apparently similar to the young succulent shoot and leaf-stalk, of the same plant; and it is as abundantly provided with vessels, or passages, for the sap; and I have proved that it is alike capable of feeding a succulent shoot, or a leaf, when grafted upon it. It appears therefore, I conceive, not improbable, that a considerable quantity of the moving fluid of the plant, passes through its tendrils: and that there is a close connection between its vascular structure and its motions.

I have proved in the Philosophical Transactions of 1806, that centrifugal force, by operating upon the elongating plumules of germinating seeds, occasions an increased growth and extension upon the external sides of the young stems, and that gravitation produces correspondent effects;

probably by occasioning the presence of a larger portion of the fluid organisable matter of the plant upon the one side, than upon the other. The external pressure of any body upon one side of a tendril will probably drive this fluid from one side of the tendril, which will consequently contract, to the opposite side, which will expand ; and the tendril will thence be compelled to bend round a slender bar of wood or metal, just as the stems of germinating seeds are made to bend upwards, and to raise the cotyledons out of the ground; and in support of this conclusion I shall observe, that the sides of the tendrils, where in contact with the substance they embraced, were compressed and flattened.

The actions of the tendrils of the pea were so perfectly similar to those of the vine, when they came into contact with any body, that I need not trouble you with the observations I made upon that plant.   An increased extension of the cellular substance of the bark upon one side of the tendrils, and a correspondent contraction upon the opposite side, occasioned by the operation of light, or the partial pressure of a body in contact, appeared in every case, which has come under my observation, the obvious cause of the motions of tendrils ; and therefore, in conformity with the conclusions I drew in my last memoir, respecting the growth of roots, I shall venture to infer, that they are the result of pure necessity only, uninfluenced by any degrees of sensation, or intellectual powers.

XV.—ON THE ACTION OF DETACHED LEAVES OF PLANTS.

[*Read before the* ROYAL SOCIETY, *June 13th,* 1816.]

SINCE I had last the honour to address a communication to the Royal Society, I have repeated great part of the experiments which formed the subjects of my former memoirs, with such additions and variations as might probably lead to the detection of any erroneous conclusions which I might have drawn ; but I have not been able to detect any errors, nor to add anything very important to my former observations. I have, however, been able to ascertain a few new facts, which I think too interesting to be lost.

I endeavoured, in my former communications, to adduce evidence that the matter, which becomes vitally united to trees, previously passes through their leaves ; and I shall now proceed to state some facts, which, I trust, will prove that a fluid possessing the power which I have attributed to the true sap actually descends through the leaf-stalks.

A slender knife was passed through some leaf-stalks of the vine, about two-thirds of an inch distant from their junction to the branch; and, down to that point, the leaf-stalks were divided longitudinally, and a transverse section, about half-an-inch long, was made through the bark opposite the middle of the leaf-stalk. A similar transverse section through the bark was made somewhat less than an inch distant below; and these sections were united by two longitudinal sections through the bark, which extended from the extremities of the upper transverse sections to the extremities of the lower; by which means pieces of bark, about half-an-inch broad and nearly an inch long, were separated from the adjoining bark. These were then detached from the alburnum, and surrounded by two folds of paper coated with wax on each side; by which all connexion and communication with the tree, except through the divided leaf-stalks, were cut off. The insulated pieces of bark, nevertheless, continued to grow, and extended downwards, and laterally, and in thickness; and thin layers of alburnum were deposited.

Leaves of the potatoe, without any portion of bark being attached to them, were taken from the plants just at the period when the tuberous roots began to be formed; and I conceived that these leaves, consistently with my former experiments and conclusions, must contain portions of the living organisable matter which would subsequently have been found in their tuberous roots. The leaves were therefore planted in pots, and placed under glass, where, being regularly and properly supplied with water, they continued to live till winter, though without emitting fibrous roots; and I then expected to find some small tubers at their bases. In this expectation I was disappointed; but the result of the experiment was not less satisfactory, the bases of the leaf-stalks themselves having swollen into conic bodies of more than two inches in circumference, and being found to consist of matter apparently similar to that which composes the tuberous roots of the plant. The enlarged parts of the leaf-stalks remained alive in the following spring; but whether they are capable of generating buds or not I have not been able to ascertain.

Leaves of mint were planted in the same manner as those above-mentioned; which grew, and continued alive through the winter, and were still living in the end of the last month, having assumed the character of the thick fleshy leaves of evergreen trees. Upon examining the mould in the pots, I found it to contain very numerous roots, which must have derived their medullary, and their cortical, and alburnous substances, from matter which had emanated and descended from the leaves.

I had frequently observed, in former experiments, that the destruction of the mature leaves of young plants not only suspended the growth of the roots, but also the growth of the immature leaves ; whence I inferred, in a former communication, that the organisable matter which composes the young leaves has always undergone a previous preparation in other leaves of the plant, either of the same or preceding season ; and I was thence led to expect that, under favourable circumstances, the mature leaves might be made to nourish and promote the growth of immature leaves, without the aid of roots. Several shoots of the vine, each about a yard long, were detached from the trees, and laid over a succession of basins of water, into which each of the mature leaves was in part depressed ; and thus circumstanced, the young leaves continued to grow, and the points of the shoots to elongate ; and all were alive, and in perfect apparent health, at the end of a month. The water necessary to preserve the young leaves must in this case have been derived from the mature leaves ; and I entertain no doubt but that the organisable matter which occasioned their growth was derived from the same source. Intersection of the bark between the mature and young leaves was not attended with any injurious consequences, and the sap must, therefore, have passed to the young leaves through the alburnum.

Consistently with the preceding circumstances, if the mature leaves be destroyed, or taken off, the fruit ceases to grow—or, if full grown, remains without richness or flavour ; and the power of feeding fruits in winter and early spring seems to be confined to evergreen plants. The orange and lemon tree, the ivy and holly, afford familiar examples of this ; and where a genus of plants consists of evergreen and deciduous species, as that of mespilus and viburnum, the evergreen species alone nourish their fruit in winter and early spring.

The probable passage of the sap from the mature to the young leaves and fruit may, I think, be easily pointed out, though decisive proof of its course will probably never be adduced. Having often detached the bark from the alburnum of the stems of young oaks, just at the period when the midsummer shoots were beginning to elongate, I observed, as others have done, that a fluid exuded from those parts of the surface of the alburnum which are called (most improperly) the medullary processes, and from correspondent points of the bark which resemble the medullary processes in organisation. This fluid has been proved, by its power of rapidly generating an organic substance, to be the true sap of the tree ; part of which, I conceive, at this period, to be passing from the bark to join the ascending current in the alburnum ; which current feeds the young succulent shoots and growing leaves. Subjecting the alburnum to a slight degree of pressure at this period, I found that a considerable quantity of liquid, being apparently the true sap of the tree, issued out laterally through the medullary processes, as well as longitudinally through the cellular substance of the alburnum; but the tubes of it continued empty, and their position was marked by depressions of the surface of the extravasated fluid.   I endeavoured to ascertain what proportion of water a given quantity of the alburnum of such oak trees contained at this period, and I found that 1000 parts lost by drying only 371 parts; which is not more than the weight of the water that the cellular substance appears capable of containing, entirely independent of the tubes.   That the tubes, nevertheless, are not always empty, but that they act at other periods of the year as reservoirs for the sap, I have given an opinion in a former communication; and I am now in possession of facts which prove them to perform this office, even in the heart wood, to a much greater extent than I had ever at any former period suspected; and which incline me to believe that the durability of the heart wood, as well as of the alburnum of the oak, will be found to depend to a great extent upon the period in which the tree is felled.

PART II

PAPERS ON PHYSIOLOGICAL HORTICULTURE,

READ BEFORE

THE HORTICULTURAL SOCIETY, IN THE YEARS 1806 TO 1838.

REPRINTED FROM THE HORTICULTURAL TRANSACTIONS.

XVI.—OBSERVATIONS ON THE MEANS OF PRODUCING NEW AND EARLY FRUITS.

[Read before the HORTICULTURAL SOCIETY, *November 4th,* 1806.]

NATURE has given to man the means of acquiring those things which constitute the comforts and luxuries of civilised life, though not the things themselves; it has placed the raw material within his reach; but has left the preparation and improvement of it to his own skill and industry. Every plant and animal, adapted to his service, is made susceptible of endless changes, and, as far as relates to his use, of almost endless improvement. Variation is the constant attendant on cultivation, both in the animal and vegetable world; and in each the offspring are constantly seen, in a greater or less degree, to inherit the character of the parents from which they spring.

No experienced gardener can be ignorant that every species of fruit acquires its greatest state of perfection in some peculiar soils and situations, and under some peculiar mode of culture: the selection of a proper soil and situation must therefore be the first object of the improver's pursuit; and nothing should be neglected which can add to the size, or improve the flavour of the fruit from which it is intended to propagate. Due attention to these points will in almost all cases be found to comprehend all that is necessary to insure the introduction of new varieties of fruit, of equal merit with those from which they spring; but the improver, who has to adapt his productions to the cold and unsteady climate of Britain, has still many difficulties to contend with; he has to combine hardiness, energy of character, and early maturity, with the improvements of high cultivation. Nature has however in some measure pointed out the path he is to pursue; and, if it be followed with patience and industry, no obstacles will be found, which may not be either removed, or passed over.

If two plants of the vine or other tree of similar habits, or even if obtained from cuttings of the same tree, were placed to vegetate, during several successive seasons, in very different climates : if the one were planted on the banks of the Rhine, and the other on those of the Nile, each would adapt its habits to the climate in which it were placed ; and if both were subsequently brought, in early spring, into a climate similar to that of Italy, the plant which had adapted its habits to a cold climate would instantly vegetate, whilst the other would remain perfectly torpid Precisely the same thing occurs in the hot-houses of this country, where a plant accustomed to the temperature of the open air will vegetate strongly in December, whilst another plant of the same species, and sprung from a cutting of the same original stock, but habituated to the temperature of a stove, remains apparently lifeless. It appears, there-fore, that the powers of vegetable life, in plants habituated to cold climates, are more easily brought into action than in those of hot climates ; or, in other words, that the plants of cold climates, are most exciteable : and as every quality in plants becomes hereditary, when the causes which first gave existence to those qualities continue to operate ; it follows that their seedling offspring have a constant ten-dency to adapt their habits to any climate in which art or accident places them.

But the influence of climate on the habits of plants, will depend less on the aggregate quantity of heat in each climate, than on the distribution of it in the different seasons of the year. The aggregate temperature of England, and of those parts of the Russian Empire that are under the same parallels of latitude, probably does not differ very considerably ; but, in the latter, the summers are extremely hot, and the winters intensely cold; and the changes of temperature between the different seasons are sudden and violent. In the spring great degrees of heat suddenly operate on plants which have been long exposed to intense cold, and in which excitability has accumulated during a long period of almost total inaction : and the progress of vegetation is in consequence extremely rapid. In the climate of England, the spring, on the contrary, advances with slow and irregular steps, and only very moderate and slowly-increasing degrees of heat act on plants in which the powers of life have scarcely in any period of the preceding winter been totally inactive. The crab is a native of both countries, and has adapted alike its habits to both; the Siberian variety introduced into the climate of England, retains its habits, expands its leaves, and blossoms on the first approach of spring, and vegetates strongly in the same temperature in which the native crab scarcely shows signs of life; and its fruit acquires a degree of maturity, even in the early part of an unfavourable season, which our native crab is rarely, or never seen to attain.

Similar causes are productive of similar effects on the habits of culti-vated annual plants ; but these appear most readily to acquire habits of maturity in warm climates; for it is in the power of the cultivator to commit his seeds to the earth at any season ; and the progress of the plants towards maturity will be most rapid, where the climate and soil are most warm.   Thus, the barley grown on sandy soils, in the warmest parts of England, is always found by the Scotch farmer, when introduced into his country, to ripen on his cold hills earlier than his crops of the same kind do, when he uses the seeds of plants, which have passed through several successive generations in his colder climate; and in my own experience, I have found that the crops of wheat on some very high and cold ground, which I cultivate, ripen much earlier when I obtain my seed-corn from a very warm district and gravelly soil, which lies a few miles distant, than when I employ the seeds of the vicinity.

The value, to the gardener, of an early crop, has attracted his attention to the propagation and culture of the earliest varieties of many species of our esculent plants; but in the improvement of these he is more often indebted to accident than to any plan of systematic culture ; and contents himself with merely selecting and propagating from the plant of the earliest habits, which accident throws in his way ; without inquiring from what causes those habits have arisen : and few efforts have been made to bring into existence better varieties of those fruits which are not generally propagated from seeds, and which, when so propagated, of necessity exercise, during many years, the patience of the cultivator, before he can hope to see the fruits of his labour.

The attempts which I have made to produce early varieties of fruit are, I believe, all that have yet been made ; and though the result of them is by no means sufficiently decisive to prove the truth of the hypothesis I am endeavouring to establish, or the eligibility of the practice I have adopted, it is amply sufficient to encourage future experiment.

The first species of fruit, which was subjected to experiment by me, was the apple ; some young trees of those varieties of this fruit, from which I wished to propagate, were trained to a south wall, till they pro-duced buds which contained blossoms.   Their branches were then, in the succeeding winter, detached from the wall, and removed to as great a distance from it, as the pliability of their stems would permit ; and in this situation they remained till their blossoms were so far advanced, in the succeeding spring, as to be in some danger of injury from frost. The branches were then trained to the wall, where every blossom I suffered to remain, soon expanded and produced fruit. This attained in a few months the most perfect state of maturity ; and the seeds afforded plants, which have ripened their fruit very considerably earlier than other trees, which I raised at the same time, from seeds of the same fruit, which had grown in the orchard. In this experiment the fecundation of the blossoms, of each variety, was produced by the farina of another kind ; from which process, I think, I obtained in this, and many similar experiments, an increased vigour and luxuriance of growth; but I have no reasons what-ever to think that plants thus generated ripen their fruit earlier than others, which are obtained by the common methods of culture. I must therefore attribute the early maturity of those I have described to the other peculiar circumstances under which the seeds and fruit ripened, from which they sprang.

I obtained, by the same mode of culture, many new varieties, which are the offspring of the Siberian crab and the richest of our apples, with the intention of affording fruits for the press, which might ripen well in cold and exposed situations. The plants, thus produced, seem perfectly well calculated, in every respect, to answer the object of the experiment, and possess an extraordinary hardiness and luxuriance of growth. The annual shoots of some of them, from newly grafted trees in my nursery, the soil of which is by no means rich, exceeded six feet and a half in height, in the last season ; and their blossoms seem capable of bearing extremely unfavourable weather without injury. In all the preceding experiments some of the new varieties inherited the character of the male, and others of the female parent in the greatest degree ; and of some varieties of fruit (particularly the golden pippin) I obtained a better copy, by introducing the farina into the blossom of another apple, than by sowing their own seeds; I sent a new variety (the Downton pippin) which was thus obtained from the farina of the golden pippin, to the Horticultural Society, last year ; but those specimens afforded but a very unfavourable sample of it; for the season, and the situation in which the fruit ripened, were very cold, and almost every leaf of the trees had been eaten off by insects. In a favourable season and situation it will, I believe, be found little, if at all, inferior, to the golden pippin, when first taken from the tree; but it is a good deal earlier, and probably cannot be preserved so long.

I proceed to experiments on the grape; which though less successful than those on the apple, in the production of good varieties, are not less favourable to the preceding conclusions. A vinery in which no fires are made during the winter, affords to the vine a climate similar to that which the southern parts of Siberia afford to the apple or crab-tree: in it a similarly extensive variation of temperature takes place, and the sudden transition from great comparative cold to excessive heat, is productive of the same rapid progress in the growth of the plants, and advancement of the fruit to maturity. My first attempt was to combine the hardiness of the blossom of the black cluster, or Burgundy grape, with the large berry and early maturity of the true sweetwater*. The seedling plants produced fruit in my vinery at three or four years old, and the fruit of some of them was very early; but the bunches were short, and ill-formed, and the berries much smaller than those of the sweetwater, and the blossoms did not set by any means so well as I had hoped.

Substituting the white chasselas for the sweetwater, I obtained several varieties, whose blossoms appear perfectly hardy, and capable of setting well in the open air; and the fruit of some of them is ripening a good deal earlier in the present year than that of either of the parent plants. The berries, however, are smaller than those of the chasselas, and with less tender and delicate skins; and, though not without considerable merits for the dessert, they are generally best calculated for the press: for the latter purpose, in a cold climate, I am confident that one or two of them possess very great excellence. I sent a bunch of one of those varieties to the Horticultural Society, in the last autumn, and I propose to send two or three others in the present year.

I have subsequently obtained plants from the white chasselas and sweetwater, whose appearance is much more promising; and the earliest variety of the grape I have ever yet seen, sprang from a seed of the sweetwater, and the farina of the red frontignac. This is also a very fine grape, resembling the frontignac in colour and form of the bunch; but I fear its blossoms will prove too tender to succeed in the open air in this country; a single bunch, consisting of a few berries, is however, all that has yet existed of this kind. The present season also affords me

---

* This grape is often confounded by gardeners, both with the white chasselas and white muscadine.

two new varieties of the vine, with striped fruit, and variegated autumnal leaves, produced by the white chasselas and the farina of the Aleppo vine : one of these has ripened extremely early, and is, I think, a good grape. When perfectly ripe, I propose sending a bunch of it for the inspection of the Horticultural Society.

In all attempts to obtain new varieties of fruit, the propagator is at a loss to know what kinds are best calculated to answer his purpose ; and therefore I have mentioned those varieties of the grape from which I have propagated with the best prospect of success. My experiments are, however, still in their infancy ; and I do not possess the means of making them on so large a scale or in so perfect a manner as I wish ; nevertheless, the facts of which I am in possession, leave no grounds of doubt in my mind, that varieties of the grape, capable of ripening perfectly in our climate, when trained to a south wall, and of other fruits better calculated for our climate than those we now cultivate, may readily be obtained ; but whether the mode of culture I have adopted and recommended be most eligible must be decided by future and more extensive practice.

I have made experiments similar to the preceding on the peach ; but I can say no more of the result of them, than that the plants possess the most perfect degree of health and luxuriance of growth, and that their leaves afford satisfactory evidence of the good quality of the future fruit. I am ignorant of the age at which plants of this species become capable of producing blossoms ; but the rapid changes in the character of the leaves and growth of my plants, which are now in their third year, induce me to believe that they will be capable of producing fruit at three or four years old.

I shall finish my paper with stating a few conclusions, which I have been able to draw in the course of many years' close attention to the subject on which I write.

New varieties of every species of fruit will generally be better obtained by introducing the farina of one variety of fruit into the blossom of another, than by propagating from any single kind. When an experiment of this kind is made, between varieties of different size and character, the farina of the smaller kind should be introduced into the blossoms of the larger ; for, under these circumstances, I have generally (but with some exceptions) observed in the new fruit a prevalence of the character of the female parent ; probably owing to the following causes. The seed-coats are generated wholly by the female

N

parent, and these regulate the bulk of the lobes and plantule : and I have observed, in raising new varieties of the peach, that when one stone contained two seeds, the plants these afforded were inferior to others. The largest seeds, obtained from the finest fruit, and from that which ripens most perfectly and most early, should always be selected. It is scarcely necessary to inform the experienced gardener, that it will be necessary to extract the stamina of the blossoms from which he proposes to propagate, some days before the farina begins to shed, when he proposes to generate new varieties in the manner I have recommended.

When young trees have sprung from the seed, a certain period must elapse before they become capable of bearing fruit, and this period, I believe, cannot be shortened by any means. Pruning and transplanting are both injurious; and no change in the character or merits of the future fruit can be effected, during this period, either by manure or culture. The young plants should be suffered to extend their branches in every direction, in which they do not injuriously interfere with each other; and the soil should just be sufficiently rich to promote a moderate degree of growth, without stimulating the plant to preternatural exertion, which always induces disease*. The periods which different kinds of fruit-trees require to attain the age of puberty, are very varied. The pear requires from twelve to eighteen years; the apple, from five to twelve, or thirteen; the plum and cherry, four or five years; the vine three or four; and the raspberry, two years. The strawberry, if its seeds be sown early, affords an abundant crop in the succeeding year. My garden at present contains several new and excellent varieties of this fruit†, some of which I shall be happy to send to the Horticultural Society.

* The soil of an old garden is peculiarly destructive.

† The hautboy strawberry does not appear to propagate readily with the other varieties, and may possibly belong to an originally distinct species. I have, however, obtained several offspring from its farina; but they have all produced a feeble and abortive blossom. If nature, in any instance, permits the existence of vegetable mules (but this I am not inclined to believe), these plants seem to be beings of that kind.

XVII.— A DESCRIPTION OF A FORCING-HOUSE FOR GRAPES; WITH OBSERVATIONS ON THE BEST METHOD OF CONSTRUCTING HOUSES FOR OTHER FRUITS.

[*Read before the* HORTICULTURAL SOCIETY, *May 3rd*, 1808.]

So much difference of opinion prevails amongst gardeners respecting the proper forms of *forcing-houses* that two are rarely constructed quite alike, though intended for the same purposes; and every gardener is prepared to contend that the form he prefers is the best, and to appeal to the test of succcessful experiment in support of his opinion. And this he is generally enabled in some degree to do, because plants, when properly supplied with food and water and heat, will succeed in houses the forms of which are very defective; and proper attention is not often paid by the gardener when his prejudices satisfy him that his labours cannot be successful. It is, however, sufficiently evident, that when the same fruit is to be ripened in the same climate and season of the year, one peculiar form must be superior to every other; and that in our climate, where sunshine and natural heat do not abound, that form, which admits the greatest quantity of light through the least breadth of glass, and which affords the greatest regular heat with the least expenditure of fuel, must generally be the best; and, if the truth of this position be admitted, it will be very easy to prove that few of our forcing-houses are at present even moderately well constructed. I therefore think that if plans and descriptions of such forcing-houses as theory and practice prove to have been properly constructed for the culture of every different species of fruit were published by the Horticultural Society, much useful information might be conveyed to the practical gardener. Under these impressions I send the following description of a vinery in which the most abundant crops of grapes have been perfectly ripened within less time, and with less expenditure of fuel, than I have witnessed in any other instance.

It is well known that the sun operates most powerfully in the forcing-house when its rays fall most perpendicularly on the roof; because the quantity of light that glances off without entering the house is proportionate to the degree of obliquity with which it strikes upon the surface of the glass; and it is, therefore, important to every builder of a forcing-house to know by what elevation of the roof the greatest quantity of light can be made to pass through it. To ascertain this point I have made many experiments, and the result of them has satisfied me that, in latitude 52°, the best elevation is about that of 34 degrees; and relative to that elevation the position of the sun, in different parts of the year, will be nearly as represented in the annexed sketch, which is taken from the vinery I have mentioned. About the middle of May, the elevation of the sun will nearly correspond with that of the asterisk A, and in the beginning of June, and again early in July, it will be vertical at B; and at Midsummer it will, at C, be only six degrees from being vertical. The asterisk D points out its position at the equinoxes, and E its position in Midwinter.

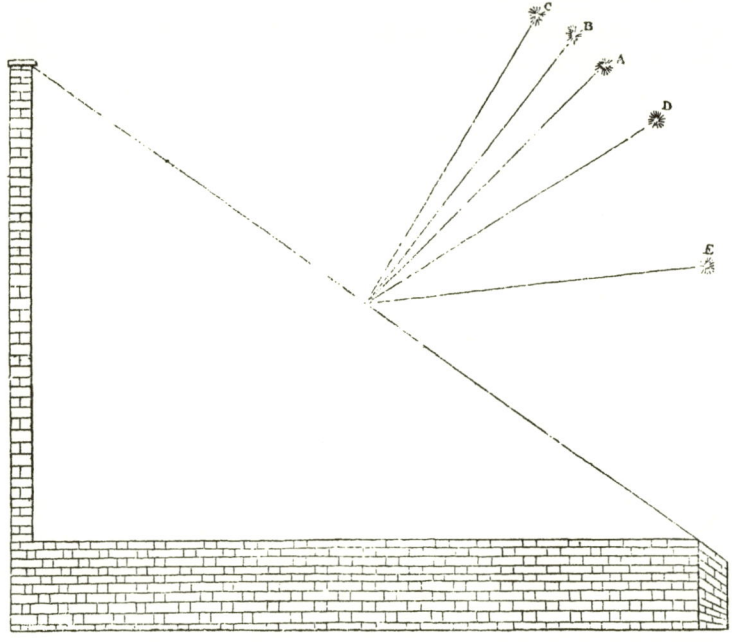

In this building, which is forty feet long, and is heated by a single fire-place, the flue goes entirely round without touching the walls; and in the front a space of two feet is left between the flue and the wall, in the middle of which space the vines, which are trained to the roofs about eleven inches from the glass, are planted; and, as both the wall and flue are placed on arches, the vines are enabled to extend their roots in every direction, whilst, in the spring, their growth is greatly excited by the heat which their roots and stems receive from the flue. Air is generally admitted at the ends only, where all the sashes are made to slide to afford a free passage of air through the house, when necessary to prevent the grapes becoming mouldy in damp seasons. About four feet of the upper end of every third light of the roof is made to lift up, (being attached by hinges to the wood-work on the top of the back wall,) to give air in the event of very hot and calm weather ; for I prefer giving air by lifting up the lights to letting them slide down, because, when the former method is adopted, no additional shade is thrown on the plants.

The preceding plan is here particularly recommended for a vinery only; but I am confident that, by sinking the front wall below the level of the ground and making a small change in the form of the bark-bed, the same elevation of roof may be made equally applicable to the pine-stove, and that no upright front glass ought, in any case whatever, to be used ; for light can always be more beneficially admitted by adding to the length of the roof, if that be properly elevated ; and much expence may be saved both in the building and in fuel. For forcing the peach or nectarine I must, however, observe that I think any house of the preceding dimensions wholly improper ; and I propose to submit a plan for the improved culture of those fruits to the Horticultural Society at a future opportunity.

The vine often bleeds excessively when pruned in an improper season, or when accidentally wounded, and I believe no mode of stopping the flow of the sap is at present known to gardeners. I therefore mention the following, which I discovered many years ago, and have always practised with success :—if to four parts of scraped cheese be added one part of calcined oyster shells, or other pure calcareous earth, and this composition be pressed strongly into the pores of the wood, the sap will instantly cease to flow; so that the largest branch may of course be taken off at any season with safety.

XVIII.—ON THE MANAGEMENT OF THE ONION.

[*Read before the* HORTICULTURAL SOCIETY, *April 4th*, 1809.]

THE first object of the Horticultural Society being to point out improvements in the culture of those plants which are extensively useful to the public, I send a few remarks on the management of one of these, the *onion :* which both constitutes one of the humble luxuries of the poor, and finds its way, in various forms, to the tables of the affluent and luxurious.

Every bulbous-rooted plant, and indeed every plant which produces leaves, and lives longer than one year, generates, in one season, the sap, or vegetable blood, which composes the leaves and roots of the succeeding spring ; and when the sap has accumulated during one or more seasons, it is ultimately expended in the production of blossoms and seeds. This reserved sap is deposited in, and composes in a great measure, the bulb; and the quantity accumulated, as well as the period required for its accumulation, varies greatly in the same species of plant, under more or less favourable circumstances. Thus the onion, in the south of Europe, acquires a much larger size during the long and warm summers of Spain and Portugal, in a single season, than the colder climate of England; but under the following mode of culture, which I have long practised, two summers in England produce nearly the effect of one in Spain or Portugal, and the onion assumes nearly the form and size of those thence imported.

Seeds of the Spanish or Portugal onion are sown at the usual period in the spring, very thickly, and in poor soil; generally under the shade of a fruit-tree; and in such situations the bulbs, in the autumn, are rarely found much to exceed the size of a large pea. These are then taken from the ground, and preserved till the succeeding spring, when they are planted at equal distances from each other, and they afford plants which differ from those raised immediately from seed, only in possessing much greater strength and vigour, owing to the quantity of previously generated sap being much greater in the bulb than in the seed. The bulbs, thus raised, often exceed considerably five inches in diameter, and being more mature, they are with more certainty preserved, in a state of perfect soundness, through the winter than those raised from seed in a single season. The same effects are, in some measure, produced by sowing the seeds in August, as is often done; but the crops often perish during the winter, and the ground becomes compressed and saddened (to use an antiquated term) by the winter rains; and I have in consequence always found that any given weight of this plant may be obtained, with less expence to the grower, by the mode of culture I recommend, than by any other which I have seen practised.

XIX.—ON POTATOES.

[*Read before the* HORTICULTURAL SOCIETY, *February* 6, 1810.]

In a paper lately read before the Society, I described a method of cultivating early varieties of the potato, by which any of those, which do not usually blossom, may be made to produce seeds, and thus afford the means of obtaining many early varieties. I also offered a conjecture, that varieties of moderately early habits, and luxuriant growth, might be formed, which would be found well adapted to field-culture, and be ready to be taken from the soil in the end of August, or the beginning of September; so that the farmer might be allowed ample time to prepare the same ground for a crop of wheat. I am now enabled to state, that the success of the experiment has in both cases fully answered every expectation that I had formed.

The facts that I have stated in the paper above referred to, and more fully in the *Philosophical Transactions*, are, I believe, sufficient to prove, that the same fluid, or sap, gives existence alike to the tuber, and the blossom and seeds, and that whenever a plant of the potato affords either seeds or blossoms, a diminution of the crop of tubers, or an increased expenditure of the riches of the soil, must necessarily take place. It has also been proved by others, as well as myself, that the crop of tubers is increased by destroying the fruit-stalks and immature blossoms as soon as they appear, and I therefore conceived that considerable advantages would arise, if varieties of sufficiently luxuriant growth and large produce, for general culture, could be formed, which would never produce blossoms.

I have since had the gratification to find that such are readily obtained, by the means which I have detailed, and I am disposed to annex more importance to the improvement of our most useful plants, than any writer on agriculture has hitherto done; because whatever increased value is thus added to the produce of the soil, is obtained without any increased expence or labour, and therefore is just so much added to individual and national wealth.

I formerly supposed that all varieties of the potato, which ripened early in the autumn, would necessarily vegetate early in the ensuing spring, and could therefore be fit for use only during winter; but I have found that the habit of acquiring maturity early in the autumn, is by no means necessarily connected with the habit of vegetating early in the spring; and therefore by a proper selection of varieties, the season of planting crops, for all purposes, may be extended from the beginning of March, nearly to the middle of May, and each variety be committed to the soil exactly at the most advantageous period.

A variety, however, which does not vegetate till late in the spring, and which ripens early in the autumn, cannot I conclude, particularly in dry soils and seasons, afford so large a produce as one which vegetates more early: I, nevertheless, obtained so large a crop from one which vegetates remarkably late in the spring, and ripens rather early in the autumn, that I was induced to ascertain, by weighing, to what the produce would have amounted had the crop extended over an acre, and I found that it would have been 21 tons, 11 cwt. 80 lb. or 48,352 lbs.

In this calculation the external rows, which derived superior advantage from air and light, were excluded. No more manure, or culture, than is usually given, had been employed, for the crop was not planted with any intention of having it weighed: the wet summer was, however, very favourable.

I am not acquainted with the ordinary amount of the weight of a good crop of potatoes, upon an acre of ground in a favourable soil, when well-manured and cultivated; but I am confident, that it may generally be made to exceed twenty tons, by a proper selection of varieties: and if four pounds of good potatoes afford, as is generally supposed, at least as much nutriment as one pound of wheat, the produce of an acre of potatoes, such as I have described, is capable of supporting as large a population, as eight acres of wheat, admitting the calculation of Mr. Arthur Young, that the average produce of an acre of wheat is $22\frac{1}{2}$ bushels or 1440 lbs.; and as an acre of wheat will certainly support as large a number of people as five acres of permanent pasture, it follows, that an acre of potatoes affords as much food for mankind, as forty acres of permanent pasture: an important subject for consideration, in a country where provisions are scarce and dear, and where so high bounties on pasture are paid in the form of taxes on tillage, that the extent of permanent pasture is certainly and consequently increasing: and it must increase, under existing circumstances; for it pays a higher rent to the landlord, and relieves the farmer from much labour, anxiety, and vexation.

To what extent a crop of potatoes will generally be increased by the total prevention of all disposition to blossom, the soil and variety being, in all other respects, the same, it is difficult to conjecture; but I imagine that the expenditure of sap in the production of fruit-stalks and blossoms alone would be sufficient to occasion an addition, of at least an ounce, to the weight of the tubers of each plant, and if each square yard were to contain eight plants, as in the crop I have mentioned, the increased produce of an acre would considerably exceed a ton, and of course be sufficient, in almost all cases, to pay the rent of the ground.

I do not know how far other parts of England are well supplied with good varieties of potatoes; but those cultivated in my neighbourhood in Herefordshire and Shropshire, are generally very bad. Many of them have been introduced from Ireland, and to that climate they are probably well adapted; for the Irish planter is secure from frost from the end of April nearly to the end of November: but in England, the potato is never safe from frost till near the end of May; indeed I have seen the leaves and stems of a crop, in a very low situation, completely destroyed as late as the 13th of June, and they are generally injured before the middle, and sometimes in the first week of September.

The Irish varieties, being excessively late, are almost always killed by the frost whilst in full blossom ; when omitting all consideration of the useless expenditure of manure, it may justly be questioned whether the tubers of such plants, being immature, can afford as nutritive, or as wholesome food, as others which have acquired a state of perfect maturity.

The preceding statement will, I trust, point out to the Horticultural Society the importance of obtaining improved varieties of the potato, and I believe no plant existing to be more extensively capable of improvement, relatively to the climate of England ; and if practical evidence were wanted to prove the extent, to which the culture of the potato is calculated to increase and support the population of a country, Ireland most amply affords it ; where population has increased amongst the Catholic poor, with almost unprecedented rapidity, within the last twenty years, under the pressure of more distress and misery, than has perhaps been felt in any other spot in Europe.

I shall conclude my present communication with some remarks upon the origin and cure of a disease, the *Curl*, which a few years ago destroyed many of our best varieties of the potato ; and to the attacks of which every good variety will probably be subject.

I observed that several kinds of potatoes, dry and farinaceous in their nature, which I cultivated, produced curled leaves, whilst those of other kinds, which were soft and aqueous, were perfectly well formed ; whence I was led to suspect, that the disease originated in the preternaturally inspissated state of the sap in the dry and farinaceous varieties. I conceived that the sap, if not sufficiently fluid, might stagnate in, and close, the fine vessels of the leaf during its growth and extension, and thus occasion the irregular contractions which constitute this disease ; and this conclusion, which I drew many years ago, is perfectly consistent with the opinions I have subsequently entertained, respecting the formation of leaves. I therefore suffered a quantity of potatoes, the produce almost wholly of diseased plants, to remain in the heap, where they had been preserved during winter, till each tuber had emitted shoots of three or four inches long. These were then carefully detached, with their fibrous roots, from the tubers, and were committed to the soil ; where having little to subsist upon, except water, I concluded the cause of the disease, if it were the too great thickness of the sap, would be effectually removed ;

and I had the satisfaction to observe, that not a single curled leaf was produced; though more than nine-tenths of the plants, which the same identical tubers subsequently produced, were much diseased.

In the spring of 1808, Sir John Sinclair informed me that a gardener in Scotland, Mr. Crozer, had discovered a method of preventing the curl, by taking up the tubers before they are nearly full grown, and consequently before they became farinaceous. Mr. Crozer, therefore, and myself, appear to have arrived at the same point by very different routes; for by taking his potatoes, whilst immature, from the parent stems, he probably retained the sap nearly in the state to which my mode of culture reduced it. I therefore conclude, that the opinions I first formed, are well founded; and that the disease may be always removed by the means I employed, and its return prevented by those adopted by Mr. Crozer.

XX.—ON THE CONSTRUCTION OF PEACH-HOUSES.

[*Read before the* HORTICULTURAL SOCIETY, *April 3rd,* 1810.]

SCARCELY any fruit can be raised in greater abundance, or with fewer chances of failure, than the peach in a forcing-house; where the insects, which often prove so formidable in the open air, are easily destroyed, and where the tree is subject to scarcely any other disease than the mildew; and I have reason to believe, that the appearance of this disease may, in general, be very easily prevented by selection of proper soil, and by proper management. But though a crop of peaches, or nectarines, is very easily obtained under glass, experience seems to have proved that neither of these fruits acquire perfection, either in richness or flavour, unless they be exposed to the full influence of the sun, during their last swelling, without the intervention of the glass. It has consequently been the practice, in some gardens, to take off the lights wholly before the fruit begins to ripen; and in warm seasons, and favourable situations, this mode of management succeeds perfectly well. But in the colder parts of England this cannot be done; and if the weather, in any part, prove cold and wet, just after the lights are taken off, the growth of the fruit is suddenly checked, and its quality greatly injured: and I have never met with the peach in so much perfection, as when it has been raised in a house where it could be conveniently exposed to the sun in

warm and bright days, and secluded from the cold night air, and rain ;
which mode of management can, I think, be adopted most conveniently
in a house constructed according to the annexed sketch and dimensions,
and the following directions.

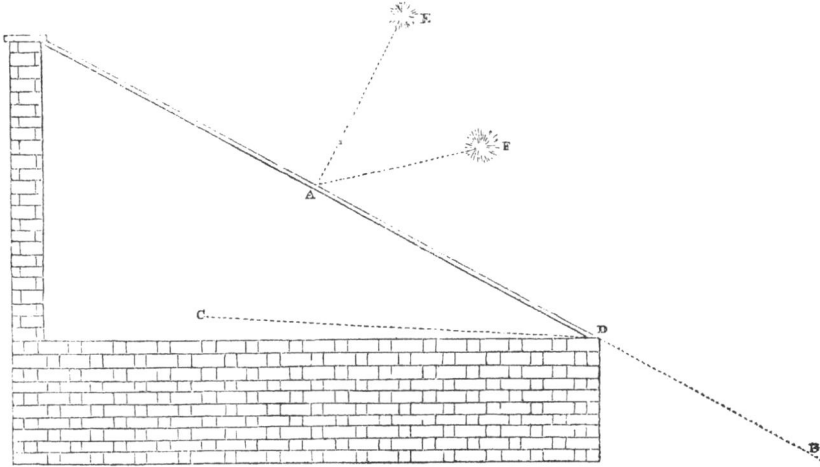

As the lights, to be moved to the required extent with facility, must
necessarily be short, the back wall of the house must scarcely extend nine
feet in height ; and this height raises the rafters sufficiently high to
permit the tallest person to walk with perfect convenience under them.
The lights are divided in the middle, at the point A, and the lower
are made to slide down to the point D, and the upper to the point
A*. The flue enters on the east or west end, as most convenient,
and passes within six inches of the east and west wall ; but not within
less than two feet of the low front wall ; and it returns in a parallel
line through the middle of the house, in the direction either east
or west, and goes out at the point at which it entered. The house
takes two rows of peach or nectarine trees, one of which is trained
on trellises, with intervals between, for the gardener to pass, parallel
with the dotted line C. These trees must be planted between the flue
and the front wall ; and the other row near the back wall, against which
they are to be trained.

If early varieties be planted in the front, and the earliest where the
flue first enters, these being trained immediately over the flue and at a
small distance above it, will ripen first ; and if the lower lights be drawn

* A bar of wood must extend from D to B, opposite the middle of each lower light, to support it when drawn down.

down in fine weather, to the point B, every part of the fruit on the trees, which are trained nearly horizontally, along the dotted line C, will receive the full influence of the sun. The upper lights must be moved, as usual, by cords and pulleys; and if these be let down to the point A, after the fruit on the front trees is gathered, every part of the trees on the back wall will be fully exposed to the sun, at any period of the spring and summer, after the middle of April, without the intervention of the glass. A single fire-place will be sufficient for a house of 50 feet long; and I believe the foregoing plan and dimensions will be found to combine more advantages than can ever be obtained in a higher or wider house.

Both the walls and flue must stand on arches, to permit the roots of the trees to extend themselves in every direction, beyond the limits of the walls; for whatever be the more remote causes of mildew, the immediate cause generally appears to be want of moisture beneath the soil, particularly if it be combined with excess of moisture, or dampness, above it. In experiments which I have made to discover the cause of mildew, in other plants, I have found that nothing so effectually prevents its appearance as abundant moisture beneath the soil; and many gardeners, who have had the misfortune to cultivate the peach in situations where the roots, at a small depth beneath the soil, were destroyed by water during winter, or where the same effect was produced by the unfavourable nature of the subsoil, must have observed the injurious effects of mildew.

I shall conclude my paper with observing, that I have never seen the peach in so great a state of perfection, as when cultivated very nearly according to the preceding directions: and I estimate so highly the advantages of bringing forward the fruit under glass, till it is nearly full-grown, and then exposing it to the stronger stimulus of sunshine, without the intervention of the glass, and excluding it from rain and dews, that I believe the peach might be thus ripened in greater perfection at St. Petersburg, in a house properly adapted to the latitude of that place, than in the open air at Rome or Naples.

XXI.—A CONCISE VIEW OF THE THEORY RESPECTING VEGETATION, LATELY ADVANCED IN THE PHILOSOPHICAL TRANSACTIONS, ILLUS-TRATED IN THE CULTURE OF THE MELON.

[*Read before the* HORTICULTURAL SOCIETY, *January* 2, 1811.]

THE council of the Horticultural Society having desired that I would send to the society a general view of my *Theory of Vegetable Physiology*, which has been published by the Royal Society, I have great pleasure in obeying their wishes ; and conceiving that I shall be able to render it more clear and useful, by making it illustrative of the proper culture of some particular plant, and by referring the reader to the papers in the *Philosophical Transactions* for evidence in support of the circumstances stated,—I have for this purpose chosen the melon.

A seed, exclusive of its seed-coats, consists of one or more cotyledons, a plumule or bud, and the caudex or stem of the future plant, which has generally, though erroneously, been called its radicle *. In these organs, but principally in the cotyledons, is deposited as much of the concrete sap of the parent plant as is sufficient to feed its offspring, till that has attached itself to the soil, and become capable of absorbing and assimilating new matter.

The plumule differs from the bud of the parent plant in possessing a new and independent life, and thence in assuming, in its subsequent growth, different habits from those of the parent plant. The organisable matter which is given by the parent to the offspring in this case, probably exists in the cotyledons of the seed, in the same state as it exists in the alburnum of trees ; and, like that, it apparently undergoes considerable changes before it becomes the true circulating fluid of the plant; in some it becomes saccharine, in others acrid and bitter, during germination †. In this process the vital fluid is drawn from the cotyledons into the caudex of the plumule or bud, through vessels which correspond with those of the bark of the future tree, and are indeed perfect cortical vessels ‡. From the point of the caudex springs the first root, which, at this period, consists wholly of bark and medulla, without any alburnous or woody matter ; and, if uninterrupted by any opposing body, it descends in a straight line towards the centre of the earth, in whatever position the seed has been placed, provided it has been permitted to vegetate at rest §.

Soon after the first root has been emitted, the caudex elongates, and taking a direction diametrically opposite to that of the root, it raises, in

* See above, No. XII.   † Above, No. V.   ‡ Above, No. XII.   § Above, No. XII.

a great many kinds of plants, the cotyledons out of the soil, which then become the seminal leaves of the young plant *. During this period the young plant derives nutriment almost wholly from the cotyledons or seed-leaves, and if those be destroyed it perishes. Gravitation, by operating on bodies differently organised and of different modes of growth, appears at once the cause why, in the preceding case, the root descends, and why the elongated plumule ascends †.

The bark of the root now begins to execute its office of depositing alburnous or woody matter : and as soon as this is formed, the sap, which had hitherto descended only through the cortical vessels, begins to ascend through the alburnum. The plumule in consequence elongates, its leaves enlarge and unfold, and a set of vessels, which did not exist in the root, are now brought into action. These, which I have called the central vessels, surround the medulla, and, between it and the bark, form a circle, upon which the alburnum is deposited by the bark, in the form of wedges, or like the stones of an arch ‡. Through these vessels, which diverge into the leaf-stalks, the sap ascends, and is dispersed through the vessels and parenchymatous substance of the leaf; and, in this organ, the fluid recently absorbed from the soil becomes converted into the true sap or blood of the plant; and as this fluid, during germination, descended from the cotyledons and seed-leaves of the plant, it now descends from its proper leaves, and adds, in its descent, to the bulk of the stem and the growth of the roots. Alburnum is also deposited in the stem of the plant, below the proper leaves, as it was previously deposited below the seed-leaves; and from this spring other central vessels, which give existence to, and feed, other leaves and buds §.

A considerable part of the ascending fluid must necessarily have been recently absorbed from the soil; but in the alburnum it becomes mixed with the true sap of the plant, a portion of which, during its descent down the bark, appears to secrete into the alburnum through passages correspondent to the anastomosing vessels of the animal economy ||. For as the cotyledons or seed-leaves first afforded the organisable matter which composed the first proper leaves, so these, when full-grown, prepare the fluid which generates other young leaves; the health and growth of which are as much dependent on the older leaves as those, when first formed, were upon the cotyledons ¶

The power of each proper leaf to generate sap, in any given species and variety of plant, appears to be in the compound ratio of its width,

---

* See above, No. VII.   † Above, No. VII.   ‡ Above, No. II.   § Above, Nos. II. and V.
|| Above, No. IX.   ¶ Above, No. V.

its thickness, and the exposure of its upper surface to light in proper temperature. As the growth of the plant proceeds, the number and width of the mature leaves increase rapidly in proportion to the number of young leaves to be formed ; and the creation consequently exceeds the expenditure of true sap. This therefore accumulates during a succession of weeks, or months, or years, according to the natural habits and duration of the plant, varying considerably according to the soil and climate in which each individual grows ; and the sap thus generated is deposited in the bulb of the tulip, in the tuber of the potato, in the fibrous roots of grasses, and in the alburnum of trees, during winter, and is dispersed through their foliage and bark during the spring and summer *.

As soon as the plant has attained its age of puberty, a portion of its sap is expended in the production of blossoms and fruit. These originate from and are fed by central vessels, apparently similar to those of the succulent annual shoot and leaf-stalk, and which probably convey a similar fluid ; for a bunch of grapes grew and ripened when grafted upon a leaf-stalk ; and a succulent young shoot of the vine, under the same circumstances, acquired a growth of many feet †.

The fruit, or seed-vessel, appears to be generated wholly by the prepared sap of the plant, and its chief office to be that of adapting the fluids, which ascend into it, to afford proper nutriment to the seed it contains‡.

I proceed to offer some observations upon the proper culture of the *melon.*

There is not, I believe, any species of fruit at present cultivated in the gardens of this country, which so rarely acquires the greatest degree of perfection, which it is capable of acquiring in our climate, as the melon. It is generally found so defective both in richness and flavour, that it ill repays the expence and trouble of its culture ; and my own gardener, though not defective in skill or attention, had generally so little success, that I had given him orders not to plant melons again. Attending, however, after my orders were given, more closely to his mode of culture, and to that of other gardeners in my neighbourhood, I thought I saw sufficient cause for the want of flavour in the fruit, in the want of efficient foliage; and appealing to experiment, I have had ample reason to think my opinions well founded.

The leaves of the melon, as of every other plant, naturally arrange themselves so as to present, with the utmost advantage, their upper

* See above, No. XII.　　† Above, Nos. III. and IV.　　‡ Above, No. II.

surfaces to the light : and if, by any means, the position of the plant is
changed, the leaves, as long as they are young and vigorous, make efforts
to regain their proper position. But the extended branches of the melon
plant, particularly under glass, are slender and feeble ; its leaves are
broad and heavy, and its leaf-stalks long ; so that if the leaves be once
removed, either by the weight of water from the watering-pot, the hand
of the gardener in pruning or eradicating weeds, or any other cause,
from their proper position, they never regain it ; and in consequence, a
large portion of that foliage, which preceded, or was formed at the same
period with the blossoms, and which nature intended to generate sap to
feed the fruit, becomes diseased and sickly, and consequently out of office,
before the fruit acquires maturity.

To remedy this defect, I placed my plants at greater distances from
each other than my gardener had previously done, putting a single plant
under each light, the glass of which was six feet long by four wide.
The beds were formed of a sufficient depth of rich mould to ensure
the vigorous growth of the plant : and the mould was, as usual, covered
with brick-tiles, over which the branches were conducted in every
direction, so as to present the largest possible width of foliage to the
light. Many small hooked pegs, such as the slender branches of the
beech, the birch, and hazel, readily afford, had been previously provided;
and by these, which passed into the mould of the bed, between the tiles,
the branches of the plants were secured from being disturbed from their
first position. The leaves were also held erect, and at an equal distance
from the glass, and enabled, if slightly moved from their proper position,
to regain it.

I, however, still found that the leaves sustained great injury from the
weight of the water falling from the watering-pot : and I therefore
ordered the water to be poured, from a vessel of a proper construction,
upon the brick-tiles, between the leaves, without at all touching them ;
and thus managed, I had the pleasure to see, that the foliage remained
erect and healthy. The fruit also grew with very extraordinary rapidity,
ripened in an unusually short time, and acquired a degree of perfection,
which I had never previously seen.

As soon as a sufficient quantity of fruit (between twenty and thirty
pounds) on each plant is set, I would recommend the further production
of foliage to be prevented, by pinching off the lateral shoots as soon as
produced, wherever more foliage cannot be exposed to the light. No
part of the full-grown leaves should ever be destroyed before the fruit is
gathered unless they injure each other, by being too much crowded

together; for each leaf, when full grown, however distant from the fruit, and growing on a distinct branch of the plant, still contributes to its support; and hence it arises, that when a plant has as great a number of growing fruit upon part of its branches, as it is capable of feeding, the blossoms upon other branches, which extend in an opposite direction, prove abortive.

The variety of melon, which I exclusively cultivate, is little known in this country, and was imported from Salonica by Mr. Hawkins. Its form is nearly spherical, when the fruit is most perfect, and without any depressions upon its surface; its colour approaching to that of gold, and its flesh perfectly white. It requires a much greater state of maturity than any other variety of its species, and continues to improve in flavour and richness, till it becomes externally soft, and betrays some symptoms of incipient decay. The consistence of its flesh is then nearly that of a water-melon, and it is so sweet, that few will think it improved by the addition of sugar. The weight of a good melon of this variety is about seven pounds.

XXII.—ON THE ADVANTAGES OF EMPLOYING VEGETABLE MATTER AS MANURE IN A FRESH STATE.

[*Read before the* Horticultural Society, *January 6th*, 1812.]

Writers upon agriculture, both in ancient and modern times, have dwelt much upon the advantages of collecting large quantities of vegetable matter to form manure; whilst scarcely any thing has been written upon the state of decomposition, in which decaying vegetable substances can be employed, most advantageously, to afford food to living plants. Both the farmer and gardener, till lately, thought that such manures ought not to be deposited in the soil till putrefaction had nearly destroyed all organic texture; and this opinion is, perhaps, still entertained by a majority of gardeners; it is, however, wholly unfounded. Carnivorous animals, it is well known, receive most nutriment from the flesh of other animals, when they obtain it most nearly in the state in which it exists as part of a living body; and the experiments I shall proceed to state, afford evidence of considerable weight, that many vegetable substances are best calculated to re-assume an organic living state, when they are least changed and decomposed by putrefaction.

I had been engaged, in the year 1810, in some experiments, from which I hoped to obtain new varieties of the plum ; but only one of the blossoms, upon which I had operated, escaped the excessive severity of the frost in the spring. The seed, which this afforded, having been preserved in mould during the winter, was, in March, placed in a small garden-pot, which was nearly filled with the living leaves and roots of grasses, mixed with a small quantity of earth : and this was sufficiently covered with a layer of mould, which contained the roots only of grasses, to prevent, in a great measure, the growth of the plants which were buried. The pot, which contained about one-sixteenth of a square foot of mould and living vegetable matter, was placed under glass, but without artificial heat, and the plant appeared above the soil in the end of April. It was three times, during the summer, removed into a larger pot, and each time supplied with the same matter to feed upon ; and in the end of October its roots occupied about the space of one-third of a square foot, its height above the surface of the mould being then nine feet seven inches.

In the beginning of June, a small piece of ground was planted with potatoes of an early variety, and in some rows green fern, and in others nettles, were employed instead of other manure ; and, subsequently, as the early potatoes were taken up for use, their tops were buried in rows in the same manner, and potatoes of the preceding year were placed upon them, and covered in the usual way. The days being then long, the ground warm, and the decomposing green leaves and stems affording abundant moisture, the plants acquired their full growth in an unusually short time, and afforded an abundant produce ; and the remaining part of the summer proved more than sufficient to mature potatoes of an early variety. The market-gardener may, probably, employ the tops of his early potatoes, and other green vegetable substances, in this way, with much advantage.

In these experiments, the plum-stone was placed to vegetate in the turf of the alluvial soil of a meadow, and the potatoes grew in ground which, though not rich, was not poor ; and, therefore, some objections may be made to the conclusions I am disposed to draw in favour of recent vegetable substances, as manures. The following experiment is, however, I think, decisive.

I received, from a neighbouring farmer, a field naturally barren, and so much exhausted by ill management, that the two preceding crops had not returned a quantity of corn equal to that which had been sowed upon it. An adjoining plantation afforded me a large quantity of fern, which I proposed to employ as manure for a crop of turnips. This was cut between the 10th and 20th of June; but as the small cotyledons of the turnip-seed afford little to feed the young plant ; and as the soil, owing to its extreme poverty, could not yield much nutriment, I thought it necessary to place the fern a few days in a heap, to ferment sufficiently to destroy life in it, and to produce an exudation of its juices; and it was then committed, in rows, to the soil, and the turnip-seed deposited, with a drilling machine, over it.

Some adjoining rows were manured with the black vegetable mould obtained from the site of an old wood pile, mixed with the slender branches of trees in every stage of decomposition, the quantity placed in each row appearing to me to exceed, more than four times, the amount of the vegetable mould, which the green fern, if equally decomposed, would have yielded. The crop succeeded in both cases; but the plants upon the green fern grew with greatly more rapidity than the others, and even than those which had been manured with the produce of my fold and stable-yard, and were distinguishable, in the autumn, from the plants in every other part of the field, by the deeper shade of their foliage.

I had made, in preceding years, many similar experiments with small trees (particularly those of the mulberry when bearing fruit in pots), with similar results : but I think it unnecessary to trespass on the time of the Society by stating these experiments, conceiving those I have mentioned to be sufficient to show that any given quantity of vegetable matter can generally be employed, in its recent and organised state, with much more advantage than when it has been decomposed, and no inconsiderable part of its component parts has been dissipated and lost, during the progress of the putrefactive fermentation.

XXIII.—ON FACILITATING THE EMISSION OF ROOTS FROM LAYERS.

[*Read before the* HORTICULTURAL SOCIETY, *February 4th*, 1812.]

IT is my custom, annually, to repeat every experiment that occurs to me, from which I have reason to expect information either in opposition to, or in favour of, the opinions I have advanced respecting the genera-tion and motion of the sap in trees ; and one of these experiments appearing to point out an improvement in the propagation of such trees by layering, as do not readily emit roots by that process, I send the following statement, under the hope that it may be acceptable to the Horticultural Society.

I have cited, in a former communication*, a part of the evidence upon which I have inferred that the sap of trees descends from their leaves through the bark; and I shall here only observe, in support of that opinion, that if a piece of bark be everywhere detached from the tree, except at its upper end, it will deposit, under proper management, as much, or nearly as much wood, upon its interior surface, as it will if it retain its natural position; and that the sap which generates the wood, deposited in the preceding circumstances, must descend through the bark, as it cannot be derived from any other source.

When a layer is prepared, and deposited in the ground, the progress of the sap, in its descent towards the original roots, is intercepted upon the side where the partially detached part, or tongue, of the layer is divided from the branch; and this intercepted sap is, in consequence, generally soon employed in the formation of new roots. But there are many species of trees which do not readily emit roots by this mode of treatment; and I suspected that, wherever roots are not emitted by layers, the sap, which descends from the leaves, must escape almost wholly through the remaining portion of bark, which connects the layer with the parent plant. I therefore attempted, in the last and the preceding spring, to accelerate the emission of roots by layers of trees of different species which do not really emit roots, by the following means, having detached the tongue of the layers from the branches in the usual manner.

Soon after Midsummer, when the leaves upon the layers had acquired their full growth, and were, according to my hypothesis, in the act of generating the true sap of the plant, the layers were taken out of the soil, and I found that those of several species of trees did not indicate any disposition to generate roots, a small portion of cellular bark only having issued from the interior surface of the bark in the wounded parts. I therefore took measures to prevent the return of the sap through the bark, from the layers to the parent trees, by making, round each branch, two circular incisions through the bark, immediately above the space where the tongue of the layer had been detached; and the bark between these incisions, which were about twice the diameter of the branch apart, was taken off. The surface of the decorticated spaces was then scraped with a knife, to prevent the reproduction of the bark, and the layers were recommitted to the soil; and at the end of a month I had the pleasure to observe that roots had been abundantly emitted by every one. In other instances, I obtained the same results by simply scraping off, at the

* Page 190.

same season, a portion of the bark, immediately at the base of the tongue of the layers, without taking them out of the ground.

By the preceding mode of management, the ascending fluid is permitted to pass freely into the layer to promote its growth, and to return till the period arrives at which layers generally begin to emit roots; the return of the sap through the bark is then interrupted, and roots are, in consequence, emitted; and I entertain little doubt that good plants of trees, of almost every species, may be thus obtained at the end of a single season. I wish it, however, to be understood, that my experiments have been confined to comparatively few species of trees; and that I am not much in the habit of cultivating trees of difficult propagation.

XXIV.—ON THE PREVENTION OF THE DISEASE CALLED THE CURL IN THE POTATOE.

[*Read before the* HORTICULTURAL SOCIETY, *February 2nd*, 1813.]

THE rough and uneven surface of the leaf, which in excess, indicates, and indeed constitutes, the disease called the curl in the potatoe, appears to exist in, and to form an essential characteristic of, every good variety of that plant; for I have never found a single variety, with perfectly smooth and polished leaves, which possessed any degree of excellence; and I have endeavoured to prove, in a former communication *, that the rough and crumpled state of the leaf probably originates in the preternaturally inspissated state of the fluid, in the firm and farinaceous potatoe. Those varieties are, however, generally most productive and grow with the greatest luxuriance, of which the leaves are smooth and polished; and this point tends to prove, that the smooth leaf is a more perfect and efficient organ than the rough one; the latter indicating some degree of approximation to disease.

I have stated, in another paper †, that I obtained a second crop of potatoes by planting those of an early variety in the same soil from which a crop of the same variety had been taken, in the month of July; and that I had employed, with success, the tops of those taken up, with green fern and nettles, as manure. But I found the tubers produced by those last planted to be much more soft and watery, when boiled, than others of the same variety, and consequently much inferior in value for every culinary purpose; and therefore, these were

* See page 185. † See page 194.

kept for the purpose of planting in the last spring. I inferred, consistently with the hypothesis I adduced in the paper last quoted, that the organisable matter these contained, being in a less firm and concrete state, would prove more disposable, and that I might therefore expect, in the succeeding season, plants of stronger growth, and more smooth and perfect foliage. The result, in every respect, coincided with my expectations; the plants presented the appearance of a different variety, and afforded a more abundant crop and larger tubers than I had ever obtained from the same variety.

This experiment was confined to a single very early kind, which had previously produced partially curled leaves; but I imagine the same mode of management will prove equally advantageous with other varieties which show similar indications of incipient disease; and as every improvement in the culture of this plant, which can add to the produce without increasing the expence, is of importance to the public, I submit the preceding account to the Horticultural Society.

A very respectable writer, in the Memoirs of the Caledonian Horticultural Society*, Mr. Dickson, has advanced an hypothesis, somewhat different from mine, respecting the curl in the potatoe: he conceives it to originate in debility arising from the too great ripeness of the tubers, and in the parent plant having too much expended itself in affording blossoms and seeds, as well as tubers. But I can scarcely accede to this hypothesis, because I do not think it probable that a plant, which is a native of Virginia, can be over-ripened in the climate of Scotland; and because those varieties, which never afford either blossoms or seeds, have, in my garden, been quite as subject to that disease as others. Mr. Dickson has stated the curious fact (and I do not entertain the slightest doubt of his perfect correctness), that a cutting taken from the extremity, which is most firm and farinaceous, of a long, or kidney-shaped potatoe, will afford diseased plants, whilst another cutting, taken from the opposite end of the same potatoe, will produce perfectly healthy plants; but I do not attribute this to the greater maturity of the buds at the extremity, than at the opposite end, for those nearest the parent plant are really the oldest, the tuber being formed by a branch, which has expanded itself laterally, instead of having extended itself longitudinally. Its buds are in consequence arranged as they would have been upon the elongated branch; and every tuber, in its incipient state of formation, will extend itself into a branch, as I have shown in the Philosophical Transactions for 1809†, provided the plant, to which it belongs, be cut off close to the

---

* See Vol. I. p. 50.   † See above, p. 157.

ground, and the current of ascending sap be in consequence diverted into, and through the tubers. Mr. Dickson, and myself, however, perfectly agree that a tuber, or part of one, which is soft and aqueous, affords a better plant than one which is firm and farinaceous ; and the trifling difference of opinion between us, being purely hypothetical, is of no importance.

I observed that the crops of potatoes, which I raised from the late ripened tubers above-mentioned, were not quite so early as others of the same variety ; but I attribute this variation in the periods of the maturity of the crops solely to different degrees of luxuriance in the plants, and to the increased size of the tubers in the one. In quality, the produce of both was the same.

XXV.—ON THE EARLY PUBERTY OF THE PEACH-TREE.

[*Read before the* HORTICULTURAL SOCIETY, *March* 2, 1813.]

IT was asserted, a few years ago, by a gentleman who had held an official situation in New South Wales, that a seedling peach-tree in that climate had produced fruit under his care when it was only sixteen months old, without having been grafted. The silence of the French writers upon gardening, respecting this earliness of puberty in the peach-tree, and the well-known circumstance that several years generally elapse between the period when a tree first springs from seed, and that in which it becomes capable of producing blossoms and fruit, appear to have induced a general disbelief of this account, which was mentioned to me, by several of my friends, as an extravagant and ridiculous falsehood ; and probably I should too readily have coincided with them in opinion, if I had not previously noticed several peculiar circumstances in the habits of seedling peach-trees. I had observed that such trees continued to grow as long as the weather continued favourable ; and that their leaves, in almost every succeeding month, assumed a more mature and improved character ; so that at the end of the first autumn, the leaves of the parent and seedling trees did not differ much from each other ; and such seedling trees, though they were retained in small pots till they were eighteen months old, and subsequently trained against a wall in the open air, and in a cold and late situation, produced fruit when only three years old.

I therefore thought it not improbable that, with the aid of glass and artificial heat, I might succeed in obtaining fruit from trees of two years old; and not impossible that, by a peculiar mode of pruning, I might obtain fruit from yearling trees, though the want of sunshine in our climate did not permit me to entertain very sanguine hopes of success.

Some peach stones, which were the produce of trees upon which I had made experiments in the year 1811, with the hope of obtaining early varieties of nectarines, were intended to have been placed in pots in a hot-house, in the beginning of January 1812; and one of my friends (I do not myself possess a hot-house) had offered me the use of his house to accelerate the germination and growth of the seedling plants. I, however, found the hot-house of my friend so much infested with insects of various kinds that I did not choose to risk my plants in it; and the seeds in consequence were not subjected to the influence of artificial heat till the middle of February, when I began to make fires in my vinery. The plants appeared above the soil early in March; and they were kept under glass during the whole summer and autumn; but without any artificial heat being applied after the end of May.

Conceiving that nature, in placing the age of puberty, in trees, so distant from the period in which they spring from seed, has intended chiefly to afford the plant, in this interval, the means of collecting a considerable store of organisable matter, before the expenditure of its sap commences in the production of blossoms and fruit, I adopted the mode of pruning and culture which, consistently with my theoretical opinions, appeared best calculated to promote that object. The leaves being the organs on which alone I believe the true sap of the tree to be generated, as many lateral shoots were suffered to remain upon each plant, as could present their foliage to the light without injuriously interfering with each other; and these were shortened, whilst very young, to the fourth or fifth leaf; and the buds in the axillæ of these leaves were destroyed as soon as they became visible; so that whatever portion of sap these leaves might generate, none might be uselessly expended. I had previously proved that leaves, under these circumstances, will promote the growth of the stem between themselves and the ground; so that any degree of taperness may be given to the stem, almost as accurately by the gardener, in proportioning the quantity and position of the foliage, as it can be subsequently given to the lifeless wood by the plane of the artificer; and I calculated that the true sap, which would be generated by the leaves upon the lower parts of the stem, and lateral shoots, would be employed in feeding the roots; whilst a portion of that, which would be generated by the foliage near the summit of the trees, might there contribute to the formation of fruit buds. The lateral shoots which were emitted near the tops of the young trees, when these had attained the height of seven or eight feet, were in consequence only shortened, the buds upon them being left, in the hope that some of them would be converted into blossoms.

The pots were filled with the green turf of the alluvial soil of a rich meadow, which substance I had previously employed with much success in similar experiments; and the pots were three times changed during the summer, and new portions of living turf added at the same periods.

The summer, however, proved so cold and cloudy, that I relinquished all hopes of success, proposing to repeat the experiment under less unfavourable circumstances; and in consequence the artificial heat, which I had intended to employ in Autumn, was not applied. I had, nevertheless, the unexpected pleasure to observe, late in the autumn, that three of the seven plants which had been the subjects of my experiment, had formed blossom buds; and these buds have subsequently presented so vigorous and healthy a character, that I do not entertain any doubt of their being capable of affording fruit.

The narrative of the planter of New South Wales was therefore, I conclude, perfectly correct; and I think it not improbable, that by shortening the lateral branches of his young plant, to give it a proper form, he incidentally adopted very nearly the same mode of pruning, which theoretical opinions pointed out to me as the best.

XXVI.—ON THE CULTURE OF THE PEAR-TREE.

[*Read before the* HORTICULTURAL SOCIETY, *May* 18*th*, 1813.]

THE pear-tree exercises the patience of the planter during a longer period before it affords fruit, than any other grafted tree which finds a place in our gardens; and though it is subsequently very long-lived, it generally, when trained to a wall, becomes in a few years unproductive of fruit, except at the extremities of its lateral branches. Both these defects are, however, I have good reason to believe, the result of improper management; for I have lately succeeded most perfectly in rendering my *old* trees very productive in every part; and my *young* trees have almost always afforded fruit the second year after being grafted; and none have remained barren beyond the third year.

In detailing the mode of pruning and culture I have adopted, I shall probably more easily render myself intelligible, by describing accurately the management of a single tree of each.

An old St. Germain pear-tree, of the spurious kind, had been trained in the fan form, against a north-west wall in my garden, and the central branches, as usually happens in old trees thus trained, had long reached the top of the wall, and had become wholly unproductive. The other branches afforded but very little fruit, and that never acquiring maturity, was consequently of no value; so that it was necessary to change the variety, as well as to render the tree productive.

To attain these purposes, every branch which did not want at least twenty degrees of being perpendicular, was taken out at its base; and the spurs upon every other branch, which I intended to retain, were taken off closely with the saw and chisel. Into these branches, at their subdivisions, grafts were inserted at different distances from the root, and some so near the extremities of the branches, that the tree extended as widely in the autumn, after it was grafted, as it did in the preceding year. The grafts were also so disposed, that every part of the space the tree previously covered, was equally well supplied with young wood.

As soon, in the succeeding summer, as the young shoots had attained sufficient length, they were trained almost perpendicularly downwards, between the larger branches, and the wall, to which they were nailed. The most perpendicular remaining branch upon each side, was grafted about four feet below the top of the wall, which is twelve feet high; and the young shoots, which the grafts upon these afforded, were trained inwards, and bent down to occupy the space from which the old central branches had been taken away; and therefore very little vacant space anywhere remained in the end of the first autumn. A few blossoms, but not any fruit, were produced by several of the grafts in the succeeding spring; but in the following year, and subsequently, I have had abundant crops, equally dispersed over every part of the tree; and I have scarcely ever seen such an exuberance of blossom as this tree presents in the present spring. Grafts of eight different kinds of pears had been inserted, and all afforded fruit, and almost in equal abundance. By this mode of training, the bearing branches, being small and short, may be changed every three or four years, till the tree is a century old, without the loss of a single crop; and the central part, which is unproductive in every other mode of training, becomes the most fruitful. I proceed to the management of young trees.

A young pear stock, which had two lateral branches upon each side, and was about six feet high, was planted against a wall early in the spring of 1810; and it was grafted in each of its lateral branches, two of which sprang out of the stem about four feet from the ground, and the others at its summit, in the following year. The shoots these grafts produced, when about a foot long, were trained downwards, as in the preceding experiment, the undermost nearly perpendicularly, and the uppermost just below the horizontal line, placing them at such distances, that the leaves of one shoot did not at all shade those of another. In the next year, the same mode of training was continued, and in the following, that is the last year, I obtained an abundant crop of fruit, and the tree is again heavily loaded with blossoms.

This mode of training was first applied to the Aston-Town pear, which rarely produces fruit till six or seven years after the trees have been grafted; and from this variety, and the colmar, I have not obtained fruit till the grafts have been three years old.

In the future treatment of my young pear-trees it is my intention to give them very nearly the form of the old tree I have described, in every respect, except that these will necessarily stand upon larger stems, which I think advantageous: and I shall not permit the existence of so great a number of large lateral branches. In both cases the bearing wood will depend wholly beneath the large branches which feed it; for it is the influence of gravitation upon the sap which occasions the early and exuberant produce of fruit.

I scarcely need add, that where, in old trees, it is not meant to change the variety, nothing more will be necessary than to take off wholly the spurs and supernumerary large branches, leaving every blossom which grows near the end of the remaining branches, or that the length of the dependent bearing wood must be different in different varieties. The Crassane, the Colmar, and Aston-Town, will require the greatest, and the St. Germain probably the least length.

XXVII.—ON THE PREVENTION OF MILDEW IN PARTICULAR CASES.

[*Read before the* Horticultural Society, *May* 4, 1813.]

The little pamphlet upon the rust, or mildew, of wheat, for which the public are indebted to the patriotic exertions of the venerable President of the Royal Society, affords much evidence in proof that this disease originates in a minute species of parasitical fungus, which is propagated, like other plants, by seeds ; and the evidence adduced would, I think, be sufficient to remove every doubt upon the subject, were the means ascertained by which the seeds of this species of fungus are conveyed from the wheat-plants of one season to those of the succeeding year. This, however, has not yet been done ; and therefore some persons still retain an opinion that the mildew of wheat consists only of preternatural processes, which spring from a diseased action of the powers of life in the plants themselves.

An hypothesis, which differs little from this, has been published in the present year respecting the dry-rot (Boletus lacrymans) of timber *. It is contended that the different kinds of fungus, which appear upon decaying timber of different species, are produced by the remaining powers of life in the sap of the unseasoned wood ; and that the same kind of living organisable matter, which, whilst its powers remained perfect would have generated an oak-branch, will, when debilitated, give existence to a species of fungus. But, if this power exists, and becomes capable, during its rapid declension, of deviating so widely from its original mode of action, the species of fungus it would produce might be expected to become successively more feeble and diminutive ; whereas the most robust and gigantic of the whole genus, the Boletus squamosus, springs from wood when that is in its last stage of decay; and the best known, and the most valuable species to mankind, of this tribe of plants, the common mushroom, appears as obviously to spring from horse-dung, under favourable circumstances, as any species of the same tribe appears to spring from decomposing wood, without the previous presence of seeds †. Yet it can scarcely be contended that any vital powers, capable of arranging the delicate organisation of a mushroom, can exist in a horse-dung ; and the admission of any such power would surely lead to the most extravagant conclusions. For if a mass of horse-dung can generate a mushroom, it can scarcely be denied that a mass of animal matter, an

* Quarterly Review, Vol. VII. page 33.
† See Nicol's Forcing, Fruit and Kitchen Gardener, 4th edition, page 119.

old cheese, may generate a mite; and if the organs of a mite can be thus formed, there could be little difficulty in believing that a larger mass of decomposing animal matter might generate an elephant, or a man.

The hypothesis therefore which supposes the various species of fungus to spring from seeds, appears to me much the least objectionable; and if the minute bodies, which are supposed to be the seeds of these plants, be really such, it will not be difficult to show that these are sufficiently numerous to account, to a great extent, for the ubiquity of the plants they are supposed to produce; particularly as such apparent seeds, owing to their excessive lightness, are capable of being everywhere dispersed by winds.

A few years ago I raised some mushrooms under glass with the intention of collecting and subsequently raising mushrooms from the seeds they might produce; and I then endeavoured to ascertain the number which would be afforded by a single fructification; for a mushroom appears to be nothing more than a fructification of the plant, though it is generally spoken of as the plant itself. I placed thin plates of talc under a very large mushroom at the period when the minute globular bodies, which are supposed to be the seeds, first began to be disengaged from its gills; and I endeavoured to count the number which fell during each successive hour, within the narrow field of a very powerful lens. The labour to my eyes was, however, so severe, that I was unable to count with any considerable degree of accuracy; but the number which fell from a single mushroom, within the succeeding ninety-six hours, exceeded, upon the lowest calculation I could make, two hundred and fifty millions. I endeavoured to raise mushrooms from these seeds, but I failed to obtain any decisive results; for though I readily procured mushroom spawn by mixing such seeds with unfermented horse-dung, I also obtained it in equal abundance, in some instances, where I had not introduced any seeds.

Immense as the number of seeds produced by a single mushroom appears, it probably is not much greater than that which a single plant of mildewed wheat would afford; and, according to this calculation, a single acre of mildewed wheat would probably afford seeds sufficient to communicate disease to every acre of wheat in the British empire, under circumstances favourable to the growth of the fungus; and I have never seen a single acre of wheat, since the publication of Sir Joseph Banks's pamphlet, so free from mildew but that it would have afforded seeds enough amply to supply the adjoining hundred acres. There is also reason to believe that the berberry-tree communicates this disease to wheat; and I have also often noticed a similar apparent parasitical fungus upon the straws of the couch-grass, in the hedges of corn-fields.

Neither the mildew of wheat, nor any other kind, can however I think, be communicated from the leaves and stems of one plant immediately to those of another : very numerous attempts made by myself to succeed in experiments of this kind having, I believe, proved wholly abortive ; though I once fancied that I had succeeded in two or three instances. I am, therefore, much inclined to believe that the parasitical fungus, which occasions every disease of this kind, enters the plant, in the first instance, by its roots, and though it may probably be transferred with the graft, and possibly by a bud, from one fruit-tree to another; and if the seeds be capable, like those of many other plants, of remaining sound a considerable time beneath the soil, or in other situations, till circumstances, which are favourable to their growth, occur, the abundant appearance of the mildew, or mushrooms, may be accounted for without supposing them to be generated wholly by the bodies from which they immediately spring.

I shall not trespass upon the time of the Horticultural Society by dwelling longer upon the primary cause of the various diseases which are comprehended under the name of mildew ; but shall proceed to the immediate object of the present memoir, which is to point out the means by which the injurious effects of the common white mildew may be, in particular cases, prevented.

The secondary and immediate causes of this disease, and of its congeners, have long appeared to me to be the want of a sufficient supply of moisture from the soil with excess of humidity in the air, particularly if the plants be exposed to a temperature below that to which they have been accustomed. If damp and cold weather in July succeed that which has been warm and bright, without the intervention of sufficient rain to moisten the ground to some depth, the wheat crop is generally much injured by mildew. I suspect that, in such cases, an injurious absorption of moisture, by the leaves and stems of the wheat plants, takes place; and I have proved, that under similar circumstances much water will be absorbed by the leaves of trees, and carried downwards through their alburnous substance ; though it is certainly through this substance that the sap rises under other circumstances. If a branch be taken from a tree when its leaves are mature, and one leaf be kept constantly wet, that leaf will absorb moisture and supply another leaf below it upon the branch, even though all communication between them through the bark be intersected ; and if a similar absorption takes place in the straws of wheat, or the stems of other plants, and a retrograde motion of the fluids be produced, I conceive that the ascent of the true sap or organisable matter into the seed-vessels must be retarded, and that it may become the food of the parasitical plants, which then only may grow luxuriant and injurious.

This view of the subject, whether true or false, led me to the following method of cultivating the pea late in the autumn, by which my table has always been as abundantly supplied during the months of September and October as in June and July; and my plants have been very nearly as free from mildew. The ground is dug in the usual way, and the spaces which will be occupied by the future rows are well soaked with water. The mould upon each side is then collected, so as to form ridges seven or eight inches above the previous level of the ground, and these are well watered; after which the seeds are sowed, in single rows, along the tops of the ridges. The plants very soon appear above the soil, and grow with much vigour, owing to the great depth of the soil, and abundant moisture. Water is given rather profusely once in every week or nine days, even if the weather proves showery; but if the ground be thoroughly drenched with water by the autumnal rains, no further trouble is necessary. Under this mode of management the plants will remain perfectly green and luxuriant till their blossoms and young seed-vessels are destroyed by frost; and their produce will retain its proper flavour, which is always taken away by mildew*.

The pea, which I have always planted for autumnal crops, is a very large kind, of which the seeds are much shrivelled, and which grows very high: it is now very common in the shops of London, and my name has, I believe, been generally attached to it. I prefer this variety because it is more saccharine than any other, and retains its flavour better late in the autumn; but it is probable that any other late and tall-growing variety will succeed perfectly well. It is my custom to sow a small quantity every ten days till midsummer, and I rarely ever fail of having my table well supplied till the end of October, though sometimes a severe frost in the beginning of that month proves fatal to my later crops.

The mildew of the peach, and of other fruit-trees, probably originates in the same causes as the mildew of the pea, and may be prevented by similar means. When the roots, which penetrate most deeply into the soil, and are consequently best adapted to supply the tree with moisture in the summer, are destroyed by a noxious subsoil, or by excess of moisture during the winter, I have observed the mildew upon many varieties

---

* One of the most experienced and close observers of our Society (Mr. Dickson) will probably recollect having seen my crops of peas in the state I have described, late in the autumn, in my garden at Elton.

of the peach to become a very formidable enemy. Where, on the contrary, a deep and fertile dry loam permits the roots to extend to their proper depth; and where the situation is not so low as to be much infested with fogs, I have found little of this disease: and in a forcing-house I have found it equally easy, by appropriate management, to introduce or prevent the appearance of it. When I have kept the mould very dry, and the air in the house damp and unchanged, the plants have soon become mildewed; but when the mould has been regularly, and rather abundantly watered, not a vestige of the disease has appeared.

It must be confessed that it is not easy to account, at first view, for the appearance of this disease under some of the preceding and various other circumstances, if it be produced by a parasitical plant which propagates by seeds; but all we ever see of the mildew is simply its fructification: the plant itself, if it be one, is wholly concealed from our senses; and it may consequently be transferred from one plant to another by the graft or bud, and never become visible till the health of the tree become affected by other causes. I could state some cases which are very favourable to this opinion, for this disease appears readily to be communicated by a graft to another tree, when that grows in the same soil, and in similar external circumstances. The different species of minute insects which feed upon the bodies of our domestic cattle are scarcely ever seen, and never injurious so long as the larger animals retain their health and vigour; but when these become reduced by famine or disease, the insects multiply with enormous rapidity, and though they are at first only symptomatic of disease, they ultimately become the chief and primary cause of its continuance. The reciprocal operation of the larger plant and the mildew upon each other may possibly be somewhat similar.

I offer the preceding opinions merely as conjectures: the hypothesis I have chosen has led me to the successful treatment of the disease in particular cases, and it may in the same way lead others: and I therefore venture to submit it to the consideration of the Horticultural Society without being very confident of its truth. If, however, the countless millions of apparently organised bodies, which are generated by the different species of fungus, be not seeds, nature appears to wander widely from its ordinary path: for amidst all its boundless profusion and exuberance, it does not ever, in other cases, appear to labour wholly in vain.

P.S. Observing that the almond-trees, round the metropolis, are likely to produce a considerable crop in the present year, I wish to recommend stocks of this species for Peaches and Nectarines to the attention of nurserymen, as likely to counteract the disposition in some varieties of the peach to become mildewed.   It has probably other qualities to recommend it, for it is obviously much more nearly allied to the peach than the plum is, if the peach and nectarine be not, as I suspect them to be, varieties only of the common almond (Amygdalus communis).   The almond stocks should be raised and retained in the nursery, in pots, as they do not transplant well.

XXVIII.—ON THE CULTURE OF THE SHALLOT, AND SOME OTHER BULBOUS-ROOTED PLANTS.

[*Read before the* HORTICULTURAL SOCIETY, *December 6th*, 1813.]

THE habits of bulbous-rooted plants of different species, relatively to the depths to which they naturally retire beneath the soil, admit of much variation, some occupying its surface, and others descending considerably beneath it.   These circumstances do not appear to have been sufficiently attended to, and injurious consequences have probably been the result, in many cases.

I have been led to adopt this opinion, and to make the experiments, which are the subject of this communication, by a complaint of my gardener, that the greater part of his crops of shallots had, during several years, generally become mouldy and perished : and I found, on enquiry, that the same thing had very often occurred in other gardens of the vicinity.   The bulbs had in all cases been planted, according to the directions of different writers upon horticulture, two or three inches beneath the soil ; and to this cause I attributed their failure.

A few bulbs of this species, which were divided, as far as practicable, into single buds, were therefore planted upon the surface of the ground, or rather above it, some very rich soil having been placed beneath them, and the mould having been raised on each side to support them, till they should become firmly rooted.   This mould was then removed by the hoe and watering-pot, and the bulbs in consequence were placed wholly out of the ground.   The growth of these plants now so closely resembled that of the common onion, as not to be readily distinguished from it ; till the irregularity of form, resulting from the numerous germs within each bulb, became conspicuous.   The forms of the bulbs, however, remained permanently different from all I had ever previously seen of the same species, being much more broad, and less long ; and the crop was so much better in quality, as well as much more abundant, that I can confidently recommend the mode of culture adopted to the attention of every gardener.

A few experiments similar to the preceding were made upon bulbs of the oriental hyacinth. Some of these were planted in the ordinary method beneath the soil, and others wholly above it, the mould being raised upon each side to cover them, and subsequently taken away ; and I found that those under the latter mode of culture flowered most strongly and in every other respect succeeded best. A compost, of great richness, formed of matter collected just without the gate of my fold-yard, and probably consisting of nearly equal parts of earth and cow-dung, by weight (if each substance had been perfectly dry), appeared to be exceedingly well adapted to this plant ; which expends much in a very short period of time in the production of leaves and flowers, and retains its foliage only a short time afterwards, and therefore probably requires more nutriment than it can generally obtain under the ordinary modes of culture. It is true that this, and some other bulbous-rooted plants, protrude their leaves and flowers as strongly, when supplied with water only, as when growing in good soil : but this growth is chiefly germination only, and during this process, in which the organs of the plant are merely formed out of matter previously assimilated, it may be questioned whether a single particle of new matter be ever vitally united to it.

A plant, of a very beautiful variety of the oriental hyacinth, which had been made to blossom with water only was, at my request, put into my hands in the last spring, just when its blossoms had begun to lose their beauty. Those were immediately taken off but the stem was suffered to remain, and the plant was removed from the bottom of water, in which it grew, into a pot sufficiently deep to receive its roots. A quantity of the rich compost above-mentioned was then, in successive portions, put into the pot, and washed in amongst the roots ; which were kept properly separated from each other. The bulb itself remained wholly out of the soil, with which it was not in contact, a thin layer of light and dry sandy loam intervening between it and the rich soil; and the bulb was also thinly covered with the same material. As the roots of the plant had been accustomed to live in water, the compost in the pot was at first kept very wet ; and the quantity of water subsequently given was lessened very gradually ; and as its leaves had been little exposed to light, it was retained under glass till the leaves perished. The bulb was then examined, and was found as solid, and apparently as perfect, as it would have been if it had germinated, as well as ultimately only grown, in a rich soil. The water in this case occasioned the extension of the roots, and the developement of the leaves, and thus was instrumental in forming organs capable of collecting and assimilating new matter; but exclusive of some impurities it contained, it probably had not given a particle of organisable matter to the plant. The formation of organs, and the action of those organs when formed, must not therefore be confounded, as has generally been done, and constantly by chemists who have endeavoured to ascertain the action of the leaves upon the surrounding air; and hence appear to have arisen the confused and contradictory results of their experiments.

I am wholly ignorant of the mode of management by which bulbous roots of different kinds, acquire so much greater perfection in the hands of the Dutch gardeners, than in those of our own countrymen : but I suspect that the Dutch gardeners employ subsoils of very great depth and richness, with which the bulbs are prevented coming into contact by the intervention of a thin layer of dry sand, with which substance they may be also thinly, or only partially, covered; and I am in part led to adopt this opinion, by observing the similarity of character in the external membranes of their bulbous roots, and of those of the shallots, which had been wholly exposed to the sun and air.

XXIX.—ON THE APPLICATION OF MANURE IN A LIQUID FORM TO PLANTS IN POTS.

[Read before the HORTICULTURAL SOCIETY, *May 17th,* 1814.]

THE quantity of earth, which the most firm and solid parts of trees afford by analysis, is well known to be very small; and even the species of these earths have been proved, by the younger Saussure, to be dependent, to a great extent, upon the component parts of the soil, in which the trees happen to have grown. A large extent and depth of soil seem therefore to be no further requisite to trees than to afford them a regular supply of water, and a sufficient quantity of organisable matter; and the rapid growth of plants of every kind, when their roots are confined in a pot to a small quantity of mould, till that becomes exhausted, proves sufficiently the truth of this position.

I have shown in a former communication*, that a seedling plum-stock, growing in a small pot, attained the height of nine feet seven inches, in a

* See page 194.

single season; which is, I believe, a much greater height than any seedling tree of that species was ever seen to attain in the open soil. But the quantity of earth, which a small pot contains, soon becomes exhausted, relatively to one kind of plant; though it may be still fertile relatively to others: and the size of the pot cannot be changed sufficiently often to remedy this loss of fertility; and if it were ever so frequently changed, the mass of mould, which each successive emission of roots would enclose, must remain the same.

Manure can therefore probably be most beneficially given in a purely liquid state; and the quantity which trees growing in pots have thus taken, under my care, without any injury and with the greatest good effect, has so much exceeded every expectation I had formed, that I am induced to communicate to the Society the particulars and the result of my experience.

I have for some years appropriated a forcing-house, at Downton, to the purposes of experiment solely upon fruit-trees; which, as I have frequent occasion to change the subjects upon which I have to operate, are confined in pots. These were at first supplied with water in which about one-tenth, by measure, of the dung of pigeons, or domestic poultry, had been infused; and the quantity of these substances (generally the latter) was increased from one-tenth to a fourth. The water, after standing forty-eight hours, acquired a colour considerably deeper than that of porter; and in this state was drawn off clear, and employed to feed trees of the vine, the mulberry, the peach, and other plants. A second quantity of water was then applied, and afterwards used in the same manner; when the manure was changed, and the same process repeated.

The vine and mulberry tree, being very gross feeders, were not likely to be soon injured by this treatment; but I expected the peach-tree, which is often greatly injured by excess of manure in a solid state, to give early indications of being over-fed. Contrary, however, to my expectations the peach-tree maintained, at the end of two years, the most healthy and luxuriant appearance imaginable, and produced fruit in the last season in greater perfection than I had ever previously been able to obtain it. Some seedling plants had then acquired, at eighteen months old, (though the whole of their roots had been confined to half a square foot of mould,) more than eleven feet in height with numerous branches, and have afforded a most abundant and vigorous blossom in the present spring, which has set remarkably well; and those trees which had been most abundantly supplied with manure have displayed the greatest degrees of health and luxuriance.

A single orange-tree was subjected to the same mode of treatment, and grew with equal comparative vigour, and appeared to be as much benefited by abundant food as even the vine and mulberry tree.

An opinion generally, though I think somewhat erroneously, prevails that many plants, particularly the different species and varieties of heath, require a very poor soil in pots; but these might, I conceive, with propriety, be said to require a peculiar soil; for I have never seen the common species of this genus spring with so much luxurance as from a deep bed of vegetable mould, which had been recently very thickly covered with the ashes of a preceding crop of heaths and other plants that had been burned upon it. And I believe, if the branches and leaves of the common species of heath were placed to decompose in water, and such water were afterwards given to the tender exotic species, that these, how heavily soever the water might be loaded with organisable matter, would be found as little capable of being injured by abundant food as the vine or mulberry tree, though the species of food which would best suit those plants might prove to every species of heath destructive and poisonous.

XXX.—ON THE ILL EFFECTS OF EXCESSIVE HEAT IN FORCING-HOUSES DURING THE NIGHT.

[*Read before the* HORTICULTURAL SOCIETY, *June 17th*, 1814.]

FEW gardeners, if any, have ever believed plants to be at all endued with powers of sensation and perception similar to those of animals; or to be, in any degree, susceptible of pleasure or pain; and yet it is very questionable whether there has ever been a single gardener, who, in the management of fruit-trees in a forcing-house, did not in some respects err by treating his trees as he would have done, if he had supposed them to possess such powers. Being fully sensible of the comforts of a warm bed in a cold night, and of fresh air in a hot day, the gardener generally treats his plants as he would wish to be treated himself; and, conse-quently, though the aggregate temperature of his house be nearly what it ought to be, its temperature during the night, relatively to that of the day, is almost always much too high. The consequences of this excess of heat during the night are, I have reason to believe, in all cases highly injurious to the fruit-trees of temperate climates, and not at all beneficial to those of tropical climates; for the temperature of these is, in many instances, low during the night. In Jamaica, and other mountainous islands of the West Indies, the air upon the mountains becomes, soon after sunset, chilled and condensed, and, in consequence of its superior gravity, descends and displaces the warm air of the valleys; yet the sugar-canes are so far from being injured by this sudden decrease of temperature, that the sugars of Jamaica take a higher price in the market than those of the less elevated islands, of which the temperature of the day and night is subject to much less variation.

During the progress of germination, in the spring, great chemical changes take place in the component parts of the sap of trees, analogous to those which have been observed in the germination of seeds. I could not detect any vestige of saccharine matter in the alburnum, either of the stem or roots of the sycamore tree in the winter; but in the spring, its sap became very sensibly sweet: and I found this sap to be much more saccharine, and of greater specific gravity, in large trees, which were prepared to nourish an abundant blossom, than in small and young trees. The sap of the same tree proved also to be subject to some variations of specific gravity, at the same period of the spring, in different years; and Duhamel has observed, that the sap of the sugar maple becomes first saccharine, and afterwards acquires an herbaceous taste; in the latter state, it probably is best calculated to feed the blossoms and unfolded buds.

At the period of the preceding chemical changes in the qualities and properties of the sap, previous to the growth of the leaves, that fluid is found to ascend during the warm part of the day, and to flow, in many species of trees, from any recent wound, and to fall again during the night, particularly if that be cold; and as variations of temperature are the apparent cause of these motions, it appears not improbable, that the chemical changes, which take place in it at this period, are promoted by the same agents.

Some experiments which I have made upon germinating seeds, have perfectly satisfied me, that these afford plants of greater or less vigour in proportion as external circumstances are favourable in promoting beneath the soil the necessary changes in the nutritive matter they contain; and I suspect that a large portion of the blossoms of the cherry and other fruit-trees in the forcing-house often proves abortive, because they are forced, by too high and uniform a temperature, to expand before the sap of the tree is properly prepared to nourish them.

I have therefore been led, during the last three years, to try the effects of keeping up a much higher temperature in the day than in the night;

and as experiments of this kind cannot be made by the common gardener, who must not risk the sacrifice of his employer's crops of fruit, I trust the following account will be honoured by the approbation of the Horticultural Society, though the experiments have been chiefly confined to the peach-tree.

As early in the spring as I wished the blossoms of my peach-trees to unfold, my house was made warm during the middle of the day; but towards night it was suffered to cool, and the trees were then sprinkled, by means of a large syringe, with clear water, as nearly at the temperature at which that usually rises from the ground, as I could obtain it; and little or no artificial heat was given during the night, unless there appeared a prospect of frost. Under this mode of treatment the blossoms advanced with very great vigour, and as rapidly as I wished them, and presented, when expanded, a larger size than I had ever before seen of the same varieties: which circumstance is not unimportant, because the size of the blossom, in any given variety, regulates, to a very considerable extent, the bulk of the future fruit. As soon as the blossoms were expanded, and the pollen began to shed, water was applied in less quantity, as a light shower, sufficient to wet the pollen, without washing it off; but when the pollen was chiefly shed, I again, to promote its absorption, sprinkled the trees abundantly with water, having previously often observed that heavy showers of rain are at this period always highly beneficial to the blossoms of the apple trees in our orchards; and almost every blossom of my peach-trees set most perfectly. The watering was regularly continued till the fruit became very nearly ripe, the roots of the trees being, at the same time, abundantly supplied with moisture and food in the manner detailed in my last paper, in which I have stated the more than ordinary size and perfection of the fruit.

My house had been previously much infested with the red spider*; but not a single one now appeared, nor scarcely an *aphis;* and the young wood became remarkable for the shortness of its joints, and the thickness, comparatively with the length of its shoots. A gardener, who is prejudiced in favour of old customs, will possibly imagine that he supplies the place of the cool evening dews of nature, and of the water in the preceding experiment, by sprinkling his flues with water, and filling his house abundantly with steam. But the effect of no two operations can be more different: in the one, the plant is suddenly chilled by cold water,

---

* I suspect, but I am no entomologist, that two distinct species of insect are confounded under this name, one of which forms a web, which the other does not. The latter kind often abounds in the open air, upon pear-trees, and appears to be, in the forcing-house, a much hardier insect than the other.

and subsequently kept cool by the evaporation of the water during the night : in the other, the steam is precipitated upon the leaves and branches of the trees, to which it necessarily communicates much heat. The former operation nearly resembles that of the shower-bath, sometimes used in this country, in which the patient is suddenly chilled by a heavy shower of cold water; the other resembles the hot steam-bath of Russia, in which he is violently heated ; and if the gardener were to try each of these processes upon himself, during a single night, I suspect he would arise in the following morning with very different feelings, unless he were blest with much peculiar hardness of constitution.　It is true, that plants do not appear to possess sensation in the ordinary sense of that term, as it is applied to animals ; but nature, in forming its whole organic creation, seems to have proceeded so much by substitutions and additions, that simple sensation, in its strict and limited sense, abstracted from all powers of perception, may not improbably be as widely diffused as organisation itself; and animal and vegetable life may be, in consequence, susceptible of similar injuries from similar external causes.　The influence of hot and damp air upon both, is greatly more powerful than that of dry air of the same temperature.　In the experiments of which Sir Charles Blagden has given an account in the Philosophical Transactions of 1775, he, with Sir Joseph Banks and others, sustained without injury a temperature of 260 degrees in dry air ; but they found damp air, at half that temperature, to be scarcely supportable : and every gardener knows, how quickly the leaves of his plants are injured by the combined action of heat and moisture.

The succulent shoots of trees, however, always appear to grow most rapidly, in a damp heat, during the night ; but it is rather elongation than growth which then takes place.　The spaces between the bases of the leaves become longer, but no new organs are added ; and the tree, under such circumstances, may with much more reason be said to be drawn, than to grow ; for the same quantity only of material is extended to a greater length, as in the elongation of a wire.

Another ill effect of high temperature during the night is, that it exhausts the excitability of the tree much more rapidly than it promotes the growth, or accelerates the maturity of the fruit : which is in consequence ill supplied with nutriment, at the period of its ripening, when most nutriment is probably wanted.　The muscat of Alexandria, and other late grapes, are, owing to this cause, often seen to wither upon the branch in a very imperfect state of maturity ; and the want of richness and flavour in other forced fruits is, I am very confident, often attributable to the same cause. There are few peach-houses, or indeed forcing-houses of any kind, in this country, in which the temperature does not exceed, during the night, in the months of April and May, very greatly that of the warmest valley in Jamaica in the hottest period of the year : and there are probably as few forcing-houses in which the trees are not more strongly stimulated by the close and damp air of the night, than by the temperature of the dry air of the noon of the following day. The practice which occasions this cannot be right : it is in direct oppo- sition to nature : and I need not point out to the intelligent members of the Horticultural Society, that the more nearly nature, in its best climates and most favourable seasons, is copied as to temperature, the more perfect will be the productions of the gardener's art.

XXXI.—ON THE MODE OF PROPAGATION OF THE LYCOPERDON CANCELLATUM*, A SPECIES OF FUNGUS, WHICH DESTROYS THE LEAVES AND BRANCHES OF THE PEAR-TREE.

[*Read before the* HORTICULTURAL SOCIETY, *December 5th*, 1815.]

I HAD the honour, two years ago, to address to the Horticultural Society some observations upon the propagation of these supposed species of parasitical plants, which, under the name of fungi†, appear as diseases upon other living plants: and of other supposed species of the same tribe, which decompose and feed upon organic substances, that have ceased to live. In the present communication, I shall endeavour to show, that one of these, at least, is a parasitical plant, which propagates like other plants, by seeds.

I observed, about seven years ago, a disease upon a few of the leaves of one of the pear-trees in my garden at Downton. Bright yellow spots, from which a small quantity of liquid exuded, appeared upon the upper surfaces of the leaves in June; and subsequently, several conic processes, about one third of an inch in length, were protruded from the same parts, but from the opposite surface, of each leaf; and from these a large quantity of brown impalpable powder, consisting of very minute globular bodies, was discharged in August and September. These minute

---

* I am indebted for the name of this species of fungus to the extensive information of Mr. Dickson, who referred me to the *Flora Danica* for a delineation of it : but Sir Joseph Banks subsequently showed me a drawing of it by Mr. Bauer, which is much more elaborate and correct.

† See page 204.

globular bodies I concluded to be seeds of a species of fungus; but as a few only of the leaves of my trees were affected, and no very injurious effects were visible, I did not take any measures to prevent their dispersion over my garden.

I did not, however, long remain ignorant of the formidable nature of my new enemy; for within two years, every pear-tree in my garden became in some degree diseased. The leaves only, at first, appeared to be injured; but the disease soon extended itself to the annual branches, in many protuberant yellow spots, beneath which the bark was found to have acquired a bright yellow colour: and as far as this colour extended, the bark, and the wood beneath it invariably perished, either in the same or following season, leaving wounds similar to these inflicted by canker, but less curable. The fruit also became diseased and worthless, and almost all the young shoots, when once attacked, perished in the following winter. These effects were not confined to my garden, but extended to the pear-trees in an orchard which was two hundred yards distant; and I cannot entertain a doubt, but that the disease was communicated to these by seeds which had been conveyed by the prevalent west winds. I endeavoured, during the summers of 1813 and 1814, to check its progress in my garden, by picking off every diseased leaf; but I found all my efforts nearly abortive, and I have been obliged to destroy the greater part of my pear-trees: those which remain have become annually more diseased, and I fear never can be ultimately preserved, unless a remedy for the disease can be discovered.

I tried the effect, in the last season, of sprinkling the leaves of different pear-trees, just at the period when eht liquid exuded from the spots upon their surfaces, with quick-lime and fresh wood ashes, in which the alkali and lime were in a caustic state; and with flowers of sulphur. The spots to which the quick-lime and ashes were applied, soon became paler; but I had not an opportunity of observing the ultimate effect of these substances: for almost all the leaves of the pear-trees upon my walls, in the last season, became covered with black and lifeless spots, and fell off prematurely. Those of a single small standard pear-tree, on which flowers of sulphur had been sprinkled, remained alive till late in the autumn; and upon these I did not observe the sulphur to operate in any degree, till the period at which the conic processes above-mentioned would have appeared; but the yellow spots then became black, and perished, without affording seeds; whence I have reason to hope, that flowers of sulphur will prevent, in some measure at least, the rapid extension of this disease.

As the existence of this species of fungus appeared, three years ago, to be confined to my garden and a few pear-trees in its vicinity, and to the hawthorn in an adjoining hedge (for it attacks the hawthorn as well as the pear-tree), I then thought that it would be practicable to ascertain decisively the means by which it transfers itself from one tree to another : and this appeared to me to be an important object; because the habits of the lycoperdon cancellatum, and of the fungus which forms the rust or mildew of wheat, are, in many respects, very similar.

I had so often tried, without success, to transfer the mildew of wheat, and other plants, from a diseased to a healthy subject, in the same season, that I had not any expectation of succeeding in an attempt of that kind; but I thought it not improbable, that I might succeed in communicating this disease to seedling plants of the pear-tree, having long ago satisfied myself that the species of fungus, which forms the mildew of wheat, always rises from the root of the plant.

I have many years been in the habit of raising annually pear-trees from seeds, with the hope and expectation of obtaining new and hardy varieties for the dessert in winter ; which may succeed without the pro-tection of a wall; and as the means I employ to obtain seeds well calculated for my purpose, necessarily cost me a good deal of time and labour, I have always planted them in pots, and in the kind of mould which long experience has pointed out to me as the best. This I have always obtained, at the period of sowing the seeds, in January or February, from the banks of a river at some distance from my garden ; and in this mould my seedling pear-trees always sprang up, and remained during the first season perfectly free from disease. In the spring of 1813, a portion of this mould, which I did not want, was intentionally placed very near some hawthorns and pear-trees, upon which the lyco-perdon cancellatum abounded, where it remained till the spring of 1814, when it was put into pots, and new seeds deposited in it. These sprang up as usual, and remained in perfect health till the end of May or begin-ning of June ; when the fungus presented itself upon almost all the first true leaves of the plants, which leaves had composed the plumules of the seeds.

That the fungus, in this case, rose from the ground, will, I think, scarcely be questioned; but it is necessary to state, that the seeds were all taken from trees which were not quite free from disease ; and that I saw in the last spring some diseased plants, in a case where every pre-caution, except that of using new pots (which had been my previous custom), had been taken ; and therefore, whilst so little is known respect-ing the habits of plants of this tribe, the preceding facts are not sufficient to support a decision, that the source of the disease might not have been in the seeds themselves. For as the fructification is probably everything which is seen of this, and many other parasitical fungous plants, the plant may extend in minute filaments through the whole body of the tree which supports it; and it appears in this view of the subject possible, that these slender filaments may extend into the seeds. The following circumstances, however, militate strongly in opposition to this conclusion. A great number of seedling pear-trees, which were very much diseased, were removed, in the last spring, from my garden to a distant situation, after having had their roots and stems carefully and repeatedly washed, and brushed, so as to remove from them every particle of the mould in which they had previously grown; and upon these not a vestige of disease has since appeared. Grafts also, which were formed of parts of diseased trees, have in all cases produced perfectly healthy foliage, even when inserted into the branches of other diseased trees; which circumstance I think interesting, because it tends to point out a further apparent similarity in the habits of this species of fungus, and that which forms the mildew of wheat: which ceases to vegetate as soon as the straw is severed from its roots, though that remains for some time green and living: whence arises the advantage of cutting mildewed crops of wheat in an immature state. Further experience can, however, alone decide these points: and the only inference I wish to draw from the facts I have stated is, that the lycoperdon cancellatum is capable, under certain circumstances, of being transferred from one plant to another in its vicinity, by means of its seeds.

I observed this disease, in the last summer, upon a few of the leaves of several pear-trees in the vicinity of London; and I fear that the fungus which occasions it is an imported species, that is likely to increase in our climate, and to become, in some situations at least, extremely injurious to one of the most valuable of our fruit-trees. I have met with several intelligent gardeners who, at first view, thought they had observed this disease some years ago; but on further inspecting its habits and injurious effects, they have always changed their opinion.

The enormous injury which the crops of wheat sustained in the year 1814 and other seasons, by mildew, attaches a great degree of interest to the investigation of the habits of parasitical plants of this tribe; and the similarity of habits of the mildew of wheat, and of the lycoperdon cancellatum, renders it probable that both are propagated in the same manner. I therefore venture to hope that the foregoing account, though very imperfect, of the apparent mode of propagation of the latter plant, may be thought deserving the attention of the Horticultural Society.

XXXII.—ON THE EFFECTS OF DIFFERENT KINDS OF STOCKS IN GRAFTING.

[*Read before the* Horticultural Society, *February 6th*, 1816.]

The practice of propagating fruits of different species, by grafting upon stocks of other species, has been so extensive, both in ancient and modern times, that the good and ill effects of it can scarcely be supposed to have escaped the observation of gardeners. Accurate information upon this subject can, however, only be acquired by experiments accurately made, and closely attended to, during many successive years, upon the comparative good and ill effects of stocks of different species, when growing in soils of the same, and of different qualities: and no such experiments, have, I believe, ever been made in this country, nor, to a proper extent, in any other. Duhamel has pointed out, with his usual ability, the erroneous opinions entertained by his countrymen upon this subject, and has given some valuable information; but he admits, that relatively to some very important points, he only details the opinions of others; and he laments that he has not himself made the experiments necessary to decide the questions, which he wishes to investigate. I also feel, that I am not, by any means, master of the subject upon which I have taken up my pen to write: but I believe that I have made and seen the result of more experiments, during the last thirty-five years, than any other person; and I venture to hope, that my experience enables me to draw a few conclusions, which may prove useful.

Whenever the stock, and graft, or bud, are not perfectly well suited to each other, an enlargement is well known always to take place at the point of their junction, and generally to some extent, both above and below it. This is particularly observable in peach-trees, which have been grafted, at any considerable height from the ground, upon plum stocks; and it appears to arise from obstruction, which the descending sap of the peach-tree meets with in the bark of the plum stock; for the effects produced, both upon the growth and produce of the tree, are similar to those which occur when the descent of the sap is impeded by a ligature, or by the destruction of a circle of bark, in the manner recommended by Mr. Williams in a former volume of the Horticultural Transactions *. The disposition in young trees to produce and nourish blossom, buds, and fruit, is increased by this apparent obstruction of the

* Vol. I. page 108.

descending sap; and the fruit of such young trees ripens, I think, somewhat earlier than upon other young trees of the same age, which grow
upon stocks of their own species; but the growth and vigour of the tree,
and its power to nourish a succession of heavy crops, are diminished,
apparently, by the stagnation, in the branches and stock, of a portion of
that sap, which, in a tree growing upon its own stem, or upon a stock of
its own species, would descend to nourish and promote the extension of
the roots. The practice, therefore, of grafting the pear-tree on the
quince stock, and the peach and apricot on the plum, where extensive
growth and durability are wanted, is wrong; but it is eligible wherever
it is wished to diminish the vigour and growth of the tree, and where its
durability is not thought important. The last remark applies chiefly to
the Moor-park apricot *.

When great difficulty is found in making a tree, whether fructiferous,
or ornamental, of any species, or variety, produce blossoms, or in making
its blossoms set when produced, success will probably be obtained in
almost all cases, by budding or grafting upon a stock which is nearly
enough allied to the graft to preserve it alive for a few years, but not
permanently. The pear-tree affords a stock of this kind to the apple;
and I have obtained a heavy crop of apples from a graft which had been
inserted in a tall pear stock, only twenty months previously, in a season
when every blossom of the same variety of fruit in the orchard was
destroyed by frost. The fruit thus obtained was externally perfect, and
possessed all its ordinary qualities; but the cores were black and without
a single seed; and every blossom had certainly fallen abortively, if it had
been growing upon its native stock. The experienced gardener will
readily anticipate the fate of the graft: it perished in the following
winter. The stock, in such cases as the preceding, promotes, in proportion to its length, the early bearing and early death of the graft.

The authority of Duhamel gives us reason to believe, that the defects
of particular soils may be remedied by a proper selection of stocks; and
that cases may occur, in which it will be eligible to bud the peach and
nectarine upon the apricot or plum. My own experience induces me to
think very highly of the excellence of the apricot stock, for the peach or
nectarine; but wherever that, or the plum stock is employed, I am
confident the bud cannot be inserted too near the ground, when vigorous
and durable trees are wanted. The opinion of Mr. Wilmot, in a former
volume of our Transactions†, is, upon this point, opposed to mine; but I

* The Abricot-Pêche, or Abricot de Nancy, of the French.

† Horticultural Transactions, Vol. I. page 216.

speak upon the evidence of long experience, and of experiments accurately and purposely made with my own hands.

The form and habit which a peach-tree of any given variety is disposed to assume, I find to be very much influenced by the kind of stock upon which it has been budded : if upon a plum or apricot stock, its stem will increase in size considerably, as its base approaches the stock, and it will be much disposed to emit many lateral shoots, as always occurs in trees whose stems taper considerably upwards ; and, consequently, such a tree will be more disposed to spread itself horizontally, than to ascend to the top of the wall, even when a single stem is suffered to stand perpendicularly upwards. When, on the contrary, a peach is budded upon the stock of a cultivated variety of its own species, the stock and the budded stem remain very nearly of the same size at, as well as above and below, the point of their junction. No obstacle is presented to the ascent, or descent, of the sap, which appears to ascend more abundantly to the summit of the tree. It also appears to flow more freely into the slender branches, which have been the bearing wood of preceding years : and these consequently extend themselves very widely, comparatively with the bulk of the stock and large branches.

When a stock of the same species with the graft or bud, but of a variety far less changed by cultivation, is employed, its effects are very nearly allied to those produced by a stock of another species, or genus : the graft, generally, overgrows its stock ; but the form and durability of the tree are generally less affected, than by a stock of a different species or genus.

Many gardeners entertain an opinion, that the stock communicates a portion of its own power to bear cold, without injury to the species, or variety of fruit, which is grafted upon it : but I have ample reason to believe, that this opinion is wholly erroneous : and this kind of hardiness in the root alone can never be a quality of any value in a stock ; for the branches of every species of tree are much more easily destroyed by frost, than its roots. Many also believe, that a peach-tree, when grafted upon its native stock, very soon perishes ; but my experience does not further support this conclusion, than that it proves seedling peach-trees, when growing in a very rich soil, to be greatly injured, and often killed, by the excessive use of the pruning-knife upon their branches, when those are confined to too narrow limits. The stock, in this instance, can, I conceive, only act injuriously by supplying more nutriment than can be expended ; for the root which nature gives to each seedling plant must be well, if not best, calculated to support it ; and the chief general conclusions which my experience has enabled me safely to draw, are, that a stock of a species, or genus, different from that of the fruit to be grafted upon it, can rarely be used with advantage, unless where the object of the planter is to restrain and to debilitate: and that where stocks of the same species with the bud, or graft, are used, it will generally be found advantageous to select such as approximate in their habits, and state of change, or improvement, from cultivation, those of the variety of fruit which they are intended to support.

XXXIII.—ON THE VENTILATION OF FORCING-HOUSES.

[*Read before the* HORTICULTURAL SOCIETY, *May 7th*, 1816.]

In a memoir which I had two years ago the honour to address to the Horticultural Society *, I stated an opinion that the gardener often erred in the application of heat, by treating his plants as he would wish to be himself treated, and consequently by keeping them much too warm during the night. Experiments, made previously and subsequently to that period, have satisfied me that he as often and as widely errs by too freely admitting the external air during the day, particularly in bright weather. Plants generally grow best, and fruits swell most rapidly, in a warm and moist atmosphere; and change of air is, to a very limited extent, necessary or beneficial. The mature leaves of plants, and, according to Saussure, the green fruits, (grapes at least), when exposed to the influence of light, take up carbon from the surrounding air, whilst the same substance is given out by every other part of the plant; so that the purity of air when confined in close vessels has often been found little changed at the end of two or three days by the growth of plants in it. But even if plants required as pure air as hot-blooded animals, the buoyancy of the heated air, in every forcing-house, would occasion it to escape, and change as rapidly, and indeed much more rapidly, than would be necessary.

It may be objected that plants do not thrive, and that the skins of grapes are thick, and other fruits without flavour, in crowded forcing-houses; but in these it is probably light, rather than a more rapid change of air, that is wanting; for in a forcing-house, which I have long devoted almost exclusively to experiments, I employ very little fire-heat; and never give air, till my grapes are nearly ripe, in the hottest and brightest

* See page 213.

weather, further than is just necessary to prevent the leaves being destroyed by excess of heat. Yet this mode of treatment does not at all lessen the flavour of the fruit, nor render the skins of the grapes thick; on the contrary, their skins are always most remarkably thin, and very similar to those of grapes which have ripened in the open air. It is always my wish to see the temperature of this house, in the middle of every bright day in summer, as high as 90°; and, after the leaves of the plants have become dry, I do not object to ten or fifteen degrees higher. In the following night the temperature sometimes falls as low as 50°; and so far am I from thinking such change of temperature injurious, I am well satisfied that it is generally beneficial.

Plants, it is true, thrive well, and many species of fruits acquire their greatest state of perfection in some situations within the tropics, where the temperature, in the shade, does not vary in the day and night more then seven or eight degrees; but in these climates the plant is exposed during the day to the full blaze of a tropical sun, and early in the night it is regularly drenched with heavy-wetting dews; and consequently it is very differently circumstanced in the day and in the night, though the temperature of the air in the shade at both periods may be very nearly the same. If the thermometer, under the above-mentioned circumstances, were to be exposed, as the plant is, to the sun, it would probably indicate, in the middle of the day, a temperature little below that of boiling water. In the forcing-house so much light and heat are repelled by the glass and wood-work of the roof, that the degree of heat to which the leaves are subjected does not greatly exceed that indicated by the shaded thermometer; and, by excess of ventilation, I have several times found the temperature of forcing-houses in the gardens of some of my friends reduced so nearly to that of the external air in the middle of a bright, but not very warm day, that the progress towards maturity of the fruit was certainly rather retarded by the shade than accelerated by the protection of the glass roof. During the night the loss, as far as related to time, was probably redeemed by the flues; but the fruit thus ripened during the night never rivals in flavour that which is chiefly ripened by confined solar heat. This kind of heat can also be made to operate in every moderately bright day without incurring either expense or increased trouble; for any observant gardener will soon discover precisely to what extent air may be confined in differently constructed forcing-houses in every different state of the atmosphere and weather, and thus guard in his absence, for a short time, against all danger of injury to the foliage of his trees; at the same time that these may be placed securely in nearly the highest temperature that can be beneficial to them.

A less humid atmosphere is more advantageous to fruits of all kinds, when the period of their maturity approaches, than in the earlier stages of their growth, and such an increase of ventilation, at this period, as will give the requisite degree of dryness to the air within the house is highly beneficial; provided it be not increased to such an extent as to reduce the temperature of the house much below the degree in which the fruit has previously grown, and thus retard its progress to maturity. The good effect of opening a peach-house, by taking off the lights of its roof during the period of the last swelling of the fruit, appears to have led many gardeners to overrate greatly the beneficial influence of a free current of air upon ripening fruits; for I have never found ventilation to give the proper flavour or colour to a peach, unless that fruit was at the same time exposed to the sun without the intervention of glass; and the most excellent peaches I have ever been able to raise, were obtained under circumstances where change of air was as much as possible prevented consistently with the admission of light (without glass) to a single tree.

XXXIV.—UPON THE PROPER MODE OF PRUNING THE PEACH-TREE, IN COLD AND LATE SITUATIONS.

[*Read before the* HORTICULTURAL SOCIETY, *May 6th*, 1817.]

THE buds of fruit-trees, which produce blossoms, and those which afford leaves only, in the spring, do not at all differ from each other, in their first state of organisation, as buds. Each contain the rudiments of leaves only, which are subsequently transformed into the component parts of the blossom, and in some species of the fruit also. I have repeatedly ascertained, that a blossom of a pear or apple tree contains parts, which previously existed as the rudiments of five leaves, the points of which subsequently form the five segments of the calyx; and I have often succeeded in obtaining every gradation of monstrosity of form, from five congregated leaves, (that is, five leaves united circularly upon an imperfect fruit-stalk), to the perfect blossom of the pear-tree. The calyx of the rose, in some varieties, presents nearly the perfect leaves of the plant, and the large and long leaves of the medlar appear to account for the length of the segments, in the empalement of its blossom. The calyx of the blossom of the plum and peach tree is formed precisely as in the preceding cases, except that the leaves, which are transmuted into the calyx, separate at the base of the fruit and become deciduous, instead of passing through and remaining a component part of it.

Every bunch of grapes commences its formation as a tendril, and it is always within the power of every cultivator to occasion it to remain a tendril.   The blossoms are all additions, the formation of which is always dependent upon other agents : and if any considerable part of the leaves be taken off the branch prematurely, or if the vine be not subjected to the influence of the requisite degree of heat and light, the tendrils will permanently retain their primary form and office; and it is very frequently observable, when much of the foliage of fruit-trees has been destroyed by insects, or when the previous season has been cold and wet, that blossoms are not formed at all, or are feeble and imperfect, and consequently abortive.   The state of the peach-trees and vines, in every part, or nearly every part of the kingdom, in the present spring, has afforded, I believe, more than sufficient evidence of the truth of the last position.

It is, I conceive, quite unnecessary to adduce arguments to prove that the buds, which are first formed in the spring, are most likely to undergo properly the necessary internal changes of structure above-mentioned, and consequently to afford more perfectly organised blossoms, than such as are not formed before the middle of the summer, or till near the approach of autumn; and if this be admitted, it will not be difficult to show, that the mode of pruning and training the peach-tree, which has been uniformly recommended, and almost as uniformly practised, is well adapted to favourable situations only.   It has been derived from the practice of the French gardeners, and is probably perfectly well suited to the climate of Paris, but by no means so well calculated (I have, I think, very good reason to believe) for the colder parts of England, as that I proceed to describe and recommend.

Every tree prepares in the summer and autumn many minute leaves, which expand and form the early foliage of the following spring, and the buds in the axillæ of these leaves are necessarily (consistent with the preceding statements,) those best calculated, in cold and unfavourable situations and seasons, to generate well organised and vigorous blossoms; and in such situations, I have often witnessed the advantage of preserving as many as practicable of these, by deviating from the ordinary mode of pruning the peach-tree.   Instead of taking off so large a portion of the young shoots, and training in a few only, to a considerable length, as is usually done, and as I should myself do to a great extent, in the vicinity of London, and in every favourable situation, I preserve a large number of the young shoots, which are emitted in a proper direction in early spring by the yearling wood, shortening each where necessary, by pinching off the minute succulent points, generally to the length of one or two inches. Spurs which lie close to the wall are thus made, upon which numerous blossom buds form very early in the ensuing summer; and upon such, after the last most unfavourable season, and in a situation so high and cold that the peach-tree, in the most favourable seasons, had usually produced only a few feeble blossoms, I observed as strong and vigorous blossoms in the present spring, as I have usually seen in the best seasons and situations; and I am quite confident that if the peach-trees, in the gardens round the metropolis, had been pruned in the manner above described, in the last season, an abundant and vigorous blossom would have appeared in the present spring. I do not, however, mean to recommend to the gardener to trust wholly, in any situation, for his crop of fruit, to the spurs produced by the above-mentioned mode of pruning and training the peach-tree. In every warm and favourable situation, I would advise him to train the larger part of his young wood, according to the ordinary method, and in cold and late situations only, to adopt to a great extent, the mode of management above suggested. A mixture of both modes, in every situation, will be generally found to multiply the chances of success; and therefore neither ought to be exclusively adopted, or wholly rejected in any situation. The spurs must not be shortened in the winter or spring, till it can be ascertained what parts of them are provided with leaf-buds.

XXXV.—OBSERVATIONS ON THE PROPER MANAGEMENT OF FRUIT-TREES, WHICH ARE INTENDED TO BE FORCED VERY EARLY IN THE ENSUING SEASON.

[*Read before the* HORTICULTURAL SOCIETY, *June 3rd*, 1817.]

THE period which any species, or variety, of fruit will require to attain maturity, under any given degrees of temperature, and exposure to the influence of light in the forcing-house, will be regulated to a much greater extent than is generally imagined, by the previous management and consequent state of the tree, when that is first subjected to the operation of artificial heat. Every gardener knows, that when the previous season has been cold, and cloudy, and wet, the wood of his fruit-trees remains immature, and weak abortive blossoms only are produced. The advantages of having the wood well ripened are perfectly well understood;

but those which may be obtained, whenever a very early crop of fruit is required, by ripening the wood very early in the preceding summer, and putting the tree into a state of repose, as soon as possible after its wood has become perfectly mature, do not, as far as my observation has extended, appear to be at all known to gardeners; though every one who has had in any degree the management of vines in a hot-house, must have observed the different effects of the same degrees of temperature upon the same plant, in October and February. In the autumn, the plants have just sunk into their winter sleep: in February they are refreshed, and ready to awake again; whenever it is intended prematurely to excite their powers of life into action, the expediency of putting these powers into a state of rest, early in the preceding autumn, appears obvious. The natural propensity of the gardener to treat his plants as in some degree sentient beings, and as he would wish to be himself treated, which sometimes misleads him (as I have remarked in a former paper)*, will in this case direct him rightly, by leading him to infer, that early rising requires early going to rest. I shall therefore state the result of a few experiments only, which will, I believe, afford satisfactory evidence of the truth of the foregoing positions.

Some vines, which grew in pots, were placed in a forcing-house, at the end of January, where they produced ripe fruit about the middle of July; and soon after that period, the pots were taken from the house and put under the shade of a north wall, in the open air. Water was subsequently given in small quantities only; and the leaves of the plants soon fell off. In August the plants were pruned; and in September they were removed to a south wall, where they soon vegetated with much vigour, and continued to grow till their young shoots were killed by frost.

Other vines, of the same varieties, were suffered to remain in the forcing-house till late in August; where they were subjected to the mode of management above described, except that they were not removed from their situation under a north wall, nor pruned, before the approach of winter. These were then placed against a south wall, where their fruit ripened well in the following season, in a climate not nearly warm enough to have ripened it at all, if the plants had previously grown in the open air.

Having raised many varieties of the peach from seed in the year 1813, I felt anxious to secure the existence of each variety till I could ascertain

* See page 213.

its merits ; and with this view, I obtained a duplicate of each by inserting a bud from every seedling plant into a stock, which I placed in the forcing-house.   Late in the autumn of the year 1815, some of the young trees, which had been obtained from these buds, were removed from the forcing-house, in which their wood had become most perfectly well ripened, in the preceding summer, to the open air, and were placed, as closely as could conveniently be done, to the seedling trees of the same varieties, which had grown wholly in the open air : and thus circum-stanced, the blossoms of the trees which had been removed from the forcing-house unfolded nine days earlier, and their fruit ripened three weeks earlier, than those upon the other trees of the same varieties.

The confinement of the roots to pots, and possibly, to a small extent, the influence of the stock (for the peach-trees in the pots grew upon apricot shoots), may have somewhat accelerated the maturity of the fruit in the experiment last mentioned ; but the chief causes of the early maturity of the fruit in both the preceding cases were, I am confident, the perfect maturity of the wood, and the high state of excitability, which had been acquired by a preternaturally long period of rest.

It is not, I believe, at all necessary that I should offer arguments to prove that a vine, which cannot be made to vegetate at all in the winter without a very high degree of heat, is not as well calculated for very early forcing as one in which the powers of life are so excitable that it is prepared to vegetate strongly in the temperature of the open air in September, and in which the power to vegetate in a low temperature will continue to accumulate progressively till spring : but it will probably be objected that as large a crop cannot be obtained from vines of which the roots are confined in pots, as from others.   This objection, however, will, I believe, prove to be wholly unfounded, whenever a very early crop is wanted ; for vines and other fruit-trees (as I have observed in former papers) when abundantly supplied with water, and manure in a liquid state, require but a very small quantity of mould.   A pot containing two cubic feet of very rich mould, with proper subsequent attention, is fully adequate to nourish a vine which, after being pruned in autumn, occupies twenty square feet of the roof of a hot-house ; and I have constantly found that vines, in such pots, being abundantly supplied with food and water, have produced more vigorous wood, when forced very early, than others of the same varieties, whose roots were permitted to extend beyond the limits of the house.

XXXVI.—UPON THE PROPAGATION OF VARIETIES OF THE WALNUT-TREE, BY BUDDING.

[*Read before the* HORTICULTURAL SOCIETY, *April 7th,* 1818.]

THE ill success of many attempts to propagate the walnut-tree by grafts, or buds, led me, in a former communication, to discourage all attempts to increase it, except by seeds, or by grafting by approach. I nevertheless continued, annually, to make a few experiments, with the hope of discovering a method of budding, which would prove successful in the culture of varieties of this fruit, and of others of equally difficult propagation ; and I have found, in ultimate success, the usual reward of patient perseverance.

The advantages of propagating varieties of the walnut-tree, by budding, will, I think be found considerable, provided the buds be taken from young, or even middle-aged healthy trees : for, exclusive of the advantage of obtaining fruit from very young trees, the planter will be enabled to select not only such varieties as afford the best fruit, but also such as endure best, as timber-trees, the vicissitudes of our climate. In this respect some degree of difference is almost always observable in the constitution of each individual seedling tree ; and this is invariably transferred with the graft or bud.

The walnut, it is true, as a fruit, contains but little nutriment, and perhaps constitutes, at best, only an unwholesome luxury : but the tree affords timber of much greater strength and elasticity, comparatively with its very low specific gravity, than any other of British growth, and it is consequently applicable to purposes for which no good substitute has hitherto been found ; the stocks of the musket of the soldier, and of the gun of the sportsman.

The buds of trees, of almost every species, succeed with most certainty, when inserted in the shoots of the same year's growth ; but the walnut-tree appears to afford an exception ; possibly in some measure because its buds contain, within themselves, in the spring, all the leaves which the tree bears in the following summer ; whence its annual shoots wholly cease to elongate soon after its buds unfold ; all its buds of each season are also, consequently, very nearly of the same age : and long before any have acquired the proper degree of maturity for being removed, the annual branches have ceased to grow longer, or to produce new foliage.

To obviate the disadvantages arising from the preceding circumstances, I adopted means of retarding the period of the vegetation of the stocks, comparatively with that of the bearing tree: and by these means I became partially successful.   There are at the base of the annual shoots of the walnut, and other trees, where those join the year-old wood, many minute buds; which are almost concealed in the bark ; and which rarely, or never vegetate, but in the event of the destruction of the large prominent buds, which occupy the middle, and opposite end of the annual wood.   By inserting in each stock one of these minute buds, and one of the large and prominent kind, I had the pleasure to find that the minute buds took freely, whilst the large all failed, without a single exception. This experiment was repeated in the summer of 1815, upon two yearling stocks which grew in pots, and had been placed during the spring and early part of the summer, in a shady situation under a north wall; whence they were moved late in July to a forcing-house, which I devote to experiments, and instantly budded.   These being suffered to remain in the house during the following summer, produced from the small buds, shoots nearly three feet long terminating in large and perfect female blossoms, which necessarily proved abortive, as no male blossoms were procurable at the early period in which the female blossoms appeared: but the early formation of such blossoms sufficiently proves that the habits of a bearing branch of the walnut-tree may be transferred to a young tree by budding, as well as grafting by approach.

The most eligible situation for the insertion of buds of this species of tree (and probably of others of similar habits) is near the summit of the wood of the preceding year, and of course, very near the base of the annual shoot; and if buds of the small kind above-mentioned, be skilfully inserted in such parts of branches of rapid growth, they will be found to succeed with nearly as much certainty as those of other fruit-trees, provided such buds be in a more mature state than those of the stocks into which they are inserted.

The advantages which may be obtained in the propagation of other species of trees by procuring buds for insertion in a more mature state than those of the stock, are sufficient to deserve some attention, and are not, I believe, at all known to gardeners and nurserymen.   The mature bud takes immediately with more certainty under the same external circumstances: it is much less liable to perish during winter; and it possesses the valuable property of rarely or never vegetating prematurely in the summer, though it be inserted before the usual period, and in the season when the sap of the stock is most abundant. I have, in different years, removed some hundred buds of the peach-tree from the forcing-house to luxuriant shoots upon the open wall;

and I have never seen an instance in which any of such buds have broken and vegetated during the summer or autumn ; but when I have had occasion to reverse this process, and to insert immature buds from the open wall into the branches of trees growing in a peach-house, many of these, and in some seasons all, have broken soon after being inserted, though at the period of their insertion the trees in the peach-house had nearly ceased to grow. The result was, in both the preceding cases, in opposition to my expectations ; but it appears necessarily to have been occasioned by the mature bud having naturally sunk into a state of repose preparatory to its long winter sleep, previously to its having been removed ; and by the more excitable state of the powers of life in the bud taken from the open wall.

If the mature buds of the peach-tree, when taken from the forcing-house, contain blossoms, these may be carried a great distance, and still afford fruit in the following spring. I have thus readily obtained fruit from blossoms sent me from the vicinity of London ; and I entertain no doubt of the practicability of obtaining fruit from blossoms sent from Paris, or even from the south of France, if properly packed. In such cases it would be necessary to pare the wood of the bud thin, instead of wholly extracting it : and this will sometimes be found expedient, when buds are to be taken from a peach-house, in which the fruit has been made to ripen early in the summer, to be inserted in the open air.

XXXVII.—UPON THE PRUNING AND MANAGEMENT OF TRANSPLANTED STANDARD TREES.

[*Read before the* HORTICULTURAL SOCIETY, *June 2nd,* 1818.]

WHEN a tree is transplanted, it loses, almost necessarily, a considerable part of its roots : and as these, in every healthy subject, are nicely proportioned to the branches, the advantages of retrenching the latter are obvious, and well known to every gardener. But relatively to the mode of retrenching the branches, and the extent of retrenchment that is beneficial, there is much discordance in the opinions and practice of different gardeners ; and often still more between the gardener and his employer ; the latter wishing to preserve the bearing branches, that he may, at an early period, obtain a crop of fruit ; and the gardener wishing to head down the tree, that he may see it shoot with vigour. Neither mode of practice is, I think, in its full extent, quite eligible to the greater number of cases; the one being too prejudicial to the growth of the tree, by occasioning the production prematurely of an unusual profusion of blossoms; and the other being, even when most successful, attended with an unnecessary loss of time: and I have found, in very extensive experience, that transplanted trees generally succeed permanently best, and as standards take the best forms, when their lateral branches, instead of being suffered to retain their whole length, or pruned off closely, are all shortened to the length of a few inches, and the top of the tree reduced to a single annual shoot. Under these circumstances the leaves become dispersed upon the stem, so as to afford nutriment to the bark of different parts of it; and the power of the wind to prevent the tree re-establishing itself is small (owing to the situation of the leaves), comparatively with the extent of the foliage which the tree exposes to light. The trees under this mode of treatment also bear as much fruit as they are capable of feeding, as soon as under any other that I have hitherto tried or seen; and within three or four years their branches generally become more widely extended than those of similar trees which are planted without being pruned. The same mode of pruning is equally well adapted to fruit and forest trees; and oaks, which I have planted when ten or twelve feet high, have not only begun immediately to grow with luxuriance, but they have within a very years wholly lost the character of transplanted trees.

The great error of modern practice is that of suffering, when the trees are not headed down, many small branches to form the summit of the transplanted tree; which branches expend its sap in the production of tufts of leaves, where those, owing to their distance from the roots, operate least beneficially in the performance of their proper office, and most injuriously by being most exposed to the influence of winds.

Whenever the roots of transplanted trees have been very much injured, or have been very long out of the ground, the number, as well as the extent of the lateral branches, should be reduced, and not more than a few inches of the leading annual shoot should be suffered to remain; but in all cases where trees are to be sent a great distance, this retrenchment of their branches should be made in the nursery from which they are to be removed; and, if it be properly executed, trees may be conveyed to great distances, under more disadvantageous circumstances than is usually supposed, without endangering life, provided they be subjected to proper subsequent management.

I received in the last spring some apple-trees from America, which were forwarded to me from London by a wrong waggon, and consequently did not arrive till near the middle of April, and many weeks after the period at which I ought to have received them. The whole of them appeared perfectly lifeless and dry, and much better fitted for fire-wood than for planting; and I scarcely entertained the slightest hope of being able to recover a single plant. I nevertheless resolved that no trouble should be spared in making the experiment.

The American nurserymen had pruned the trees much in the way I wished (though in a very rough and careless manner, and obviously without any other object than convenience in packing them); and I had therefore little more to do in pruning them than to take away such branches as were broken and wholly dead. The trees, which were about four feet high, were then planted in a situation where they were perfectly screened from the morning sun, and just as much water was given as was sufficient to close the moulds to the roots. Their stems were then sprinkled with water, by an engine, sufficiently to wet the bark; and this was repeated at six o'clock every morning through the months of May, June, and July; but no water was given immediately to the roots, previous experience having led me to believe that excess of moisture is, in such cases, generally injurious, and often fatal.

About midsummer a few of the trees began to exhibit some feeble symptoms of life; several subsequently shot vigorously, some to the length of eighteen inches; and out of sixty-four trees, I lost only three. They succeeded, in the aggregate, better than other trees of nearly the same age, which were only removed from a contiguous nursery, but which were not sprinkled with water; the season having proved cold and dry, and consequently extremely unfavourable to transplanted trees.

I had previously seen in other instances, though never in so apparently hopeless a case, the good effects of sprinkling the stems and branches of transplanted trees before the sun began to shine upon them in the morning, both in the forcing-house and in the open air. In the forcing-house I have found that water may be also thus applied with advantage in the evening as well as in the morning; but, in the open air, I have had reason to think its operation injurious, when the succeeding night has proved cold.

XXXVIII.—ON THE CULTURE OF THE GUERNSEY LILY.

[*Read before the* HORTICULTURAL SOCIETY, *August 3rd,* 1819.]

A WISH has been expressed by the Council of the Society, that a method of cultivating the *Amaryllis Sarniensis,* or *Guernsey Lily,* should be discovered, by which the bulbs of that plant might be made to afford blossoms, regularly, through successive seasons : and I, in consequence, address the following communication upon that subject ; believing, that I can satisfactorily account for its sparing production of blossoms in our climate, and point out a mode of cultivating it, by which it may be made to blossom, much more freely than it usually does, though I have not attained the object desired by the Society.

Bulbous roots increase in size, and proceed in acquiring powers to produce blossoms, only during the periods in which they have leaves, and in which such leaves are exposed to light ; and these organs always operate most efficiently when they are young, and have just attained their full growth.   The bulb of the Guernsey Lily, as it is usually cultivated in this country, rarely produces leaves till September, or the beginning of October, at which period, the quantity of light afforded by our climate is probably quite insufficient for a plant, which is said to be a native of the warm and bright climate of Japan ; and before the return of spring, its leaves are necessarily grown old, and nearly out of office, even when they have been safely protected from frost through the winter. It is, therefore, not extraordinary, that a bulb of this species, which has once expended itself in affording flowers, should but very slowly recover the power of blossoming again.   The operation also of a cold climate, in retarding its period of vegetation, must have led the plant into late habits, like those of the vines, described by Mr. Arkwright, in our Transactions * ; and, consequently, instead of being naturalised, and adapted to our climate as plants become, which propagate by seeds, it is, probably, now less capable of producing a regular annual succession of blossoms, than a similar variety of the same species of plant, immediately imported from Japan, would be.

Considering, therefore, the deficiency of light and heat, owing to the late period of its vegetation, as the chief cause, why this plant so fails to produce flowers, I infer that nothing more would be required to make it blossom, as freely, at least, as it does in Guernsey, than such a slight degree of artificial heat, applied early in the summer, as would prove

* See Horticultural Transactions, Vol. III. p. 95.

sufficient to make the bulbs vegetate a few weeks earlier than usual in the autumn.

Early in the summer of 1816, a bulb, which had blossomed in the preceding autumn, was subjected to such a degree of artificial heat, as occasioned it to vegetate six weeks, or more, earlier than it would other-wise have done. It did not, of course, produce any flowers ; but in the following season, it blossomed early, and strongly, and afforded two offsets. These were put, in the spring of 1818, into pots, containing about one-eighth of a square foot light and rich mould, and were fed with manured water, and their period of vegetation was again accelerated by artificial heat. Their leaves, consequently, grew yellow from matu-rity, early in the present spring, when the pots were placed in rather a shady situation, and near a north wall, to afford me an opportunity of observing to what extent, in such a situation, the early production of the leaves in the preceding seasons had changed the habit of the plant. I entertained no doubt but that both the bulbs would afford blossoms, but I was much gratified by the appearance of the blossoms in the first week in July. Wishing to obtain seeds, I then removed the plants to a forcing-house, in which they have flowered very strongly ; and the appearance of the seed-vessels gives much reason to suppose that I shall succeed in obtaining seeds, though I am not at present able to speak decisively.

From the success of the preceding experiment, I conclude that if the offsets, and probably the bulbs, of this plant which have produced flowers, be placed in a moderate hot-bed, in the end of May, to occasion the early production of their leaves, blossoms would be constantly afforded in the following season : but it will be expedient to habituate the leaves, thus produced, gradually to the open air, as soon as they are nearly full grown, and to protect them from frost till the approach of spring.

Should seedling plants be obtained, the powers of life in those, will probably prove more alert : and I think it probable, that, with a mode-rate degree of care, these may be made to afford blossoms in successive seasons ; though it should be found impracticable to give that habit to the offsets of the individual seedling plant, now in cultivation.

XXXIX.—UPON THE EFFECTS OF VERY HIGH TEMPERATURE ON SOME SPECIES OF PLANTS.

[*Read before the* HORTICULTURAL SOCIETY, *December 7th*, 1819.]

HAVING constructed a forcing-house for the purpose of attempting the culture of the mango, and a few other species of tropical fruits, I have endeavoured to ascertain, with accuracy, the advantages and disadvantages, of employing very high temperature during the day in bright weather, and of comparatively low temperature during the night, and in cloudy weather; and I communicate the following account of my experiments, considering the results to have been generally very favourable, and where unsuccessful, not wholly uninteresting.

A fire of sufficient power, only, to preserve in the house a temperature of about 70°, during summer, was employed, but no air was ever given, nor its escape facilitated, till the thermometer, perfectly shaded, indicated a temperature of 95°; and then only two of the upper lights, one at each end, were let down about four inches. The heat of the house was consequently sometimes raised to 110°, during the middle of warm and bright days, and it generally varied, in such days, from 90° to 105°, declining during the evening to about 80°, and to 70° in the night.

Late in the evening of every bright and hot day, the plants were copiously sprinkled with water, nearly of the temperature of the external air; and the following were the effects produced upon the different species.

The Melon. Plants of this species were trained upon a trellis near the glass, which was of the best quality, and these exhibited a greater degree of health and luxuriance, than I had ever before seen; but not a single flower ever unfolded; a great profusion of minute blossoms, nevertheless, appeared in succession at the points of the shoots, all of which perished abortively. I was much disappointed at the result of this experiment; from which I confidently expected to obtain fruit of the greatest excellence.

The Water Melon. A plant of this species, treated in the same manner as the melon plants above mentioned, grew with equal health and luxuriance, and afforded a most abundant blossom; but all its flowers were male. This result did not, in any degree, surprise me; for I had many years previously succeeded, by long continued very low temperature, in making cucumber plants produce female flowers only; and I entertain but little doubt, that the same fruit-stalks might be made, in this and the preceding species, to support either male or female flowers, in obedience to external causes.

The Guernsey Lily. I transferred plants of this species, from the open air to the hot-house, in the summer, with the hope of obtaining seeds, in which I was wholly disappointed. The flowers expanded very beautifully; but their pollen never shedded. The plants have, nevertheless, subsequently grown with more than ordinary vigour; and I entertain scarcely any doubt that the same roots which afforded flowers in the present season, will blossom strongly in the next. It appears therefore from this, and the two preceding experiments, that the same degree of temperature, which may promote the growth, and exuberant health of the plant, may, at the same time, render it wholly unproductive of fruit or offspring.

The Fig Tree. Several varieties of this species were subjected to experiment; but the trees, although planted in pots, grew with so much luxuriance, and afforded me so little prospect of fruit, that I removed all except those of the large white variety, from the house. The white fig-tree succeeded perfectly, first ripening its spring-figs, (those which usually ripen in the open air in this country,) and afterwards its summer figs. The trees then produced new leaves and branches: and the fruit, which would have appeared in the next spring, ripened in high perfection in September. Subsequently also a few of those, which, in the ordinary course of the growth of the tree, would have appeared as the summer crop of next year, have ripened, and these, though far inferior to those of the preceding crops, have not been without merit.

The Nectarine. A seed of this species of fruit was planted in a hot-bed, in January last, and it vegetated in the succeeding month. It was subsequently removed to the hot-house, in which it continued to grow through the summer, without being in the smallest degree drawn by the high temperature in which it was placed: its wood, on the contrary, is remarkably short-jointed, and is covered with blossom-buds; from which I think it will be practicable to obtain ripe fruit, within sixteen months of the period, at which the plant first sprang from the ground.

The Orange and Lemon. A very high temperature appeared peculiarly favourable to plants of these species, or, I believe, more properly of this species; for I consider both, with the citron and shaddock, to be varieties only of the lime. A plant which sprang from seed in March, had, in the end of August, attained the height of more than four feet, with proportionate strength; when wanting the place it occupied for another purpose, it was removed from the house. I obtained in April a plant of the China orange, with one very small fruit upon it, which has ripened in much apparent perfection, and the tree exhibits every appearance of the most exuberant health.

The Mango. (Mangifera Indica.) This species of fruit-tree appears to possess great peculiarity of constitution; for, although a native of a very hot and bright climate, and capable of bearing, with apparent benefit, the hot drying winds of Bengal, it vegetates freely, and retains its health in comparatively low temperature, and under a cloudy atmosphere. The plants I possess sprang from seeds in October 1818; and the leaves acquired, during winter, their proper dark colour, and remained in perfect health till spring; although, not possessing at that period, a hot-house, I was very ill prepared to preserve them. In March they began to shoot a second time, without having been, I believe, at any period subjected to a higher temperature than 60°, and some of them are now shooting strongly; although the temperature of my house during the last five weeks, except once or twice in very bright days, has rarely been so high as 60°. The mode of growth of this plant appears also to be very singular; it extends a few inches, and then closes its terminal buds, as if its growth for the season were ended. One of my plants has done so nine times within the last thirteen months, without having acquired a greater height than two feet seven inches. I am much inclined to believe that the mango might be raised in great abundance and considerable perfection in the stove in this country, for it is a fruit which acquires maturity within a short period. It blossoms, in Bengal, in January, and ripens in the end of May; and Mr. Turner, in his journey to Thibet, states that he found the mango growing in latitude 27° 50 in Boutan, in the same orchard with the apple-tree; the apples ripening in July, and the mangoes in September. And another Eastern traveller of credit (I think it is Mr. Barrow), mentions an instance in which a frost, sufficiently severe to have injured the crops of barley, had proved fatal to the blossoms (only) of the mango-trees.

The Alligator, or *Avocado pear. (Laurus Persea.)* The plants of this species have grown with rather troublesome luxuriance in my house, though they have been generally confined to small pots; one plant to which a larger pot was given is more than six feet high, with branches extending five feet wide; and a stem, the growth of a single year, exceeding, at its base, an inch in diameter. To obtain fruit of this species within the narrow limits of a forcing-house, it would be necessary to propagate from buds or grafts taken from the extreme branches of trees of considerable age.

The Mammee-tree. (Mammea Americana.) Very contrary to my

expectations, this plant, a native of Jamaica, proved extremely impatient of heat and light, and its young leaves always required to be shaded when the temperature of the house exceeded 90°. But with proper attention to screen the leaves from the mid-day sun, till they acquired maturity, the young trees of this species have succeeded as well as those of any of the preceding species.

Several other plants, part of them natives of temperate climates, grew in my house through the whole summer, without any one of them being drawn, or any way injured, by the very high temperature to which they were occasionally subjected; and from these, and other facts, which have come within my observation, I think myself justified in inferring, that, in almost all cases in which the object of the cultivator is to promote the rapid and vigorous growth of his plants, a very high temperature, provided it be accompanied by bright sunshine, may be employed with great advantage; but it is necessary that the glass of his house should be of good quality, and that his plants be placed near it, and be abundantly supplied with food and water. In the preceding experiments, water was made the vehicle of food to the roots of the plants, in the manner I have described in a former communication *, and with similar good effects.

My house contains a few pine-apple plants, in the treatment of which I have deviated somewhat widely from the common practice; and, I think, with the best effects; for their growth has been exceedingly rapid, and a great many gardeners, who have come to see them, have unanimously pronounced them more perfect than any which they had previously seen. But many of the gardeners think that my mode of management will not succeed in winter, and that my plants will become unhealthy, if they do not perish, in that season; and as some of them have had much experience, and I very little, I wish at present to decline saying more relative to the culture of that plant.

* See above, page 211.

XL.—UPON THE CULTURE OF THE PINE-APPLE, WITHOUT BARK, OR OTHER HOT-BED.

[*Read before the* HORTICULTURAL SOCIETY, *March 7th*, 1820.]

IN a communication which I had the honour to send to the Horticultural Society in the last autumn, upon the effects of very high temperature, when accompanied by very bright sunshine, upon some species of plants, I mentioned that I had made a few, apparently very successful, experiments upon the culture of the pine-apple: but I declined, at that period, to describe the means I had used; because several experienced gardeners in the vicinity were of opinion that my plants could not be made to survive, in health at least, the winter. The same gardeners have since frequently visited my hothouse, and they have unanimously pronounced my plants more healthy and vigorous than any they had previously seen : and they are all, I have good reason to believe, zealous converts to my mode of culture.

I had no intention whatever to attempt to raise pine-apples till the autumn of 1818, when I received from one of my friends in this vicinity, Mr. Ricketts, of Ashford Hall, some seeds of the mango, and soon afterwards some more seeds of that, and other tropical fruit-trees, from one of our members, Mr. Pallmer. I then resolved to erect a hothouse, chiefly for the purpose of attempting to cultivate the mango ; but I had long been much dissatisfied with the manner in which the pine-apple plant is usually treated, and very much disposed to believe the bark bed, as Mr. Kent has stated it in our Transactions*, " worse than useless," subsequently to the emission of roots by the crowns or suckers. I therefore resolved to make a few experiments upon the culture of that plant ; but as I had not at that period, the beginning of October, any hothouse, I deferred obtaining plants till the following spring. My hothouse was not completed till the second week in June, at which period I began my experiment upon nine plants, which had been but very ill preserved through the preceding winter by the gardener of one of my friends, with very inadequate means, and in a very inhospitable climate. These, at this period, were not larger plants than some which I have subsequently raised from small crowns, (three having been afforded by one fruit,) planted in the middle of August, were in the end of December last ; but they are now beginning to blossom, and, in the opinion of every gardener who has seen them, promise fruit of great size and perfection. They are all of the variety known by the name of Ripley's Queen Pine.

Upon the introduction of my plants into the hothouse, the mode of management, which it is the object of the present communication to

* Horticultural Transactions, Vol. III. page 288.

describe, commenced. They were put into pots of somewhat more than a foot in diameter, in a compost made of thin green turf, recently taken from a river side, chopped very small, and pressed closely, whilst wet, into the pots; a circular piece of the same material, of about an inch in thickness, having been inverted, unbroken, to occupy the bottom of each pot. This substance, so applied, I have always found to afford the most efficient means for draining off superfluous water, and subsequently of facilitating the removal of a plant from one pot to another, without loss of roots. The surface of the reduced turf was covered with a layer of vegetable mould obtained from decayed leaves, and of sandy loam, to prevent the growth of the grass roots. The pots were then placed to stand upon brick piers, near the glass; and the piers being formed of loose bricks (without mortar), were capable of being reduced as the height of the plants increased. The temperature of the house was generally raised in hot and bright days, chiefly by confined solar heat, from $95°$ to $105°$, and sometimes to $110°$, no air being ever given till the temperature of the house exceeded $95°$; and the escape of heated air was then, only in a slight degree, permitted. In the night the temperature of the house generally sunk to $70°$, or somewhat lower. At this period, and through the months of July and August, a sufficient quantity of pigeon's dung was steeped in the water, which was given to the pine-plants, to raise its colour nearly to that of porter, and with this they were usually supplied twice a day in very hot weather; the mould in the pots being kept constantly very damp, or what gardeners would generally call wet. In the evenings, after very hot days, the plants were often sprinkled with clear water, of the temperature of the external air; but this was never repeated till all the remains of the last sprinkling had disappeared from the axillæ of the leaves.

It is, I believe, almost a general custom with gardeners, to give their pine-plants larger pots in autumn, and this mode of practice is approved by Mr. Baldwin*. I nevertheless cannot avoid thinking it wrong; for the plants, at this period, and subsequently, owing to want of light, can generate a small quantity only of new sap; and consequently, the matter which composes the new roots, that the plant will be excited to emit into the fresh mould, must be drawn chiefly from the same reservoir, which is to supply the blossom and fruit: and I have found that transplanting fruit-trees, in autumn, into larger pots, has rendered their next year's produce of fruit smaller in size, and later in maturity. I therefore would not remove my pine-plants into larger pots, although those in which they grow are considerably too small.

* Baldwin's Practical Directions for the Culture of the Ananas, page 16.

As the length of the days diminished, and the plants received less light, their ability to digest food diminished. Less food was in consequence dissolved in the water, which was also given with a more sparing hand ; and as winter approached, water only was given, and in small quantities.

During the months of November and December, the temperature of the house was generally little above 50°, and sometimes as low as 48° *. Most gardeners would, I believe, have been alarmed for the safety of their plants at this temperature ; but the pine is a much hardier plant than it is usually supposed to be; and I exposed one young plant in December to a temperature of 32°, by which it did not appear to sustain any injury. I have also been subsequently informed by one of my friends, Sir Harford Jones, who has had most ample opportunities of observing, that he has frequently seen, in the East, the pine-apple growing in the open air, where the surface of the ground, early in the mornings, showed unequivocal marks of a slight degree of frost.

My plants remained nearly torpid, and without growth, during the latter part of November, and in the whole of December ; but they began to grow early in January, although the temperature of the house rarely reached 60° ; and about the 20th of that month, the blossom, or rather the future fruit, of the earliest plant became visible ; and subsequently to that period their growth has appeared very extraordinary to gardeners who had never seen pine-plants growing, except in a bark-bed or other hotbed. I believe this rapidity of growth, in rather low temperature, may be traced to the more excitable state of their roots, owing to their having passed the winter in a very low temperature comparatively with that of a bark-bed. The plants are now supplied with water in moderate quantities, and holding in solution a less quantity of food than was given them in summer.

In planting suckers, I have, in several instances, left the stems and roots of the old plant remaining attached to them ; and these have made a much more rapid progress than others. One strong sucker was thus planted in a large pot upon the 20th of July ; and that is beginning to show fruit. Its stem is thick enough to produce a very large fruit ; but its leaves are short, though broad and numerous ; and the gardeners, who have seen it, all appear wholly at a loss to conjecture what will be the value of its produce. In other cases, in which I retained the old stems and roots, I selected small and late suckers, and these have afforded me the most perfect plants I have ever seen ; and they do not exhibit any

* Subsequently to the time this paper was sent to the Society, I have been informed, that the thermometer was once, in the last winter, so low as 40 degrees.

symptoms of disposition to fruit prematurely. I am, however, still ignorant whether any advantage will be ultimately obtained by this mode of treating the queen-pine, but I believe it will be found applicable with much advantage in the culture of those varieties of the pine which do not usually bear fruit till the plants are three or four years old.

I shall now offer a few remarks upon the facility of managing pines in the manner recommended, and upon the necessary amount of the expense. My gardener is an extremely simple labourer, he does not know a letter or a figure; and he never saw a pine-plant growing, till he saw those of which he has the care. If I were absent, he would not know at what period of maturity to cut the fruit; but in every other respect he knows how to manage the plants as well as I do; and I could teach any other moderately intelligent and attentive labourer, in one month, to manage them just as well as he can : in short, I do not think the skill necessary to raise a pine-apple, according to the mode of culture I recommend, is as great as that requisite to raise a forced crop of potatoes. The expense of fuel for my hothouse, which is forty feet long by twelve wide, is rather less than seven-pence a day here, where I am twelve miles distant from coal-pits; and if I possessed the advantages of a curved iron roof, such as those erected by Mr. Loudon, at Bayswater, which would prevent the too rapid escape of heated air in cold weather, I entertain no doubt, that the expense of heating a house forty-five feet long and ten wide, and capable of holding eighty fruiting pine-plants, exclusive of grapes or other fruits upon the back wall, would not exceed four-pence a day. A roof, of properly curved iron bars, appears to me also to present many other advantages : it may be erected at much less cost, it is much more durable, it requires much less expense to paint it, and it admits greatly more light.

I have not yet been troubled with insects upon my pine-plants, and have not, of course, tried any of the published receipts for destroying them. Mr. Baldwin recommends the steam of hot fermenting horse-dung* : I conclude the destructive agent, in this case, is ammoniacal gas; which Sir Humphrey Davy informed me he had found to be instantly fatal to every species of insect; and if so, this might be obtained at a small expense, by pouring a solution of crude muriate of ammonia upon quicklime; the stable or cow-house would afford an equally efficient, though less delicate fluid. The ammoniacal gas might, I conceive, be impelled, by means of a pair of bellows, amongst the leaves of the infected plants, in sufficient quantity to destroy animal, without injuring vegetable life : and it is a very interesting question to the gardener, whether his hardy enemy, the red spider, will bear it with impunity.

* Baldwin's Practical Directions, &c. page 30.

XLI.—PHYSIOLOGICAL OBSERVATIONS UPON THE EFFECTS OF PAR-TIAL DECORTICATION, OR RINGING THE STEMS OR BRANCHES, OF FRUIT-TREES.

[*Read before the* Horticultural Society, *June 6th,* 1820.]

It has not, I think, been sufficiently explained by what means the obstruction, or prevention, of the passage of the fluids of trees through their bark operates in occasioning an increased production of blossom, and a more rapid growth, and more early maturity, of the fruit: the gardener is in consequence, in many cases, unable to foresee whether he is likely to obtain benefit, or to sustain injury, from the operation ; and he is wholly without the means of knowing how to adopt his mode of operating, with any degree of precision, to the object which he has in view. I therefore address the following observations under the impression that the hypothesis which I have advanced in different papers in the Philosophical Transactions will afford a satisfactory explanation of the cause of all the above-mentioned effects.

According to that hypothesis, the true sap of trees is wholly generated in their leaves, from which it descends through their bark to the extremities of their roots, depositing in its course the matter which is successively added to the tree ; whilst whatever portion of such sap is not thus expended sinks into the alburnum, and joins the ascending current, to which it communicates powers not possessed by the recently-absorbed fluid. When the course of the descending current is intercepted, that necessarily stagnates, and accumulates above the decorticated space ; whence it is repulsed, and carried upwards, to be expended in an increased production of blossoms and of fruit ; and, consistently with these conclusions, I have found that part of the alburnum which is situated above the decorticated space to exceed in specific gravity, very considerably, that which lies below it. The repulsion of the descending fluid therefore accounts, I conceive, satisfactorily for the increased produce of blossoms, and more rapid growth of the fruit, upon the decorticated branch ; but there are other causes which operate in promoting its more early maturity. The part of the branch which is below the decorticated space is ill supplied with nutriment, and ceases almost to grow: it in consequence operates less actively in impelling the ascending current of sap, which must also be impeded in its progress through the decorticated space. The parts which are above it must therefore be less abundantly supplied with moisture ; and drought, in such cases, always operates very powerfully in accelerating maturity.

When the branch is small, or the space from which the bark has been taken off is considerable, it almost always operates in excess; a morbid state of early maturity is induced, and the fruit is worthless.

If this view of the effects of partial *decortication*, or *ringing*, be a just one, it follows that much of the success of the operation must be dependent upon the selection of proper seasons, and upon the mode of performing it being well adapted to the object of the operator. If that be the production of blossoms, or the means of making the blossoms set more freely, the ring of bark should be taken off early in the summer, preceding the period at which blossoms are required; but if the enlargement and more early maturity of the fruit be the object, the operation should be delayed till the bark will readily part from the alburnum in the spring. The breadth of the decorticated space, as Mr. Sabine has justly observed, must be adapted to the size of the branch *; but I have never witnessed any except injurious effects whenever the experiment has been made upon very small or very young branches; for such become debilitated and sickly long before the fruit can acquire a proper state of maturity. I have found a tight ligature, applied in the preceding summer, in such cases, to answer, in a great measure, all the purposes of ringing, with far less injurious consequences to the tree; and if such were applied to the stems or principal branches of cherry-trees which are to be forced very early in the following year, I believe the blossoms would be found to set more freely, and the fruit to attain an early maturity. I have also succeeded in preserving, to a great extent, the health of a ringed branch by instantly covering the exposed surface of the alburnum with a tight bandage of coarse thread coated with bees-wax, if the branch were small; or of fine packthread, if it were large; so as wholly to fill the space from which the bark had been taken. By such means the desiccation and consequent death of the external surface of the alburnum have been prevented; and I consequently think it not improbable that the operation might be performed with advantage upon the cherry-tree, and some other fruit-trees, to which it has hitherto been found destructive. I have tried, with the most ample success, in the present spring, the application of such a bandage upon a ringed branch of a fig-tree; and the evidence I have obtained of its mode of operation has not been confined to a recent period, for I applied such a bandage in the first experiment I ever made upon a plant, and at the distance (I have particular reasons for knowing) of precisely half a century from the present time;—when I was a school-boy of ten years old.

* See Horticultural Transactions, Volume IV. page 124.

I am not friendly to the process of ringing, in whatever manner it may be performed ; and I think it never should be adopted unless in cases where blossoms cannot be otherwise obtained, or where, in very early forcing, the value of a single crop of fruit exceeds the value of the tree. For it is a process which promotes the expenditure, whilst it diminishes the creation, of the vital fluid of the tree, which must also suffer in all subsequent periods, from the organic injuries it sustains.

XLII.—UPON THE CULTURE OF THE FIG-TREE, IN THE STOVE.

[*Read before the* HORTICULTURAL SOCIETY, *July* 18*th*, 1820.]

IN a communication respecting the effects of very high temperature upon certain species of plants, which was addressed by me to the Horticultural Society in the last autumn *, I stated that fig-trees of one variety had afforded four successive crops in the same season. The fourth crop, at that period, was only beginning to ripen, and I thought the fruit somewhat inferior in quality to that which had ripened early in the season ; but the subsequent portion of it proved most excellent ; and some figs, which were gathered upon Christmas-day, were thought by myself, and a friend who was with me, much the best we had ever tasted. The same plants have since ripened four more crops, being eight within twelve months ; and upon a ringed branch of one year old, and about an inch in diameter, a ninth crop, consisting of sixty figs, will ripen within the next month. I possess only two plants, each growing in a pot, which contains something less than fourteen square inches of mould, and occupying together a space equal to about sixty-four square feet of the back wall of my pine-stove ; from which space the number of figs that have been gathered within twelve months has been little, if any, less than three hundred : and I see every prospect of a succession of crops till winter. I therefore send the following account of the mode of culture which has been employed, in the hope that it may prove useful to those who are sufficiently admirers of the fig to think it deserving a place in the forcing-house.

My trees grow, as I have stated in the communication to which I have above alluded, in exceedingly rich mould, and are most abundantly supplied with water which holds much manure in solution. They consequently shoot with great vigour, notwithstanding the small space

* See above, page 239.

to which their roots are confined ; and they require some attention to restrain them within the limits assigned to them ; but I have found the following mode of treatment perfectly efficient and successful.

Whenever a branch appears to be extending with too much luxuriance, its point, at the tenth or twelfth leaf, is pressed between the finger and thumb, without letting the nails come in contact with the bark, till the soft succulent substance is felt to yield to the pressure. Such branch in consequence ceases subsequently to elongate ; and the sap is repulsed to be expended where it is more wanted. A fruit ripens at the base of each leaf, and during the period in which the fruit is ripening, one or more of the lateral buds shoots, and is subsequently subjected to the same treatment, with the same result. When I have suffered such shoots to extend freely to their natural length, I have found that a small part of them only became productive either in the same or the ensuing season, though I have seen that their buds obviously contained blossoms. I made several experiments to obtain fruit in the following spring from other parts of such branches, which were not successful ; but I ultimately found that bending such branches, as far as could be done without danger of breaking them, rendered them extremely fruitful ; and in the present spring thirteen figs ripened perfectly upon a branch of this kind within the space of ten inches. In training, the ends of all the shoots have been made, as far as practicable, to point downwards.

When I made my former communication upon this subject, I supposed that the variety which had succeeded so well in my hothouse was the large white fig, the cuttings from which I raised my plants having been sent to me as such ; and that its size had been somewhat diminished by the confinement of the roots to pots, and the exuberant produce of fruit. I have, however, recently seen a private letter of the late Mr. Speechley's (the well-known author of Treatises on the Culture of the Pine-Apple and Vine), in which he speaks of a white fig that he had found to succeed perfectly in high temperature, but the name of which he does not appear to have known ; and I believe that which I am cultivating to be the one he has described. The form of the fruit, in its most perfect state, is an oblate spheroid of nearly two inches in width ; but its length often exceeds its breadth, and it then tapers to the point next the stalk.

XLIII.—ON THE CULTIVATION OF THE COCKSCOMB.

[Read before the HORTICULTURAL SOCIETY, *Dec.* 19, 1820.]

THE flower of the *cockscomb*, which I sent to the meeting of the Society on the 17th of October, may be considered a fair sample of all that I grew this year; two of six having been larger, and two somewhat smaller *.

In cultivating these plants, I have treated them precisely as I do my pine-apple plants, having in some respects a similar object in view; for in both a single fruit-stalk of great strength is requisite, the protrusion of which should be retarded as long as possible, consistently with the rapid growth of the plant. The compost I employed was the most nutritive and stimulating that I could apply, consisting of one part of unfermented horse-dung fresh from the stable and without litter, one part of burnt turf, one part of decayed leaves, and two parts of green turf, the latter being in lumps of about an inch in diameter, to keep the mass so hollow that the water might have free liberty to escape, and the air to enter. Manure was also given in a liquid state by steeping pigeon-dung in the water, which was given very freely. The plants were put, whilst very small, into pots of four inches diameter, and three inches deep; as soon as their roots had reached the sides of the pots, and before they had become in any degree matted, they were transplanted into pots of a foot in diameter, and about nine inches deep. Particular attention was paid to the state of the roots, for I have reason to think that the compression of them in the pot has, under all circumstances, a tendency to accelerate the flowering of plants.

Under this mode of treatment, the plants became large and strong before they showed a disposition to blossom; they usually divide into many branches (as the pine-apple plant will also do), which will greatly injure them, if due attention be not paid to remove the side branches when very young. My plants were at all times so placed that their leaves reached within a few inches of the glass, and they were subjected to the same heat (from 70° to 100°), during the summer, as my pine-apple plants.

The seeds of the plants which I raised in the present season were not sown till too late in the spring; and if I were to repeat the experiment, I entertain no doubt of producing much larger flowers than the one I sent you; for the variety, I believe, is of superior excellence. It affords seeds very sparingly, as you would perceive by the specimen sent.

* The flower sent by Mr. Knight measured eighteen inches in width and seven inches in height from the top of the stalk; it was thick and full, and of a most intense colour. A very accurate drawing of it has been executed by Mrs. Pope, and placed in the library of the Society. (*Note by Mr. Sabine.*)

XLIV.—OBSERVATIONS ON HYBRIDS.

[*Read before the* HORTICULTURAL SOCIETY, *February* 6, 1821.]

MUCH difference of opinion appears to exist between my friend the Hon. William Herbert and myself, relatively to the production of hybrid plants ; he supposing that many originally distinct species are capable of breeding together, without producing mules (that is, without producing plants incapable of affording offspring) ; and I considering the fact of two supposed species having bred together, without producing mules, to be evidence of the original specific identity of the two. Our difference of opinion is, however, I believe, apparently much greater than it really is : for I readily concede to Mr. Herbert, that great numbers, perhaps more than half, of the species enumerated by botanical writers, may be made to breed together, with greater or less degrees of facility : but upon what sufficient evidence the originally specific diversity of these rests, I have never been able to obtain anything like satisfactory information ; and I cannot by any means admit that plants ought to be considered of originally distinct species, merely because they happen to be found to have assumed somewhat different forms or colours in an uncultivated state. The genus Prunus contains the P. Armeniaca, P. Cerasus, P. domestica, P. insititia, P. spinosa, P. sibirica, and many others. Of these, I feel perfectly confident that no art will ever obtain offspring (not being mules) between the Prunus Armeniaca, P. Cerasus, and P. domestica : but I do not entertain much doubt of being able to obtain an endless variety of perfect offspring between the P. domestica, P. insititia, and P. spinosa ; and still less doubt of obtaining an abundant variety of offspring from the P. Armeniaca and P. sibirica. The former, the common apricot*, is found, according to M. Regnier (for a translation of whose account we are indebted to Mr. Salisbury)†, in a wild state in the Oases of Africa. It is there a rich and sweet fruit, of a yellow colour. The fruit of the P. sibirica, seeds of which came to me last year from Dr. Fischer of Gorenki, is, on the contrary, I understand, black, very acid, and of small size : but nevertheless, if these apparently distinct

* The early period at which the apricot unfolds its flowers leads me to believe it to be a native of a cold climate : and I suspect the French word abricot, the English apricock, and the African Berrikokka, to have been alike derived from the Latin word præcocia, which the Romans (there is every reason to be believe) pronounced praikokia, and which was the term applied to early varieties of peaches, which probably included the apricot. The Greeks also wrote the Latin word, as I suppose the Romans to have pronounced it, Πρακοκια. Hardouin's edition of Pliny, lib. 15. sec. xi.

† See Horticultural Transactions, Vol. III. Appendix, page 23.

species will breed together, and I confidently expect they will, without giving existence to mule plants, I shall not hesitate to pronounce these plants of one and the same species; as I have done relatively to the scarlet, the pine, and Chili strawberries. Botanists may nevertheless, if they please, continue to call these transmutable plants, species; but if they do so, I think they should find some other term for such species as are not transmutable, and which will either not breed together at all, or which, breeding together, give existence to mule plants. I do not, however, feel any anxiety or wish to defend my own hypothetical opinions upon this subject; on the contrary, I shall be most happy to see them proved erroneous; and my chief object in addressing the present communication to the Horticultural Society is to point out a circumstance which is more favourable to Mr. Herbert's opinions than any other which has come under my observations.

I sent to the Society, some years ago, a fruit which sprang from a seed of the sweet almond and the pollen of a peach blossom, and which in every respect presented the character of a perfectly melting peach. When the tree which afforded that fruit first produced blossoms, I introduced into them the pollen of another peach-tree, with the view of obtaining more improved varieties of the peach of this family, and the necessary preparation of such blossoms prevented my noticing an imperfection which I have since observed in them. Little or no pollen is ever produced in them; and though the tree has borne well subsequently upon the open wall, and has produced perfect seeds without any particular attention being paid to it, I suspect that its blossoms have been fecundated by those of some adjoining nectarine trees. Having, however, often observed that varieties of the same acknowledged identical species, when one was in a highly cultivated and the other in a perfectly wild state, did not readily succeed when grafted upon each other, owing probably to the very different qualities of their circulating fluids, I conceived it possible that the same causes might have prevented a perfect union at once taking place between the almond and peach tree. I therefore waited till I had an opportunity of observing, in the last summer, the blossoms of a second generation, which proved in every respect as imperfect as those of the first tree, and, like those, afforded fruit and perfect seeds with the pollen of an adjoining nectarine tree. This result, which I did not anticipate, appears interesting; but I hesitate in drawing, at present, any inferences from it*.

* Since the foregoing observations were addressed to the Horticultural Society, a tree which sprang from a seed of a sweet almond and pollen of the early violet nectarine has produced a profusion of perfectly well organised blossoms, with abundant pollen, after having, in the three

The vegetable and animal worlds present so much similarity in almost everything which respects the generation of offspring, that the extent to which mules are permitted to exist in the animal world might have been expected to point out the utmost limits of their existence amongst plants ; for every animal is driven by its instinctive feelings to seek its proper mate, whilst an unrestrained and unlimited intercourse between plants is carried on by the incidental operation of wind and insects. But if the fruit-tree obtained from the almond and pollen of the peach be a mule, nature has already permitted it to propagate offspring to an extent rarely, if at all, known in the animal world. I have, however, heard it asserted, that female mule birds have been known to breed under similar circumstances ; that is, with a male of the same species as the male parent of the mule : but upon trying the experiment, it did not succeed at all in my hands. The mule birds laid eggs, apparently well organised, upon which they sat ; but the eggs soon became putrid ; and I had good reason to believe, that the first pulse of life had never beaten in any of them.

If hybrid plants had been formed as abundantly as Linnæus and some of his followers have imagined, and such had proved capable of affording offspring, all traces of genus and species must surely long ago have been lost and obliterated ; for the seed-vessel even of a monogynous blossom often affords plants which are obviously the offspring of different male parents ; and I believe I could adduce many facts which would satisfactorily prove that a single plant is often the offspring of more than one, and, in some instances, of many male parents. Under such circumstances, every species of plant which, either in a natural state or cultivated by man, has been once made to sport in varieties, must almost of necessity continue to assume variations of form. Some of these have often been found to resemble other species of the same genus, or other varieties of the same species, and of permanent habits, which were assumed to be species ; but I have never yet seen a hybrid plant, capable of affording offspring, which had been proved, by anything like satisfactory evidence, to have sprung from two originally distinct species ; and I must therefore continue to believe, that no species capable of propagating offspring, either of plant or animal, now exists, which did not come as such immediately from the hand of the Creator.

Having spoken, in the preceding account, of mule birds, I will take this opportunity of recording a very singular circumstance which came

preceding years, afforded imperfect blossoms only. If such pollen prove efficient, which I see no reason to doubt, either the specific identity of the peach and almond, or the transmutability of the two species, will be proved. But if the peach be an originally distinct species, where could it have lain concealed from the Creation to the reign of Claudius Cæsar ?

under my observation, whilst I was engaged in the experiments which I have stated. A person informed me that a farmer, who resided a few miles distant from me, possessed a mule bird, which was bred between the common hen and the wood-pigeon; and which my informant had seen, and described with accuracy : I took, in consequence, the earliest opportunity of seeing the farmer, and the supposed mule bird; because I thought that nature had strictly prohibited the production of mules between species so distinct, and had usually made the death of the female the price of the attempt. The information I obtained was, that the children in his house (his infant brothers and sisters) had reared a young wood-pigeon and a motherless chicken together; that these became much attached to each other, and appeared to have paired, the wood-pigeon constantly paying court to the young hen, as he would have done to a female of his own species. The hen subsequently laid eleven eggs, which she sat upon, and produced one offspring, the bird in question. It was wholly without comb, and it had soft turgid nostrils, extremely similar to those of a wood-pigeon; and the whole profile of its head, exclusive of the point of the beak, bore a most striking resemblance to that of its supposed male parent. It, however, certainly was not the offspring of a wood-pigeon, nor a mule; for it bred freely. I ought to have preserved the bird, which was offered me, and perhaps I convict myself of an act of unpardonable stupidity in not having done so. But it was a great favourite with the children who possessed it; and I did not like to deprive them of it. The animal physiologist will draw his own conclusions respecting these singular facts; I do not feel qualified to give an opinion.

XLV.—UPON THE MANAGEMENT OF FRUIT-TREES IN POTS.

[Read before the HORTICULTURAL SOCIETY, May 8th, 1821.]

I HAVE more than once mentioned, in the Transactions of this Society, the importance of giving to fruit-trees, from which a crop of fruit is required very early in the season, a high degree of excitability, or the power to vegetate very strongly in moderately low temperature, at the period when they are first subjected to artificial heat*: and I have pointed out the advantages of retaining all trees, which are intended to afford such very early crops, in pots†. In the present season, I have

---

* See paper on the Culture of the Pine-apple, p. 242; also paper on the Proper Management of Fruit-trees which are intended to be forced very early, p. 228.

† See paper on Culture of Pine-apple, p. 242.

endeavoured to ascertain within how short a period, in the ordinary temperature of my pine-stove, plants of the Chasselas and Verdelho vine could be made to yield more mature fruit.

The subjects of this experiment had produced a crop of fruit previously to midsummer 1820, and in the following month of July they had been taken from the stove, after having been for some time sparingly supplied with water, and placed under a north wall; in which situation they remained nearly torpid till autumn, when they were pruned. Early in the winter, I observed in them strong symptoms of a disposition to vegetate, though they remained in the cold and shaded situation in which they were first placed, when removed from the stove; and on the 12th of January, I found the buds so much swollen, that I feared the exposure to frost would prove fatal to them, and the pots were consequently removed to the stove.

In this, the sudden increase of temperature occasioned every visible bud to unfold itself within a very few days; and on the 17th of the following month, being thirty-six days after the pots were brought into the stove, the berries of some bunches of the Verdelho grape were so far grown, that I could have thinned them with advantage. In the end of March, the Chasselas grapes became soft and transparent, and in the middle of April some bunches were as mature, and much more yellow, than those of the same kind usually are when first brought to the London market in the spring; though the weather had been, during the early part of the spring, dark and cloudy, and consequently unfavourable. The wood of these vines appeared nearly mature in the end of the last month (April); and by removing them from the stove for a short time to a cold and shaded situation, and subsequently replacing them in the stove, I do not doubt the practicability of obtaining another crop from them within the present year.

A pot which contains a quantity of mould equal to a cube of fourteen inches has been found large enough for a vine whose foliage occupied a space of twenty square feet; water holding manure in solution being abundantly given: and I have seen grapes acquire a larger size, and other fruits a higher flavour, under such management than under any other.

The supposed necessity of frequently removing fruit-trees which grow in pots, to other pots of larger dimensions, appears to present a good deal of inconvenience; but I have readily obviated this necessity by means which I can confidently recommend to the attention of gardeners. When the plant or fruit-tree is first placed in the pot in which it is long to remain, I mix with the compost some material, in greater or less quantity, which is capable of ultimately affording nutriment, but which will decompose slowly. In some cases I have used with success slender half-decayed branches from my wood pile ; and in others I have employed sound chips, chiefly of apple-tree, mixed with mould, and in sufficient quantity to occupy at least one-fourth of the space afforded by the pot. As the roots of the plant increase, the lifeless wood gradually decomposes, at the same time giving food and space to the roots, which consequently do not become injuriously compressed in the pot. I possess a nectarine-tree which has grown nine years in the same pot, and which vegetated more strongly in the present spring than I can recollect it previously to have done. Several successive crops of fungi usually appear upon the surface of the pots under the preceding circumstances ; but I have had no reason to think these injurious.

The trouble of conveying water to numerous pots, in hot weather, would be very considerable; but a simple mode of applying the very ingenious contrivance of Mr. Loddiges, by which water is dispersed as in showers upon the foliage of his plants, and which has been described in the Society's Transactions*, would reduce this labour to the act of turning a cock : and if it were desirable to diminish or wholly take away the supply from any particular spot, this might easily be effected, by partially or wholly closing the apertures through which the water is made to escape from the pipe.

XLVI.—AN ACCOUNT OF AN IMPROVED METHOD OF RAISING EARLY POTATOES IN THE OPEN GROUND.

[*Read before the* HORTICULTURAL SOCIETY, *June 5th*, 1821.]

THE destruction, in the present season, of early crops of potatoes by frost in this vicinity, (particularly in the gardens of those who could ill bear the loss they have sustained,) has led me to address to the Society the following account of some deviations from the ordinary modes of practice in the culture of that plant, which I have found successful in not only affording plants which more effectually recover when impeded by frost, but also in furnishing a larger and more early produce under ordinary circumstances.

It has long been known that abundant crops of late and luxuriant varieties of early potatoes may be obtained by planting very small pieces only

---

* See Vol. III. page 14 of the Transactions of the Horticultural Society.

of their tuberous root : for the plants of those varieties always acquire a considerable age before they begin to generate tubers, and therefore do not too soon begin to expend themselves in the production of tubers ; and the size which these acquire within any given period in the spring will be to a great extent regulated by the strength of the plants, at the period when they first spring from the soil ; and strong plants of such varieties can be afforded only by sets of considerable size. I have, in consequence, for some years past, selected in the autumn the largest tubers, and these nearly of an equal size, for planting in the spring ; and I have found that these not only uniformly afford very strong plants, but also such as readily recover when injured by frost : for being fed by a copious reservoir beneath the soil, a reproduction of vigorous stems and foliage soon takes place, when those first produced are destroyed by frost, or other cause.

When the planter is anxious to obtain a crop within the least possible time, he will find the position in which the tubers are placed to vegetate by no means a point of indifference ; for these being shoots, or branches, which have grown thick instead of elongating, retain the disposition of branches to propel their sap to their leading buds, or points most distant from the stems of the plants of which they once formed parts. If the tubers be placed with their leading buds upwards, a few very strong and very early shoots will spring from them ; but if their position be reversed, many weaker and later shoots will be produced ; and not only the earliness, but the quality of the produce in size, will be much affected.

In the spring, when the young plants are just beginning to appear in the rows, I have often found it very advantageous to raise the mould over them in ridges by an operation perfectly similar to moulding the plants. Protection has been thus given against frost, and I have not found the period of maturity of the crop to have been in any degree retarded.

It has been contended that there is much waste in the practice above described of planting large sets ; because the old tuber is often found to have lost little in weight, when an early crop is taken up in an immature state : and it has thence been inferred, that a very small part only of the matter of the old tubers enters into the composition of the new. But I believe a false inference has in this case been drawn, and that, under ordinary circumstances, a very large portion of the soluble matter of the old tubers is employed in the formation of the new : for I have proved by experiments purposely made, that the vital union, and community of circulating fluid, between the old tuber and the plant which has sprung from it, is not so soon dissolved.

s

Some potatoes of rather large size and early habit were placed in such situations that the fibrous roots only of the plants entered into, or were in contact with, the soil. Thus circumstanced, an abundant blossom appeared, and seeds would have been produced; but both the blossoms, and the runners which would have formed young tubers, were alike removed.

The old tubers, though fully exposed to the sun and air, still retained life, and were obviously supplied with moisture by the stems, which had sprung from them; and the result was ultimately just what I had anticipated. The plants, after many frustrated efforts to produce blossoms and tubers upon every part of their branches, at last threw their sap back into the old tubers; and a numerous crop of young tubers was suspended from the buds, or eyes, of the old. This did not occur till autumn; and therefore the vital union must have subsisted through the whole summer; and I entertain but very little doubt, that such a union subsists under ordinary circumstances, till almost the whole of the soluble and organisable matter of the old tubers has been absorbed by the new. To what extent this occurs is, however, a point of little consequence : the important fact of the crop being increased by the employment of large sets has been proved by accurate experiments, in many successive seasons.

XLVII.—ON GRAFTING THE VINE.

[*Read before the* HORTICULTURAL SOCIETY, *September* 18*th*, 1821.]

THE practice of grafting the vine appears to be very ancient; for it is mentioned both by Cato and Columella* in a way which shows that it was common in the vineyards of Italy at the period in which they wrote. It must, consequently, have been an operation of easy execution, though it is rarely seen to succeed well in the hands of the modern gardener ; who is, nevertheless, certainly much better provided with instruments, and can scarcely be supposed to be inferior in skill or science to the cultivators of that period. It is therefore probable, that the ancients were acquainted with some mode of operating, of which the modern gardener is ignorant. It is well known that the ancients, in propagating the vine, employed cuttings, which consisted partly of year-old, and partly of two-year-old wood ; and the modern gardener, in deviating from this mode of practice, has adopted one which does not possess a single advantage,

* Cato, cap. 42.   Columella, lib. IV. c. 29.

and which is in every respect worse. I conceived it probable, in the last spring, that the success of the Roman cultivators in grafting their vines might have arisen from the selection of grafts similar to their cuttings; and the result of the following experiment leads me to believe my conjecture to be well founded. I selected three cuttings of the black Hamburgh grape, each having at its base one joint of two years old wood. These were inserted in, or rather fitted to, branches of nearly the same size, but of greater age; and all succeeded most perfectly. The clay which surrounded the base of the grafts was kept constantly moist; and the moisture thus supplied to the graft operated very beneficially at least, if it was not essential to the success of the operation. A very skilful gardener in my vicinity, to whom I mentioned my intention of trying the foregoing experiment, was completely successful by a somewhat different method. He used grafts similar to mine; but his vine grew under the roof of the hot-house, in which situation he found it difficult to attach such a quantity of clay as would supply the requisite degree of moisture to the graft, and he therefore supported a pot under each graft, upon which he raised the mould in heaps sufficiently high to cover the grafts, and supply them with moisture.

Some very intelligent gardeners have asserted, that they have seen the berries of some of the smaller varieties of grape enlarged by the use of stocks of larger or more luxuriant varieties.

I possess no information relative to this statement; and the object of this communication is merely to point out the means by which new varieties may be introduced into the forcing-house without loss of time or produce.

The grafts which I used consisted of about two inches of old wood, and five of annual wood, by which means the junction of the new and old wood, at which point cuttings most readily emit shoots and receive nutriment, was placed close to the head of the stock, and a single bud only was exposed to vegetate.

XLVIII.—FURTHER OBSERVATIONS ON THE CULTIVATION OF THE PINE-APPLE.

[Read before the Horticultural Society, *March* 5, 1822.]

THE following circumstances, relative to the habits of the pine-apple plant, appear to me so interesting and singular, that I am induced now to send an account of them to the Horticultural Society, though I have so recently addressed a communication* upon nearly the same subject. In that communication I mentioned the extraordinary growth of a pine-apple, which had passed the whole of the last summer and autumn in very low temperature, and which then, in the beginning of November, continued to increase in size, four months having at that time elapsed, since the period of its blossoming. I saw the same fruit in the first week of the last month (February), when it still continued perfectly green, and apparently growing rapidly. Our member Mr. Mearns, who has had not only the advantages of long and very attentive experience, but who has also visited the stoves of a very great number of the most celebrated cultivators of the pine-apple in different parts of the kingdom, has been to view the fruit above mentioned; and he assures me that he has never seen a queen pine-apple growing upon so small a plant, so perfectly well swelled out, in any season of the year, under any circumstances. He was of opinion, when he saw it, which was early in the last month, that it would probably ripen about the end of the present month, or early in April. It had passed the winter in the temperature which is usually given to common green-house plants, and it had certainly not had the advantages in any degree of judicious management, having been very irregularly, and at times much too profusely, supplied with water. What will be the merits of it when ripe, time alone can show; but I shall here observe, that I have found all fruits (and particularly the melon) to acquire their highest state of excellence when their growth has been slow—provided it has been regularly progressive, and that the fruit has ultimately attained its proper size and perfect maturity; and I believe, that no fruit has ever been seen perfect, either in taste or flavour, the growth and maturity of which had been greatly accelerated by much fire-heat, and of necessity, abundant water. I am, therefore, much inclined to believe, that the pine-apple will be found to acquire its highest state of excellence, when a considerable time elapses between the period of its blossom and that of its maturity.

Should it be found easily practicable, as I very confidently believe

* See above, page 242.

it will, to retard the ripening of the fruit of those plants of the pine-apple which blossom late in the summer, or early in the autumn, such fruit might be made to supply our tables abundantly in the spring or early summer months.

Since my last paper upon the management of the pine-apple plant was written, I have placed a few plants, which have blossomed in autumn, in very high temperature (generally above that of 80°), and very near to white glass of good quality; and so circumstanced, even the queen pine-apple has swelled nearly, if not quite, as rapidly, as it usually does in the best seasons of the year, and its taste and flavour have been quite as good as those of that kind usually are in winter. Other varieties have succeeded better, and one which I received without a name from the West Indies, and which I am informed is the St. Vincent's pine-apple, acquired, in the last month, a degree of excellence both in taste and flavour which I have rarely found equalled in any season*.

XLIX.—DESCRIPTION OF A MELON AND PINE PIT.

[*Read before the* HORTICULTURAL SOCIETY, *July* 16*th*, 1822.]

I SENT an account to the Horticultural Society, in the last spring†, of a pine-apple, which, having blossomed in the month of July, did not ripen till the following month of April, owing to its having passed the autumn and winter in low temperature ; and I thence inferred that pine-apples might easily be so managed as to supply the market abundantly in seasons when few species of fruit can be obtained. In the present spring I erected a small pine-pit upon a new construction, for the purpose of ascertaining, by experiment upon a few plants, whether my opinions were well founded ; but not having more plants than my houses could

NOTE BY THE SECRETARY, JOS. SABINE, ESQ.

   * A few days after this paper was read to the Society, being on a visit to the president, at Downton, I had the gratification of observing the condition and appearance of the pine-apple plants described by him in the communication above referred to ; the plants, which were expected to begin showing their fruit in the next month, though young, were remarkable for their vigour and strength. They were grown in pots of much larger size than usual, which were raised so as to bring the upper leaves nearly in contact with the glass. The plants themselves were firmly rooted in the mould, their leaves were of peculiar breadth and substance, the stems were short and of unusual thickness, and the whole had the appearance of extraordinary health.

   † See the preceding paper. The experiment has since been, in many instances, repeated, with similar results.

conveniently contain, I have applied the structure erected for pine-apple plants to the culture of melons only, during the summer.

These having succeeded most admirably, and a great number of gardeners having examined my machinery, and given their unqualified approbation of it, I send the following description of it to the Horticultural Society, flattering myself that it will be found, in the aggregate, superior to any now employed, which can be erected at so small an expense, and managed with so little cost and trouble. It consists of a hollow wall, similar in every respect, as to construction, to that described by Mr. Silverlock in the Transactions of the Society*; and I cannot describe it better than by using his words: "It is built nine inches thick, with sound even-sized bricks, placed edgeways, the joints being carefully made, and laid with the very best mortar. The bricks are placed with their faces and ends alternately to the outside, so that those which have their ends exposed become ties to the surfaces of the wall. In each succeeding course, as the wall is built, the bricks with their ends outwards are placed on the centre of the bricks which are laid lengthways in the course below. Thus a hollow space is formed in the middle of the wall, of four inches in width, which is only interrupted where the tying bricks cross it, but there is a free passage for air from top to bottom of the wall."

My front wall is four feet, and my back wall five feet six inches high, enclosing a space of six feet wide and fifteen feet long, and the walls are covered with a wall-plate, and with sliding lights, as in ordinary hotbeds.

The space included may be filled to a proper depth with leaves, or tan, when it is wished to promote the rapid growth of plants; but at present it contains only nine large pots, in which the melon plants grow, and the stems of these are supported by a trellis at a proper distance from the glass. The wall is externally surrounded by a hot-bed composed of leaves and horse-dung, by which it is kept warm; and the warm air contained in its cavity is permitted to pass into the inclosed space through many small perforations in the bricks. At each of the lower corners is a passage, which extends along the surface of the ground, under the fermenting material, and communicates with the cavity of the wall, into which it admits the external air to occupy the place of that which has become warm and passed into the pit. The entrances into these passages are furnished with grates, to prevent the ingress of vermin of every kind. The hot-bed is moved and renewed in small successive

---

* See Horticultural Transactions, vol. IV. page 224.

portions, so that the temperature may be permanently preserved, the ground being made to descend a little towards the wall on every side, that the bed in shrinking may rather fall towards than from the walls; and I entertain no doubt but that the perpetual ingress of warm air, even without an internal leaf-bed, will prove sufficient to preserve pine-apple-plants without the protection of mats, except in very severe weather. I have nothing further to add, but that the melon plants are the most healthy and luxuriant that I ever possessed, and that their fruit is swelling with more than ordinary rapidity. I annex (plate 6) a sketch of a section and plan of the pit, without which, I fear, the preceding account would scarcely prove intelligible.

The perforations in the interior of the wall, are from eighteen to nearly twenty inches distant from each other, and they do not begin till the fifth row of bricks from the bottom. When the pit is intended for early cucumbers or melons, and the lower part is consequently to be filled with leaves or tan, the holes in the bricks should only be made above the surface of whatever may be put into the pit, or, if previously made below, must be closed.

REFERENCES TO PLATE 6.

A. Sliding lights.   D. Hollow wall.
B B. Wall plates.   E. Dung linings.
C. Water groove.   F. Air funnels.

L.—UPON THE ADVANTAGES AND DISADVANTAGES OF CURVILINEAR IRON ROOFS TO HOT-HOUSES.

[*Read before the* HORTICULTURAL SOCIETY, *October 1st*, 1822.]

A WISH has more than once been expressed to receive from me an account of my opinion of the comparative advantages and disadvantages of the iron curvilinear, and common hot-house roofs of sliding lights, in the culture of the pine-apple, as soon as experience should have enabled me to give it. I am now, I believe, in possession of sufficient information to enable me to give an opinion with some degree of confidence, having had the experience of three summers, in which I have nearly sacrificed more than two hundred very fine fruiting pine-apple plants in my curvilinear roofed hot-house. I have, however, ultimately succeeded to the full extent of my hopes and expectations, and I give a decided preference to the curvilinear roof. I must nevertheless admit, that it has some defects, which I shall endeavour to point out, and set in opposition to its perfections.

The curvilinear iron roof certainly transmits heat more rapidly than one of wood of the ordinary construction, but not to any considerable extent, I think, more rapidly than a roof composed of wood and glass would do, if the wood were employed in as small quantities as the iron is, and not nearly to as great an extent as a roof composed wholly of glass would do, if such could be constructed.

My house is fifty feet long, and ten feet wide, and it is heated by a single fire of moderate size; and I have found that single fire fully sufficient to keep pine-apple plants in a healthy growing state, in all seasons of the year, without the aid of bark or hot-bed of any kind, and without the protection of any kind of covering *.   I have always used it as a fruiting-house, and my plants, after being placed in it, have grown admirably, and have shown fruit well; but the fruit has never till the present year, except in one instance, when the plants stood close to the door, swelled properly.   Its taste and flavour have nevertheless been good, and it constantly ripened in a singularly short time.

The fruit which appeared in September and October, in the last autumn, became ripe in January, and whenever one fruit became ripe, its aroma appeared to accelerate the maturity of all in its vicinity.

The queen pine-apples were generally very similar to those I have usually seen at the shops in London in the months of April and May; and with imperfections arising, I believe, from the same source, the want of efficient ventilation.

In houses of ordinary construction, with roofs of sliding lights, air enters and escapes at all times with much rapidity; and the consequent change of air is very nearly, if not wholly, sufficient to enable the pine-apple to acquire maturity and perfection at all seasons; provided the flues operate with sufficient power to give the requisite temperature. But in my house, with a curvilinear roof, I acquired the power of almost wholly preventing any change of air whatever; and I exercised that power too extensively, after the fruit was shown, and particularly after a part of it had nearly acquired maturity.   In the last spring I adopted a mode of ventilation, from which I expected to derive all the advantages of change of air, without materially lowering the temperature of the house; and the success of it has greatly exceeded the expectations I had entertained.   I shall best be able to show the advantages of this mode of

---

* A much higher temperature than my machinery enables me to give, and varying from 75° to 90° in winter, and from 80° to 105° in summer, would, however, be highly beneficial : and I feel quite confident that in a dry stove of such temperature pine-apples might, under appropriate management, be abundantly ripened, and in considerable perfection, in any part of the year.

ventilation, by giving (plate 7) a slight sketch of the form of a section of my house, in which D marks the position of cylindrical passages of nearly two inches diameter through the front wall. Through these, which are placed eighteen inches distant from each other, along the whole front wall of the house, the air, whenever the weather is warm, is suffered to enter freely, and its entrance is at other times more or less obstructed in proportion to its coldness : but it is never wholly excluded, except during the nights in very severe weather.

The passages through the front wall are placed at just such a distance from the ground, as will occasion them to direct the air, which enters, either into contact with, or to pass closely over, the heated covers of the flue. It consequently becomes heated, and is impelled amongst the pine-apple plants, which stand in rows behind each other, each row of plants being so far elevated above that before, as to place every plant at nearly an equal distance from the glass roof. A thermometer was placed at H, being equally distant from each end of the house, and I had the satisfaction to observe, that the temperature of that part of the house in which the thermometer stood was raised between two and three degrees, when the external air was at 40°. This effect was, I conclude, produced by the heated air being impelled into the body of the house amongst the plants, instead of being permitted to rise, as it had previously done, and to come instantly into contact with the roof : and by suspending light bodies amongst the plants, I ascertained that the previously confined air was thus constantly kept in a state of rapid motion. The air is suffered to escape through passages of four inches wide and two inches and a half high, at E, which passages are placed at the same equal distances as those in the front wall, and, like those, are opened or closed as circumstances require. The trouble of opening or closing such passages, after substances of proper form are prepared and suspended for the purpose, is very small, much less, I think, than that of moving the lights of any house of ordinary construction ; and the effect of the kind of ventilation obtained upon the growth of my plants and fruit, is everything I wish it to be.

I have stated that my whole house is heated by a single flue : this enters at the west end of it, and thence passes along the whole front within sixteen inches of the wall. It then returns twenty feet towards the middle of the house and back again, the smoke escaping at the end opposite to that which it enters. The flue is consequently single at the end of the house, which adjoins the fire place, and triple in the last twenty feet of the opposite end ; by which means a nearly equal temperature is everywhere given.

It has been objected, that the water which drops from bars of iron is extremely noxious to pine-apple plants; but I have not found this to be in any degree the case: for having placed a plant in such a situation that the water from a cast-iron rafter dropped upon it, in summer, and removing it only as soon as the mould became sufficiently moist, I could not discover that the plant had, during a month, sustained the slightest injury. Another objection made to iron roofs is, that the metal is very subject to rust. This is perfectly true, provided they be not kept well painted; but if one-third of the sum requisite to keep a wooden roof properly painted be expended upon the iron roof, no injury will ever be sustained from the liability of that to suffer from rust. I must, however, take this opportunity of observing, that the bars of all the iron roofs I have yet seen have been exceedingly ill-formed. The metal, instead of being rolled thin with grooves, and made to descend into the house far below the level of the glass, should be compressed into the least compass consistent with sufficient strength; and its lower surface, instead of being brought to a thin edge, should be hemicylindrical in form. None of the edges or angles which are now presented, and which are most subject to rust, would then exist; less shade would be thrown upon the plants in the mornings and evenings; and the condensed steam would be less subject to drop from the bars upon the plants; though this, in a house constructed as mine is, can never do any injury.

I have remarked, in a former communication, that I suspected pine-apple plants might suffer under the influence of a bright sun during the whole length of an English summer's day, in a hot-house with a curvilinear roof such as mine, if the glass were of good quality. I am not prepared positively to say whether such apprehensions are well or ill-founded: but I have thought it best to be provided with a net, such as those usually employed to protect fruit-trees, of proper form to cover my house, if necessary; and I am satisfied that I could have used it with advantage, if I had possessed it, in some very hot days in the beginning of June.

The ends of my house are of brickwork; but I think the end opposite the door ought to contain a window of about two feet square, to permit a free passage of air through, upon the door being opened in very hot weather: my own house is, however, without one.

In conclusion, I wish to observe, that a curvilinear iron roof may be erected at much less expense than one of wood: two shillings and six-pence a foot being, I conceive, a fully remunerating price to the builder of such a house as mine, the glass being white, and of the quality called *best seconds.*   Green glass might be afforded on much lower terms ; but I do not recommend it, being confident that in our climate pine-apple plants suffer a hundred days by want of light, for one in which, with proper care, they sustain injury by excess of it.

LI.—A NEW AND IMPROVED METHOD OF CULTIVATING THE MELON.

[*Read before the* HORTICULTURAL SOCIETY, *November 15th,* 1822.]

I HAVE described, in a preceding paper*, a new kind of hot-bed, into which, by means of a hollow wall, a heated current of air is made at all times to enter, without any mixture of the vapour arising from the fermenting material ; and in which the temperature is raised and supported by a rapid change of air, instead of being lowered, as it is in every other kind of hot-bed with which I am acquainted.

My object in the construction of this hot-bed, was the culture of the pine-apple ; but I employed it in the last summer in raising melons ; and I succeeded so much more perfectly than I had ever previously done, that I am led to hope the following account of the mode of culture adopted, will be honoured by the approbation of the Horticultural Society.

Before I began to raise my melon plants, I calculated, as I think every gardener ought to do, who cultivates this fruit, the amount in weight which I might expect to obtain in perfection, from a given extent of glass roof.   The heaviest crop of good grapes, which I had ever seen growing in a forcing-house, did not appear to me to exceed a pound to every fifteen inches square of glass roof, taking into the admeasurement every part of such roof.   The vines had, in such cases, lived through many successive seasons, and possessed a large extent of roots and branches, everywhere amply stored with the true sap, or living blood, of the plant generated in a preceding season, and possessing powers relative to vegetable life analogous to those of the blood of torpid animals. Their blossoms and minute leaves had also been the product of the labour of a past season.   The melon plants had, on the contrary, everything to accomplish, not only in a single season, but in a small part of such season ; and therefore I considered a pound of fruit to every fifteen inches of glass roof, to be the largest amount of perfect fruit upon which I could

* See above, page 262.

venture to calculate.   The variety of melon, which I proposed to cultivate, was a Persian kind, chiefly grown in the vicinity of Ispahan, whence
it takes its name.   Its form is nearly that of a cucumber, acquiring
frequently more than a foot in length, and weighing about seven pounds.
It possesses, in my estimation, very great excellence as a fruit ; but it is
of very difficult culture, the blossoms not setting freely, and the fruit,
on account of the excessive thinness of its skin, being very subject to
decay prematurely in the damp atmosphere of an ordinary hot-bed : and
I had, on these accounts, for some years wholly ceased to cultivate it.

Having already described, with sufficient minuteness, the mode of construction and plan of my hot-bed, I need not, at present, do anything
more than describe the manner in which my plants were managed in the
last season ; they were not planted till late in the spring, and therefore
did not produce blossoms capable of affording fruit till the second week
in July ; and it had consequently, in the last season, to grow and ripen
under a very cloudy sky.   Each plant was placed by itself in a pot of
about eighteen inches in diameter in its widest part, and of about a foot
deep, inside measure, the mould in them being very rich and light, and
constantly kept sufficiently moist with manured water ; and the number
of pots was equal to the number of melons, which I proposed that my
hot-bed should contain at one time.   These pots were supported at the
south and lowest side of the bed about fourteen inches below the glass
roof ; and the plants were trained upon a trellis at the same distance
from the roof, and parallel to it.   By these means, and by giving to each
plant a similar extent of space, I expected to see each melon swell, and
be equally well fed and ripened ; and I calculated upon the further
advantages of being able to give or to withhold water from each plant
according to the state of growth, or approaching maturity of its fruit ;
and also upon that of being able to introduce other pots and plants, as
soon as I had gathered the produce of each plant.   My success in every
respect wholly exceeded my expectations, the bed proving an instrument
of much greater powers than I had calculated upon ; and I was assured
by Sir Harford Jones, who first supplied me with seeds of the variety,
(which he had brought from Persia,) that he had never seen plants of
more healthy growth, nor with fruit better swelled, even in its native
climate.   The only enemy with which the gardener will, I believe, have
to contend, is the red spider ; and against the attacks of this he must
guard his plants, by frequently sprinkling their leaves lightly with clear
warm water.

I had a singular opportunity in this experiment of obtaining evidence of the truth of an opinion, which I gave many years ago*, that every leaf, even the most distant, of a melon-plant, contributes to feed its fruit. One of my plants exhibited appearances which led me to conclude that a fruit was set, and was swelling rapidly upon it. My gardener, on the contrary, was very positive that no such fruit existed ; and having myself searched in vain to find it, I was compelled to relinquish my opinion ; this however I resumed upon observing the habit of the plant two days afterwards, when I ordered the lights to be taken off, and every branch to be minutely examined. It was then discovered, that a melon, at the extremity of a straggling branch, had fallen through the trellis, and was hanging half a yard below it. In this situation, it had been entirely shaded by the crowded foliage of another plant ; but nevertheless it had grown in less than fourteen days to be nearly a foot long, and it weighed at least four pounds. That it had derived the material necessary to its rapid growth from the sap of the parent plant cannot, I think, be doubted : and the evidence that the most distant part of the plant contributed to feed it, is certainly extremely strong ; for the fruit grew at the distance of at least six feet from those parts of the plant which led me to infer its existence.

By what means the sap generated in the distant foliage was carried to this fruit in sufficient quantity, is a very interesting question to the physiologist, and not less so to the scientific gardener.

I have at different periods made an immense variety of experiments to ascertain by what organs, and under what circumstances, the lifeless inorganic matter, which is absorbed by the roots of plants, becomes converted into their true sap, or living vegetable blood ; and the result of every experiment has led me to believe, that in all cases where plants possess leaves, as distinct organs, it is in such organs alone, and under the influence of light, that this process takes place. The powers which roots of various forms and cuttings, and other detached parts of plants, possess of emitting foliage have appeared to me to be wholly, in all cases, dependent upon the presence of true sap previously deposited within them. Like the cotyledons of seeds, they appear to be reservoirs only, which contain, but never create : and it has been long ascertained that seedling plants perish, or at best scarcely retain life, if deprived of their cotyledons, even after the radicle has penetrated deeply into the soil, and the elongated plumule has reached its surface ; a discovery which appears to be universally given to Bonnet, but which belongs to Malpighi.

The following experiment, with many others which I could adduce,

* See above, n. xxi. p. 189, 191.

appears to prove that powers have been given to the mature leaf, which have been denied to the roots and branches of plants, and to the cotyledons of their seeds, unless the latter expand into and assume, as they in many cases do, the form and office of leaves.   In an early part of the summer some leaves of mint, (Mentha piperita,) without any portion of the substance of the stems upon which they had grown, were planted in small pots, and subjected to artificial heat, under glass.   They emitted roots and lived more than twelve months, having assumed nearly the character of the leaves of evergreen trees : and upon the mould being turned out of the pots, it was found to be everywhere surrounded by just such an interwoven mass of roots, as would have been emitted by perfect plants of the same species.   These roots presented the usual character of those organs, and consisted of medulla, alburnum, bark, and epidermis ; and as the leaf itself, during the growth of these, increased greatly in weight, the evidence that it generated the true sap, which was expended in their formation, appears perfectly conclusive.

Supposing the leaves of the melon plant to possess (as I do not entertain a shadow of doubt that they do) powers similar to those of the mint above mentioned, and of other plants, and that all the foliage may be made to contribute to feed a single fruit, it is not easy to conceive by what means this can be done, without the circulation of a very large portion of the true sap of the plant (even of that generated in its most distant foliage) through such single fruit, be assumed.   And it appears difficult upon any other grounds to account for the extremely rapid growth which, under such circumstances, takes place in a single fruit, with the influence of the fruit upon the most distant parts of the plant, and the dependence of the ultimate weight and perfection of the fruit upon the extent of the foliage of the plant.   In an experiment which I made some years ago, a single melon, of the Rock Canteloup variety, grew upon a plant which occupied more than thirty feet of the surface of a hot-bed, but under green glass of ordinary quality ; where it acquired the weight of thirteen and a half pounds, having during its growth given the whole plant full employment, and apparently put the services of every leaf in requisition, though some of them grew at nearly six feet distance from it.

The disadvantages of leaving too numerous a crop on any plant are sufficiently well known,and every skilful gardener is able to calculate, from the extent and vigour of his melon-plants, what number of fruits, of any given variety, each plant is capable of supporting ; but when a melon-plant has many fruits to support, it is often a partial parent, by which one offspring is very abundantly fed whilst another starves ; and hence often arises the great disparity in the quality of fruit of the same plant.

This cannot occur when each plant has a single fruit only to support, and is given a sufficient extent of foliage ; and, under this mode of culture, the most shy and the most free bearer become equally productive ; for every plant will readily offer all that is wanted—a single fruit.

I have already stated that I think a melon-plant of any saccharine variety will require about fifteen inches square of glass roof for every pound of fruit ; and in this calculation I include glass of good quality. There may possibly be varieties of the melon which will afford a larger produce than that above-mentioned ; but whatever variety be cultivated, I feel confident that quite as large a produce may be obtained by the mode of culture above recommended as by any other ; and I cannot but believe a larger produce of good fruit, owing to the advantages of a constant supply of warm air, and the power of giving, and of permanently maintaining, in the bed a high and regular temperature, without the introduction of steam, and the power of securing to each fruit its due share of nutriment. I am also of opinion that great advantages might be thus obtained in the very early culture of the cucumber. The cavity of the bed might be filled with leaves, or other material which would afford a temperate and permanent heat ; whilst a current of warm and dry air would be made to flow constantly into the bed above the level of the mould in which the plants were placed. When the bed is intended for this purpose, the perforations through the bricks should be confined to those which stand above the level of the mould.

As soon as the crop of melons in my bed was expended, the pots were removed, and others of smaller size, and containing pine-apple plants, were introduced and supported upon a frame of wood at proper distance from the glass, a new lining being at the same time given to the bed. These plants have subsequently thriven exceedingly, and I entertain no doubt of their continuing to thrive through the winter ; for the powers of a constant, though small, current of heated air to sustain a high temperature are very great, operating not only by introducing heat, but also in opposing the ingress of the cold external air,—a circumstance to which I particularly wish to attract the attention of the gardener.

I will take this opportunity of suggesting an improvement in the construction of the common pine-stove. If the wall which surrounds the bark-bed were made hollow, and its cavity given a communication beneath the soil (as in the hot-bed I have described), at its lower corners, with the external air ; that would pass into the cavity of the wall, and escape into the house through passages immediately beneath the coping of such walls ; and warm air might be thus at all times freely introduced with much advantage to the plants, and in winter with a very considerable diminution of the expenditure of fuel ; and indeed I feel perfectly confident that, by the proper application of hollow walls in a shed behind a hot-house, every kind of forcing culture might be successfully carried on without the use of a particle of fuel, and with a moderate quantity only of bark, or leaves, or other fermenting material.

LII.—AN ACCOUNT OF THE INJURIOUS INFLUENCE OF THE PLUM-STOCK UPON THE MOORPARK APRICOT.

[*Read before the* HORTICULTURAL SOCIETY, *April 1st*, 1823.]

IN the selection of stocks for the reception of grafts or buds of different species of fruit-trees, the English gardeners and nurserymen generally suppose that, when a stock is employed upon which the inserted graft, or bud, will grow freely and permanently, everything which is expedient or beneficial is done. It is even supposed that cases exist in which much advantage is obtained by the use of a stock of a different species, and even of a different genus. The peach and nectarine trees are thus generally believed to succeed better upon the plum than upon the native stock ; and some varieties of the pear have been pronounced by Miller to acquire their highest state of perfection upon quince-stocks ; but I suspect that Miller formed his opinion rather upon the external colour and size of the fruit than upon its intrinsic qualities, and decided, as every gardener who had honestly sent the best produce of his garden to his employer's table would probably have done, that the sample of his fruit which exhibited the finest colour and the largest size was the best ; and it is well known that a young pear-tree, when growing upon a quince-stock, affords fruit of brighter colours, and, in some varieties, of larger size ; and that the tree is rendered more governable, and therefore more productive, when trained to a wall. Taking off a circular ring of bark, or what is called ringing the stock, gives a similar increase of size to the fruit, and of brilliancy to its colour ; but its pulp is rendered much less succulent and melting ; and I suspect that the effects of a quince-stock, and of ringing, will be found very nearly similar,—each operating to interrupt the free and proper course of the sap. Some varieties of pears are known to be spoiled by the quince-stock; and I entertain little doubt but that the quality of every species of fruit, to some extent, suffers when grown upon a stock of another species or genus.

I have been led to these conclusions by the following circumstances, which have within the last two years come under my observation. I have stated, in a former communication, that the Moorpark apricot succeeds much better upon its native stock than upon a plum-stock. I had observed that its foliage acquired a deeper shade of colour, and that it retained its verdure very considerably later in the autumn; and its fruit appeared to me to be singularly excellent. I had not, however, at that period an apricot-tree growing upon a plum-stock, upon quite the same aspect; and I therefore hesitated to ascribe the superiority of the fruit to any operation of the native stock. But I have subsequently planted two trees, growing upon plum-stocks, and two upon apricot-stocks, upon the same aspects, and in a similar soil; giving those upon the plum-stocks the advantage of some superiority in age, and I have found the produce of the apricot-stocks to be in every respect greatly the best. It is much more succulent and melting, and differs so widely from the fruit of the other trees, that I have heard many gardeners, who were not acquainted with the circumstances under which the fruit was produced, contend against the identity of the variety. The buds were, however, taken from the same tree.

I have also some reasons for believing that the quality of the fruit of the peach-tree is, in some cases at least, much deteriorated by the operation of the plum-stock. My garden contains two peach-trees of the same variety, the Acton Scott, one growing upon its native stock, and the other upon a plum-stock,—the soil being similar, and the aspect the same. That growing upon the plum-stock affords fruit of a larger size, and its colour, where it is exposed to the sun, is much more red; but its pulp is more coarse, and its taste and flavour so inferior, that I should be much disposed to deny the identity of the variety, if I had not inserted the buds from which both sprang with my own hand.

Having tried experiments only in one soil, and in the same situation, I, of course, have stated the foregoing circumstance chiefly with the view of exciting other horticulturists to make similar experiments; and it is particularly desirable that such should be tried in the garden of the Society.

I think it probable that the quality of the nectarine will be still more

T

affected, its pulp being less succulent than that of the peach ; but I have not at present any facts worth adducing in support of this opinion.

One valid objection to the use of peach-stocks must be admitted : trees budded upon them certainly cannot be transplanted with an equal certainty of success ; and particularly trained trees : but those I am very much disposed to call spoiled trees, which appear calculated to gratify the impatience of the planter, but which often ultimately disappoint his hopes.   I have never found any difficulty in transplanting young budded peach trees with perfect success.

The peach stones, having been protected from severe frost through the winter, may be planted in drills, at about eight inches distant from each other, and a space of about two feet was left between the rows.   The plants will spring up in April, and in August and September will be of proper age and size to be budded about two inches from the ground. The nurseryman therefore will have the advantage of taking his buds from the trees whilst the fruit is upon them, and he can in consequence easily guard against errors, which much too frequently occur ; and he may be quite certain that none of his buds will break prematurely· Buds may be inserted in the early part of October ; and in the last autumn, I introduced some with perfect success in November.   Late in the autumn, I generally shorten the roots of my young peach-stocks, particularly those roots which descend perpendicularly into the soil, by introducing a spade into the ground on two sides of each plant, but without moving it, or further disturbing its roots.   Thus managed, the buds shoot very freely ; and with proper attention to preserve their fibrous roots, and to pack them properly, they may, I am certain, be sent to the most distant parts of the island without danger of their being killed by their removal.   Older trees possibly cannot be removed without danger of their failing ; but I transplanted a peach tree in the last autumn of ten years old, which grows upon its own roots, and was more than ten feet high ; and it is this spring emitting its blossoms as freely as those trees which have not been transplanted.   Its roots were, however, well preserved, and its branches properly retrenched.

Peach and nectarine trees, particularly of those varieties which have been recently obtained from seed, may be propagated readily by layers, either of the summer or older wood; and even from cuttings, without artificial heat ; for such strike root freely.   But the most eligible method appears to be that of sowing the stones, and budding the young plants in the same season ; and I will venture to assert, that peach and nectarine trees may be thus raised with much less expense and trouble, than by the ordinary method of budding upon plum-stocks; and that the rapidity of their growth will amply compensate for the small size at which it will be expedient to plant them.   An opinion prevails amongst gardeners, that such trees will prove very short-lived; in opposition to this, I have nothing further to say, than that I have plants of more than twelve years old, one of fourteen years old, which certainly show no disposition to die, nor any appearance of having grown old.

LIII.—AN ACCOUNT OF SOME MULE PLANTS.

[*Read before the* HORTICULTURAL SOCIETY, *May 6th,* 1823.]

THE excessive rarity of mule plants in a perfectly wild state (if in such they exist at all), and the facility with which they are in many cases obtained in the garden, seem to countenance the opinion which is entertained by many botanists, that plants of different species do not readily breed with each other, till their natural habits have been broken and changed by the operation of culture through some successive generations. Vegetable mules are, however, never produced except under circumstances which rarely, if ever, occur in a perfectly natural state; for experiment has satisfied me, that not only the pollen of the alien species must be introduced at the proper period, but also, that the natural pollen must be kept away not only at that precise period, but generally, for several succeeding days afterwards: also, and even under the most favourable circumstances, I have never succeeded in obtaining mules, unless the plant, or a considerable branch of a fruit tree, has been reduced to the necessity of nourishing mule offspring, or none.   When the later blossoms on a fruit tree were suffered to remain, such branch either threw off the fruit which would have afforded mule plants, or the natural pollen was found to have been subsequently introduced by insects or winds, and to have annihilated the operation of that obtained from the plant of another species.   Not improbably some erroneous conclusions may also have been drawn, owing to varieties of permanent habits into which different species of plants have sported, under the influence of different soils and climates, in a perfectly natural state, having been mistaken for originally distinct species; for I perfectly agree with Mr. Herbert*, in thinking that the number of species of plants, which came immediately from the hand of

* Horticultural Transactions, Vol. IV. page 16.

nature, is probably much smaller than that now found in the catalogues of botanical writers : and it is also wholly impossible to distinguish such natural varieties from originally distinct species, by any peculiarities in their external character. In the present imperfect and limited state of our information, it is therefore, in many cases, difficult to decide whether plants are or are not mules; it being still questionable whether mere natural varieties, after they have through successive generations assumed very widely different forms and characters, are found to breed with each other as readily as other varieties of the same species, of similar habits ; and that real mule plants have, in some instances, and under certain circumstances, produced offspring, (mules like themselves, I suspect,) cannot, I believe, be questioned.

The principal object of the present communication is to describe two new kinds of mule plants, which have recently come within my observation. One of these presents the singularity of being, though certainly a mule, in some degree deserving the attention of the fruit-gardener ; and the other affords me the means of pointing out a new species of fruit, in the Morello cherry, to the improvement of which I wish particularly to invite the attention of the experimental gardener.

The results of many experiments upon the different kinds of strawberries which are cultivated in our gardens, led me, some years ago, to conclude that we possess three distinct species of that genus : the wood or Alpine, the scarlet in many states of variation, and the hautbois. I failed to obtain mule plants between the Alpine and the scarlet, and hautbois, which I inferred to be of distinct species; because they did not, under favourable circumstances, breed at all with each other. But I have subsequently seen, in the possession of my friend Mr. Williams of Pitmaston, mule plants obtained from the seeds both of the scarlet and hautbois, and the pollen of the Alpine strawberry. One of these, which sprang from the seed of the hautbois, presents in its foliage and habit the character of its female parent, without any perceptible variation. It blossoms very freely, and its blossoms set well ; but the growth of the fruit subsequently remains very nearly stationary during the whole period in which the hautbois strawberry grows and ripens ; after which it swells and acquires maturity. It is then rich and high-flavoured, but of less size than the hautbois, and without seeds. Mr. Williams, however, informed me that he had once obtained a single seed, which afforded a mule plant in every respect similar to its parent. I have sent a few plants of each kind to our garden, and I believe the varieties will be thought to deserve culture by those who are admirers of the flavour of the hautbois, and wish to prolong its season. The plants in my garden afford a second blossom in autumn.

Not entertaining any doubt of the specific identity of the Morello and common cherry, I made experiments upon a large scale, confidently anticipating the production of some very valuable new varieties; and I had in consequence not less than twenty trees, which afforded blossoms in the last season. Buds of many of these had been inserted into the bearing branches of old cherry trees, which were trained to walls of different aspects; and blossoms, which were all apparently well organized and perfect, were everywhere abundantly produced, but very nearly all proved abortive. From a south wall I obtained five cherries from nearly as many thousand blossoms, and four of these did not contain seeds. One variety was very large, and nearly similar in colour to its male parent, the Elton cherry; but its colour was somewhat deeper. Its flesh was white and melting, with very abundant juice; but containing only a small portion of saccharine matter. The others were worthless, and all the plants are, I believe, unquestionably mules.

As a species of fruit, I consider the Morello cherry to present very strong claims to the attention of the horticulturist. The hardiness of its blossoms, which I have found to be alike patient of heat and cold; the large size of the fruit, with its abundant juice, and power of retaining its soundness and perfection long after it has become mature; and the exuberant produce of the tree in situations where the common cherry succeeds but ill, render it, with all its present imperfections, most valuable: and there appears to be no reasonable ground for doubt, but that richer and possibly larger varieties of it may be generated by proper culture through a few successive generations. Should the fruit become rich, a less exuberant produce must however be expected; for sugar appears to be an article, the production of which requires a large expenditure of the vital juices of the tree.

We possess, I believe, in the Flemish and Kentish cherry, two varieties of the same species with the Morello; and the Toussaint, and one or two others described by Duhamel in his *Traité des Arbres Fruitiers*, appear to belong to the same family. The Morello cherry-tree is obviously the "Cérisier très-fertile" of this author.

I have seen the blossoms and fruit of the Morello cherry-tree bear, in the forcing-house, the temperature of seventy and even of eighty degrees, without any injurious or peculiar effects, except that the plumules of the seeds produced in such high temperature expanded with something very like blossoms upon the points. Small white leaves, in every respect similar to the petals of blossoms, were in many instances arranged as in a perfect blossom, which withered and died, whilst a bud upon the lower part of the stem vegetated, and the period of puberty in the plants did not subsequently appear to be at all accelerated by the operation of the high temperature in which the seeds had been ripened.

I do not offer plants of the mule varieties above-mentioned of the cherry to the Society, because I feel quite confident of their being wholly useless.

LIV.—SOME REMARKS ON THE SUPPOSED INFLUENCE OF THE POLLEN, IN CROSS BREEDING, UPON THE COLOUR OF THE SEED-COATS OF PLANTS, AND THE QUALITIES OF THEIR FRUITS.

[*Read before the* HORTICULTURAL SOCIETY, *June 3d*, 1823.]

IT has been long ago ascertained by physiologists, that the seed-coats, or membranes which cover the cotyledons of the seeds of plants, with the receptacles which contain such seed-coats, are visible some time before the blossoms acquire their full growth ; and the existence of these organs is, therefore, obviously independent of the influence of the pollen upon the growth of the internal and essential parts of the future seeds. The seed-coats also, and the fruit of some species of plants, acquire nearly, if not wholly, their perfect growth when the pollen has been entirely withheld, or when, from other causes, it has not operated ; and from these circumstances, and other observations, it has been inferred, that neither the external cover of the seeds, nor the form, taste, or flavour of fruits, are affected by the influence of the pollen of a plant of a different variety or species.   There exists, however, some difference of opinion upon these points ; and the experiments of Mr. Goss upon the pea, of which an account is given in a paper recently printed in the Transactions of the Horticultural Society*, appear strongly to countenance the opinion, that the colour of the seed-coats, at least, may be changed by the influence of the pollen of a variety of a different character ; and hence he infers, with apparent reason, the probability that the taste and flavour of fruits may be also affected.

The narrative of Mr. Goss is unquestionably quite correct ; but I believe that there is an error in the inference which he has drawn ; and I am anxious that such error, if it exist, should be pointed out ; because

* See Vol. V. page 234.

it may occasion many experiments to be made to prove that which I conceive to have been already sufficiently proved ; and, consequently, cause the useless expenditure of time and labour, which might be advantageously employed in similar investigations upon other plants in the wide and unexplored field which lies open to the experimental Horticulturist.

The numerous varieties of strictly permanent habits of the pea, its annual life, and the distinct character in form, size, and colour of many of its varieties, induced me, many years ago, to select it for the purpose of ascertaining, by a long course of experiments, the effects of introducing the pollen of one variety into the prepared blossoms of another. My chief object in these experiments was to obtain such information as would enable me to calculate the probable effects of similar operations upon other species of plants ; and I believe it would not be easy to suggest an experiment of cross breeding upon this plant, of which I have not seen the result, through many successive generations. I shall, therefore, proceed to give a concise account of some of these experiments, or rather (as I wish not to occupy more than necessary of the time of the Society), to state the results of a few of them, believing that I shall be able to explain satisfactorily the cause of a coloured variety of the pea having been apparently changed into a white variety, by the immediate influence of the pollen in the experiment of Mr. Goss.

When, in my experiments, the pollen of a gray pea was introduced into the prepared blossoms of a white variety, no change whatever took place in the form, or colour, or size of the seeds ; all were white, and externally quite similar to others which had been produced by the unmutilated blossoms of the same plant. But these when sown in the following year uniformly afforded plants with coloured leaves and stems, and purple flowers ; and these produced gray peas only. When the stamens of the plants which sprang from such gray peas were extracted, and the pollen of a white variety, of permanent habits, was introduced, the seeds produced were uniformly gray ; but many of these afforded plants with perfectly green leaves and stems, and with white flowers, succeeded, of course, by white seed. In these experiments, the cotyledons of all the varieties of peas employed or produced were yellow ; and, consequently, the peas with white seed-coats retained their ordinary colour, though they contained the plumules and cotyledons of coloured pea plants. The cotyledons of the blue Prussian pea, which was the subject of Mr. Goss's experiments, are, on the contrary, blue ; and the colour of these being perceptible through the semi-transparent seed-coats, occasioned those to appear blue, though they are really white ;

the whole habits of that plant are those of a white pea.   The colour of the cotyledons only were, I therefore conceive, changed ; whilst the seed-coats retained their primary degree of whiteness.   I must consequently venture to conclude, that the opinions of Mr. Salisbury, quoted by Mr. Goss, which have also very long been mine, viz. that neither the colour of the seed-coats, nor the form, taste, or flavour of fruits, are ever affected by the immediate influence of the pollen of a plant of another variety or species, are well-founded.

I need not add, that Mr. Seton's experiment mentioned in the note to Mr. Goss's paper, is also most perfectly accurate ; though the results differed from those obtained by Mr. Goss, owing, I imagine, to the greater permanence of colour in the cotyledons of the green Imperial pea, which was the subject of his experiments.

LV.—ON THE PREPARATION OF STRAWBERRY PLANTS FOR EARLY FORCING.

[Read before the HORTICULTURAL SOCIETY, *March* 16*th,* 1824.]

THE method of preparing strawberry plants for early forcing, by putting the plants into pots a year, or longer, before they are intended to afford fruit, is generally perfectly successful, and is in every respect eligible, except that it requires a good deal of time and trouble.   For if the pots be not regularly watered during the summer after the plants are put into them, the size of the future fruit will be considerably reduced ; and if during the following winter the pots be not carefully protected from excess of moisture and frost, a great part of the fibrous roots, which lie in contact with the internal surface of the pots, will be found lifeless in the spring; and many of the pots, if their quality be not very good, will be broken by the expansion of the frozen water.

The minute fibrous roots of trees (the *chevelu* of the French writers) have been pronounced by them, and by all the naturalists of this country, who have written upon the subject, to be, like the leaves of deciduous plants, annual productions only : and such is the opinion of Duhamel, or rather his decision respecting facts within his own observation ; for he rarely, if ever, favours his readers with his opinions.   If the fibrous roots of plants, which have, like the strawberry plant, the whole habits of trees, be annual productions only, any effort to preserve them through the winter must be useless ; but I deny the fact of their being annual productions only ; and I contend that whenever they are found wholly lifeless round the surface of the mould of the pots, as they often are after unfavourable winters, the growth and produce of the plants in the succeeding season will be much diminished.

The mode of management which I have adopted, and which it is the object of the present communication to recommend, is the following.

I manure a small piece of ground very highly, but very superficially, just covering the manure with mould ; thus deviating widely from my ordinary practice of putting the manure deep in the soil to occasion the roots to descend deep, that they may be enabled to supply proper moisture in dry weather. The ground being prepared, the strongest and best rooted runners of the preceding year are selected and planted in rows, one foot apart, in the beginning of March. The distance between each plant is eight inches in one half the rows, and four inches only in the other half, the thickly and thinly planted rows occurring alternately. In July all the plants of the thickly planted rows are removed to ground that has produced an early crop of peas or potatoes ; and these, having their roots well preserved, always afford me an abundant crop of fruit in the following summer. The other plants remain unnoticed till the end of November, when the mould between the rows is removed with the spade, and the most widely extended lateral roots detached from it. The spade is also made to pass under each plant, and between it and the next adjoining, so that each plant becomes capable of being removed at a subsequent period without having any of its roots ruptured ; and the whole of these should be preserved as entire as is practicable. As each plant becomes detached from the surrounding soil, the ground is closed around it, and it remains till it is wanted ; but it should be placed in its pots as early as the middle of February, if it be not sooner removed. At this period innumerable radicles will be seen to spring from the sides of the older roots, and these readily extend themselves into any proper soil that is placed in contact with them. I always employ soil of the richest quality, and very finely reduced ; and a good deal of water, holding manure in solution, is employed to occasion the newly introduced soil to occupy all space previously vacant in the pots. The plants are then in a state to be subjected immediately to artificial heat.

Having denied, in opposition to the generally received opinion, that the slender fibrous roots of trees and plants, having the habits of trees, are of annual duration only ; and the subject being of much importance to the gardener ; I will state a few facts in support of my opinion. That many of the fibrous roots usually perish in winter I admit ; but under favourable circumstances I have seen a very large portion perfectly alive and growing in the spring; and in the last year I tried the following experiment, the evidence of which is, I think, conclusive. Having observed that fig-trees of some varieties are capable of ripening their fruit in much higher temperature than others, I thought it expedient to try whether the same variation of power to bear different degrees of temperature did not exist in varieties of other species of fruits. Young plants of different new varieties of nectarines were therefore placed in the stove in the spring of 1823, where they grew well till Midsummer, after which all, except one, indicated, by shedding prematurely their full grown young leaves, the presence of excess of temperature. One tree, whether owing to any peculiarity of the constitution of the variety, or other cause, remained in full health till the end of the summer; when its wood and foliage, having become perfectly mature, and the latter beginning to turn yellow and fall off, it was removed, in September, to the open wall. In this situation it remained till the middle of December, its roots having been purposely carefully guarded from injury either from excess of moisture, or of frost. In December, owing to the high excitability the plant had acquired by the treatment to which it had been previously subjected, its buds showed much disposition to vegetate; and it was consequently taken from the pot to the situation it was intended permanently to occupy.

Supposing the minute fibrous roots of a plant, thus treated, to be, like its leaves, organs of annual duration only, they ought in this case to have wholly ceased to live; but on the contrary, I found them all alive, and all in the act of elongating. The evidence in this, and in many other cases, of the fibrous roots continuing to live and vegetate in a second season is positive; that of my opponents is wholly negative; and a little positive evidence in this, as in all other cases, is more than equivalent to a great deal of negative evidence. I must therefore conclude, in opposition to the opinion of those whom I am much disposed to treat with deference, that the preservation of the minute fibres of plants is important; and I believe almost every experienced gardener will coincide with me.

LVI.—ON THE CULTIVATION OF STRAWBERRIES.

[*Read before the* HORTICULTURAL SOCIETY, *December* 21*st*, 1824.]

MR. KEENS has published, in the Transactions of the Horticultural Society*, some excellent observations upon the proper modes of managing different varieties of the strawberry ; in conjunction, however, with some opinions which I do not think well founded : and as I rarely see in the gardens of my friends that which is, in my opinion, even a moderately good crop of strawberries, I shall proceed to state some conclusions which theory and practice have conjointly led me to draw, relatively to the most advantageous modes of culture of those species and varieties of fruit.

I perfectly coincide in opinion with Mr. Keens, that the spring is the only proper season for planting. At that season of the year, the ground, having been properly worked and manured, will long continue light and permeable to the roots, which will consequently descend during the summer deeply into the soil. Abundant foliage will be produced, which will be fully exposed, through the summer, to the light ; and much true sap will be generated, whilst very little, comparatively, will be expended ; for if any fruit stalks appear, those should be taken off. In the following season, as Mr. Keens has justly observed, a superior crop will be borne than by plants of greater age, or differently cultivated.

When plantations of strawberries are made, as they usually are, in the month of August, the plants acquire sufficient strength before winter to afford a moderate crop of fruit in the following year : but the plants will not have formed a sufficient reservoir of true sap to feed even such a crop, without being too much impoverished ; their spring foliage will be also exhausted in feeding the fruit, and will continue, through the summer, to shade the leaves subsequently produced. The aggregate produce in two seasons will, in consequence, generally be found to be less in quantity, and very inferior in quality, to that afforded in one season by a plantation of equal extent, made in the spring.

Mr. Keens suffers his beds to continue three years, though he admits that the produce of the first year is the most abundant, and of the best quality ; and in order to afford his plants sufficient space, when they are three years old, he places them at too great distances, in my opinion, from each other, to obtain the greatest produce from the smallest extent of ground. He places his hautbois and pine strawberry plants at eighteen

* Vol. II. page 392.

inches apart in the rows, with intervals of two feet between the rows; each square yard consequently contains three plants only. I have placed Downton strawberry plants, which require as much space as those of the hautbois, or pine, in rows at sixteen inches distance from each other, and with only eight inches distance between the plants; which is nearly nine to each square yard; and I have found each plant at such distances nearly, if not quite, as productive, as when placed with much wider intervals. The old scarlet strawberry I have also found to bear admirably when plants have been placed in rows of one foot distance from each other, with spaces of half that distance between the plants; and I think I have obtained more than twice the amount of produce from the same extent of ground which I should have obtained, if my plants had been placed at the distances recommended by Mr. Keens. My beds are, however, totally expended at the end of sixteen or seventeen months from the time of their being formed, and the ground is then applied to other purposes. I have consequently the trouble annually of planting; but I find this trouble much less than that of properly managing old beds; and I am quite certain that I obtain a much larger quantity of fruit, and of very superior quality, than I ever did obtain, by retaining the same beds in bearing during three successive years, from the same extent of ground.

There is a very large strawberry of most luxuriant growth raised from seed by Mr. Williams of Pitmaston, called the yellow Chili, which will alone, of those varieties which I have cultivated, require, in my opinion, wider intervals than those I have mentioned; and the distances recommended by Mr. Keens will, I think, be found expedient, where that variety is cultivated. It is a variety of much merit, and of most extraordinary size, a single fruit, raised in my garden, in the last season, having weighed 558 grains. Some plants of it were sent by Mr. Williams to the Society's garden in the last spring.

I perfectly approve of, and have long practised, the mode of management recommended by Mr. Keens, of placing some long dung between the rows, where it has all the good effects which he ascribes to it; but to his practice of digging between the rows I object most strongly; for by shortening the lateral roots in autumn, the plants not only lose the true sap, which such roots abundantly contain; but the organs themselves, which the plants must depend upon for supplies of new food in the spring, must be, to a considerable extent, destroyed. This mode of treating strawberry-plants is much in use amongst country gardeners, and I have amply tried it myself, but always with injurious effects; and I do not hesitate to pronounce it decidedly bad.

The wide intervals recommended by Mr. Keens certainly permit the fruit to be gathered with much convenience ; but spaces to receive the feet of the gatherers of the fruit may be easily made ; and it is much better that a small number of strawberries should be destroyed, than that a large quantity should fail to be produced, owing to more than necessarily wide, void spaces.

Taking off the runners is not expedient in the mode of culture I recommend, and, under all circumstances, this must be done with judgment and caution; for every runner is, in its incipient state of formation, capable of becoming a fruit stalk, and if too great a number of the runners be taken off in the summer, others will be emitted by the plants, which would, under other circumstances, have been transmuted into fruit stalks.   The blossoms, consequently, will not be formed till a later period of the season, and the fruit of the following year will thence be defective alike in quantity and quality : and, under the mode of culture recommended, a large part of the runners, when these are taken off in the spring, will be required to form the new beds.

I have found the alpine strawberries to succeed best, when seedling plants, raised very early in the spring, or those obtained from runners of the preceding year, have been planted in the beginning of April, at one foot apart, in beds of about four or five feet wide, with intervals between the beds. It is expedient, in the culture of these varieties, that the superficial soil should be extremely rich ; because much the most valuable part of their produce is obtained from runners of the same season, and these require to be well nourished.   If a good alpine variety be planted, the blossoms of all the runners will rise with the third leaf.   The best which I have seen affords a white fruit, similar in form to the red variety ; and the old plants of this, as well as the runners, continue to bear till the blossoms are destroyed by frost : and both the white wood and the white alpine strawberries, appear to me to retain their flavour more perfectly in autumn than the red.   The habits of the white alpine variety above-mentioned, of which I have sent plants to the garden of the Society, are permanent in the seedling plants ; provided the seed be grown at some distance from plants of the coloured varieties of the same species.

Mr. Keens supposes the alpine strawberry-plants to be incapable of producing blossoms till they are a year old; but I have shown that they afford fruit in a very few months after they have sprung from seeds. He also supposes that the seedling plants of other species of strawberries do not produce fruit till they are two years old. I entertain no doubt but that he is correct, when the plants are raised in the open ground; but when I have employed, as I have always done, artificial heat early in the spring, I have obtained abundant crops from yearling plants of every species.

LVII.—UPON THE BENEFICIAL EFFECTS OF PROTECTING THE STEMS OF FRUIT-TREES FROM FROST IN EARLY SPRING.

[*Read before the* HORTICULTURAL SOCIETY, *February 1st,* 1825.]

THE blossoms of fruit trees fall off abortively in some seasons, and produce much fruit in others, in which the weather, relatively to tempera-ture and moisture, has been nearly the same during the flowering season of such trees; and it is in very favourable, or very unfavourable seasons only, that the gardener can, with any degree of precision, pronounce what portion of his blossoms will afford fruit. If a larger part of it than he has been led to anticipate prove abortive, he generally attributes its falling off to something which he calls a blight, and which he supposes to be the operation of some unknown noxious quality in the atmosphere, during the season in which his trees have been in blossom.

Many circumstances have at different periods come under my observa-tion, which have led me to draw a different conclusion, and to believe that whenever a very large portion of the well organized blossom of fruit trees falls off abortively, in a moderately favourable season, the cause of the failure may generally be traced to some previous check which the motion and operation of the vital fluid of the tree has sustained.

It is well known that the bark of oak trees is usually stripped off in the spring, and that in the same season the bark of other trees may be easily detached from their alburnum, or sap-wood, from which it is at that season separated by the intervention of a mixed cellular and mucilaginous substance; this is apparently employed in the organization of a new layer of fibre, or inner bark, the annual formation of which is essential to the growth of the tree. If, at this period, a severe frosty night, or very cold winds occur, the bark of the trunk or main stem of the oak tree becomes again firmly attached to its alburnum, from which it cannot be separated till the return of milder weather. Neither the health of the tree, nor its foliage, nor its blossoms, appear to sustain any material injury by this sudden suspension of its functions; but the crop of acorns invariably fails. The apple and pear tree appear to be affected to the same extent by similar degrees of cold. Their blossoms, like those of the oak, often unfold perfectly well, and present the most healthy and vigorous character ; and their pollen sheds freely. Their fruit also appears to set well; but the whole, or nearly the whole, falls off just at the period when its growth ought to commence. Some varieties of the apple and pear are much more capable of bearing unfavourable weather than others; and even the oak trees present in this respect some dissimilarity of constitution.

It is near the surface of the earth that frost in the spring operates most powerfully; and the unfolding buds of oak and ash trees, which are situated near the ground, are not unfrequently destroyed, whilst those of the more elevated branches escape injury; and hence arises, I think, a probability that some advantages may be derived from protecting the stems or larger branches of fruit trees, as far as practicable, from frost in the spring; and the following facts appear strongly to support this conclusion.

Mr. Williams of Pitmaston pointed out to me, two or three years ago, an apple tree which, having had its stem and part of its larger branches covered by evergreen trees, had borne a succession of crops of fruit; whilst other trees of the same variety, and growing contiguously in the same soil, but without having had their stems protected, had been wholly unproductive. I subsequently saw, in the garden of another of my friends, Mr. Arkwright of Hampton Court, in Herefordshire, a nectarine tree, which having sprung up from a seed accidentally in a plantation of laurels, had borne as a standard tree three successive crops of fruit. The possessor of it, with the intention of promoting its growth and health, cut away the laurel branches which surrounded its stems, in the winter of 1823-4, and in the succeeding season not a single fruit was produced. Never having known an instance of a standard nectarine tree bearing fruit in a climate so unfavourable, or nearly so unfavourable, I was led to expect that the variety possessed an extraordinary degree of hardiness: but having inserted some buds of it into bearing branches upon the walls of my garden at Downton in the autumn of 1822, I have not had any reason to believe that its blossoms are at all more patient of cold than those of other seedling varieties of the nectarine.

I planted some years ago in my garden, under a wall, in a north-east aspect, and shaded by a contiguous building, a common Chinese rose tree (Rosa indica) and a plant of Irish ivy. Both have risen considerably above the top of the wall, which is thirteen feet high; and the rose tree, of which the stem is wholly covered by the branches and foliage of the ivy, has annually produced more abundant flowers, and exhibited symptoms of more luxuriant health, than any other tree of the same kind in my possession. The soil in which it grows is poor and unfavourable; and I am unable to discover any cause, except the protection it receives, from which it has derived its luxuriant health and growth.

Ivy is generally, I believe, known to gardeners as a creeping dependent plant only: but when the trees have acquired a considerable age, and have produced fruit-bearing branches, these exhibit an independent form of growth, which they retain when detached, and form very hardy evergreen shrubs of low stature. If these were intermixed with plants of the more delicate varieties of the Chinese rose, or other low deciduous and somewhat tender flowering shrubs, so that the stems of the latter would be covered in the winter, whilst their foliage would be fully exposed to the light in summer, I think it probable that those might be successfully cultivated in situations where they would perish without such protection: and the evergreen foliage of the ivy plants in winter would be generally thought ornamental. Detached fruit-bearing branches of ivy readily emit roots, and the requisite kind of plants would therefore be easily obtained.

LVIII.—AN ACCOUNT OF A METHOD OF OBTAINING VERY EARLY CROPS OF THE GRAPE AND FIG.

[*Read before the* HORTICULTURAL SOCIETY, *March 1st,* 1825.]

MR. ARKWRIGHT* has proved that vines, of which the wood and fruit have ripened late in one season, will vegetate late in the following season, under any given degree of temperature; and I have shown the converse of this proposition to be equally true†; the plants under each different mode of treatment requiring a period of rest, during which they regain their expended excitability. The following statements will show that Mr. Arkwright and myself have met at the same point, like navigators who have continued to proceed east and west in diametrically opposite courses, the one with an apparent loss and the other with an apparent gain of time.

A Verdelho vine, growing in a pot, was placed in the stove early in the spring of 1823, where its wood became perfectly mature in August.

---

* Horticultural Transactions, Vol. III. page 95.   † See above, p. 228.

It was then taken from the stove and placed under a north wall, where it remained till the end of November, when it was replaced in the stove; and it ripened its fruit early in the following spring. In May it was again transferred to a north wall, where it remained in a quiescent state till the end of August. It then vegetated strongly, and showed abundant blossom, which upon being transferred to the stove set very freely; and the fruit, having been subjected to the influence of a very high temperature, ripened early in the present month, February. The plant will retain its foliage till April, and will not be prepared to vegetate again till late in the spring, and it is at the present period very nearly in the same inexcitable state with those described by Mr. Arkwright. This experiment will probably succeed well with those varieties of the vine only which produce blossoms somewhat freely, and are of hardy habits; but abundant crops of fruit of these may be obtained at any period of the winter or spring by proper previous management of the plants, and by the application of a higher or lower degree of temperature.

The white Marseilles fig, and the other white variety of Duhamel, the *Figue blanche*, which very closely resemble each other*, succeed most perfectly under similar treatment ; and if the trees be taken from the stove in the end of May or beginning of June, and placed under a north wall till September, and be then again transferred to the stove, they will begin to ripen their fruit in January or February, and continue to produce it till the end of May or the beginning of June, when they should be again removed from the stove. The figs which ripen in January and February are not so good as those ripened in more favourable seasons: but they are nevertheless very good fruit, and valuable in mid-winter ; and the trees, if the temperature be proper (and they are extremely patient of heat), grow equally well in all seasons, if the roof of the stove be properly constructed, and the glass be of good quality.

So small a quantity of the fruit which is formed in the preceding autumn, of either of those varieties of the fig, sets in any climate, that it will rarely be found to deserve much attention; and I usually prune off as much of the annual wood as is necessary to reduce the trees to such forms and sizes as I think most convenient, without paying any regard to their blossom buds. It appears probable that many of those varieties of the fig which will not at all bear the high temperature of a stove in summer, may succeed well in winter and early spring; but I have not yet had sufficient experience to enable me to decide.

* Traité des Arbres Fruitiers, Tom. I. page 211.

U

LIX.—ON THE CULTURE OF STRAWBERRIES.

[*Read before the* Horticultural Society, *May 17th*, 1825.]

At the period when, in the last year, I addressed to the Horticultural Society some observations upon the culture of different species and varieties of strawberries *, I had seen the successful result of other experiments ; but as my experience had then been chiefly confined to a single season, I thought it better to wait for the further evidence which the present spring has afforded me.

It is, I believe, the general practice of gardeners to select the early runners of one season to place in pots for forcing in the following spring. Instead of these, I selected, as soon as their fruit had been gathered, the roots, which in the mode of culture recommended in my last communication * upon the subject, had borne one crop of fruit ; but which had been planted too closely in their beds to be retained there long with advantage. The roots of these, to which a good deal of mould remained attached, were retained as perfect as was practicable ; but their branches, which in some varieties were become very numerous, and which in all were too abundant, were reduced to three at most in the large varieties, and to four in the smaller ; and the plants were all placed so deeply in the soil, after their old and decaying leaves had been taken off, that their buds alone remained above it. Soil of extremely rich quality had been chosen for the purpose, and water holding manure in solution was rather abundantly given to the pots ; the plants I by these means obtained, apparently owing to their possessing a more copious reservoir of sap beneath the soil, afforded me a more abundant crop of fruit, and of superior quality, to that which I believe I could have obtained from younger plants. A single plant of this kind will be found sufficient for a pot, the size of which must be regulated by the habits of the variety of strawberry.

Summer planting is, I think, always in some degree objectionable ; because the plants can never have time enough to extend their roots to a sufficient depth beneath the soil to save themselves from being injured by drought in the following spring. But as the whole extent of the soil which is allotted to produce strawberries becomes, under this mode of management, every year productive of fruit, it may in some situations be the most eligible. Whenever this mode of culture is adopted, I would

* See page 283.

recommend the kind of plants above mentioned to be selected, and to be treated in every respect as if they were to be placed in pots for forcing ; except that their roots should be made to extend as deeply as practicable into the soil in which they are planted. In summer planting I have also found great advantage in using the runners of the preceding year : these had been planted with a dibble within three inches of each other, in rows, and with intervals of only six inches between the rows, till the ground in summer was ready to receive them ; a very small space was thus found to afford plants enough for a large plantation ; and these having acquired greater strength, with more strong and more numerous roots, afforded a much more copious produce in the following season than could possibly have been obtained from younger plants. By placing the plants ulti-mately near each other—those of the large varieties within six inches of each other in the rows, and with intervals of fourteen inches between the rows ; and those of the smaller varieties within four inches of each other in the rows, and with intervals of a foot only between the rows—as large, or nearly as large, a weight of fruit may be obtained, I think, from any given extent of ground, as by planting early in the spring, provided water be supplied in the spring in sufficient quantity ; but the fruit will rarely rival that which will be produced by plantations made early in the preceding spring either in quality or size ; it will, nevertheless, excel both in quantity and quality the produce of the preceding year's runners either in the open air or forcing-house.

Whenever strawberry-plants are wanted for very early forcing, it is advantageous that their roots should have been well established in their pots in the preceding autumn, and well preserved through the winter ; but for late forcing I have obtained very good subjects by the following means :—Plants which had produced one crop of fruit were taken up as soon as all their fruit had acquired maturity, and were planted at nine inches apart in soil which had been manured superficially only, and their roots were spread horizontally near the surface of the soil ; late in the autumn the roots were as much detached from the soil as would have been requisite if they had then been to be planted in pots, but they were replaced in the soil till the end of February ; being at that period placed in pots, they produced an abundant crop of very fine fruit. I found, under this mode of management, pans without any apertures to permit the escape of the water to be preferable to pots, apparently owing to the finely-reduced mould having more perfectly closed round the fibrous roots in the form of mud in the pans than in pots of the ordinary construction. In giving water to plants which grow in vessels from which it cannot escape, the gardener will avoid supplying it in excess ; but strawberry-plants whilst growing are not easily injured by any degree of moisture in the soil.   It is scarcely necessary to mention that it will be advantageous in the first. as well as in the second transplantation, not to detach the roots more than necessary from the soil in which they have grown.

LX.—ON THE CULTIVATION OF THE AMARYLLIS SARNIENSIS, OR GUERNSEY LILY.

[*Read before the* HORTICULTURAL SOCIETY, *December* 20*th*, 1825.]

So many splendid species and varieties of Crinum, and other plants of the Liliaceous tribe, have within a few years been introduced into our gardens, that the culture of the Amaryllis Sarniensis, or Guernsey Lily, notwithstanding the unrivalled splendour of its blossoms when closely inspected, has to some extent ceased to interest the modern gardener. I should consequently think the matter of my present communication scarcely worth sending to the Horticultural Society, if I were not per-fectly confident that the same mode of culture is applicable to bulbous roots of every kind which do not flower freely (exclusive of those which grow in water), and with but little variation to plants of every kind. Wishing, however, at the present time, to confine myself to very narrow limits, I shall simply relate the experiments which I have made upon the Guernsey Lily, with the conclusions which I have drawn from the result of those experiments ; and my narrative will, I think, be most plain and intelligible, if I confine it to treatment, through successive seasons, of a single root of that plant.

The gardener possesses many means of making trees produce blossoms; by ringing, by ligatures, and by depressing their branches; and the increasing thickness of the bark of these necessarily obstructs the course of the descending fluid, and thus tends to render them productive of blossoms.   But none of these mechanical means can be made to operate upon the habits of bulbous-rooted plants; and I thence inferred, that in the culture of these I should best succeed by adopting such measures as would first occasion the generation of much true sap, and subsequently promote in it such chemical changes as would cause it to generate blossoms ; and under these impressions I made, amongst others, the fol-lowing experiments, the results of which have in every respect answered my expectations and wishes.

A bulb of the Guernsey Lily, which had flowered in the autumn of

1822, was placed in a stove as soon as its blossoms had withered, in a high temperature, and damp atmosphere. It was planted in very rich compost, and was amply supplied with water, which held manure in solution. Thus circumstanced, the bulb, which was placed in the front of a curvilinear-roofed stove, emitted much luxuriant foliage, which continued in a perfectly healthy state till spring. Water was then given in smaller and gradually reduced quantities till the month of May, when the pot in which it grew was removed into the open air. In the beginning of August the plant flowered strongly, and produced several offsets. These, with the exception of one, were removed; and the plant, being treated precisely as in the preceding season, flowered again in August 1824. In the autumn of that year it was again transferred to the stove, and subjected to the same treatment; and in the latter end of the last summer, both bulbs flowered in the same pot with more than ordinary strength, the one flower-stem supporting eighteen, and the other nineteen large blossoms. One of these flowered in the beginning of August, when its blossoms were exposed to the sun and air during the day, and protected by a covering of glass during the night, by which mode of treatment I hoped to obtain seeds; but the experiment was not successful. The blossoms of the other bulb appeared in the latter end of August, and were placed in the same situation in the stove which the bulb had occupied in the preceding winter; and I by these means obtained three apparently perfect seeds. One of these, the smallest, and seemingly the least perfect, was placed immediately in a pot in the stove, where it has already produced a plant. The old bulbs have been again placed in the stove, where they have emitted abundant foliage, and where I do not doubt they will again generate blossoms.

In the foregoing experiments, I conceive myself to have succeeded in occasioning the same bulbs to afford blossoms in three successive seasons; by having first caused the production of a large quantity of true sap, and subsequently, by the gradual abstraction of moisture, having caused that sap to become inspissated, and in consequence adapted to the production of blossom-buds. Some gardeners entertain an opinion that bulbs may be excited to produce blossom-buds by being kept very dry, after their leaves have withered: but I believe this opinion to be wholly unfounded, and that the blossoms are always generated whilst the living foliage remains attached to the bulb.

I have made nearly similar experiments upon some fibrous-rooted plants, without the aid of artificial heat, with similar, and, to me, with more interesting results, an account of which I shall reserve for a future communication.

LXI.—UPON THE CULTURE OF CELERY.

[Read before the HORTICULTURAL SOCIETY, *December 5th,* 1826.]

THAT which can be very easily done, without the exertion of much skill or ingenuity, is very rarely found to be well done, the excitement to excellence being in such cases necessarily very feeble. The practice of a very large number of British gardeners, in the management and culture of exotic plants and fruits, and in every difficult department of their professions, probably approximates to, if it have not in many instances attained, perfection ; whilst the culture of many of the common esculent plants is still capable of much improvement. I shall at present confine my observations to one of these, the Apium graveolens, or celery. This plant, under the name of smallage, a worthless and almost poisonous weed, is found in its wild state growing most luxuriantly in rank soils by the sides of wet ditches, where it can obtain at the same time abundant food and moisture. Without being very well supplied with food, it will not thrive at all in our gardens, and therefore it rarely fails to obtain a proper quantity of manure ; but as with this it is in most seasons found to grow moderately well, the gardener has not paid due attention to the circumstance of its being naturally almost an aquatic plant. I have during several seasons supplied my celery plants much more copiously with water than is usually done, and always with the best effects ; but in the last excessively dry season, I gave water so profusely that the ground was constantly kept wet ; and before the plants were moulded up above the common level of the ground, that to some extent round their roots was so perfectly saturated with moisture as to wholly preclude the probability of the plants suffering by want of it during the remaining part of the summer. My gardener had not raised his plants at the usual and proper season in the last spring, the seeds not having been sown till nearly the end of April ; but nevertheless the plants had acquired in the middle of September nearly the height of five feet. Not the quantity only, but the quality also of the produce, was greatly improved by the abundant supply of water ; for it became, as might have been inferred, more crisp and tender. The rows were five feet distant from each other ; but those spaces were not sufficiently wide to permit the plants to be moulded up to the proper height ; and this circumstance, joined to the preternatural tenderness of the leaf-stalks, caused those to be broken and beaten down so much by the first windy weather, that my crop, though very excellent, was not nearly as perfect as it might have been.

The plants also were placed within about eight inches of each other in the rows; and their foliage was so injuriously crowded, that I believe I might have obtained as large, if not a larger quantity of marketable produce, if only half as many plants had been used.

I have little more to add to the excellent directions * which Mr. Judd has given in our Transactions for the culture of this plant, except that I believe wide intervals between the rows, and between the plants in the rows, when food and water are abundantly given, will be found beneficial. I also think that in preparing the bed into which the plants are first removed from the seed-bed, considerable advantages will be obtained by covering a thin layer of dung, not in a very rotten state, with about two inches deep of mould; for under these circumstances, whenever the plants are removed, the dung will adhere tenaciously to their roots; and it will not be necessary to deprive the plants of any part of their leaves. Younger and smaller plants may therefore be used; for their growth, under the preceding circumstances, will not be at all checked; and I need not point out to the experienced gardener, that the younger his plants are, the less subject they will be to run to seed, or pipe, as it is called, in the autumn.

LXII.—UPON THE CULTURE OF THE PRUNUS PSEUDO-CERASUS, OR CHINESE CHERRY.

[*Read before the* HORTICULTURAL SOCIETY, *February* 20*th*, 1827.]

THE Prunus Pseudo-Cerasus, or Chinese cherry, has been so recently † introduced into Europe, and has been hitherto so little propagated or cultivated, that probably not even its name is known to the greater

---

* See Horticultural Transactions, Volume III. page 45.

† This cherry was introduced from China by Mr. Samuel Brookes, of Ball's Pond, in 1819, and he presented a plant of it in 1822 to the Horticultural Society. It has since, in two instances, been imported from China by the Society, through the assistance of Mr. Reeves. In the year 1824, it produced a crop of fruit in one of the houses in the Chiswick garden, which ripened within fifty days from the time the blossoms opened. In that year, a figure of the plant in flower was published by Mr. Bellenden Ker, in the Botanical Register, tab. 800, with the name of Prunus paniculata, under the impression that it was the species so named by Thunberg. It received its present name of Prunus Pseudo-Cerasus from Mr. Lindley, in his report on the New and Rare Plants (see Horticultural Transactions, Vol. IV. page 90) which had flowered in the garden at Chiswick, previously to March 1824. It is readily distinguished as a distinct species from the common cherry and the morello cherry, by its bearing its flowers in racemes, and by the peduncles being hairy. It is known in China by the name of Yung Fo, but is only cultivated as an ornamental plant at Canton, where it rarely produces fruit.

number of gardeners. It has, however, properties and qualities which will render it an acquisition of considerable value; and I am perfectly confident that it has not yet been seen, in this country, nearly in the greatest state of excellence which it is capable of acquiring. I have therefore addressed to the Horticultural Society the following observations upon the propagation and culture of it, believing that I am better acquainted with the means of propagating it than any other person is, though I am sensible that I am but ill prepared to execute the task which I have undertaken.

I received a plant of the Chinese cherry from the garden of the Horticultural Society in the summer of 1824, after it had produced its crop of fruit; and it was preserved under glass, and subjected to a slight degree of artificial heat, till the autumn of that year. It appeared very little disposed .to grow, but produced one young shoot, which afforded me a couple of buds for insertion in stocks of the common cherry. Soon after Christmas the tree was placed in a pine-stove, where it presently blossomed abundantly, and its fruit set perfectly well, as it had previously done in the garden of the Society, and it ripened in March. The cherries were middle-sized, or rather small compared with the larger varieties of the common cherry; they were of a reddish amber colour, very sweet and juicy, and excellent for the season in which they ripened. The roots of the tree were confined to rather a small pot, and the plant was not even in a moderately vigorous state of growth; I therefore infer that the fruit did not acquire either the size or state of perfection which it would have attained if the tree had been larger, and in a vigorous state of growth, and the season of the year favourable.

I inserted the two buds which I had obtained into stocks of the common cherry; and they seemed to take well, but both appeared lifeless in the spring, though one vegetated late in the summer, and is now bearing a few cherries in the pine-stove.

During the last spring and early part of the summer, the old tree retained in the stove put out very numerous roots from the bases of its young branches, similar to those emitted, under similar circumstances, by the vine; and I thence inferred that the species might be readily propagated by cuttings; and having planted some cuttings in the pine-stove this year, in January, I have proved that plants may be thus raised with perfect facility.

I endeavoured to obtain seedling plants in the present spring; but a single seed only has vegetated. The remainder decayed without vegetating, but owing to what cause I am at present ignorant. I do not however doubt of better future success, or that numerous varieties of this species of cherry will be readily obtained from seedling plants.

I intended to have obtained a very early crop of cherries from the old tree in the present year, and for that purpose I had placed it in the open air, to winter, in the autumn ; proposing to introduce it into the stove in November. But unfortunately going from home for a few days just before the time when I proposed to introduce it into the pine-stove, two very severe frosty nights occurred, which so much injured the blossom-buds, which were very far advanced, that they all fell off abortively, as those of a peach-tree would certainly have done under similar circumstances. The tree, however, did not sustain further injury, and I believe that the species will be found quite hardy enough to succeed in the open air, if trained to a wall. It is much disposed to vegetate very early in the spring; and thence its blossoms, like those of the apricot-tree, will probably require some protection. This highly-excitable habit seems to indicate a plant of a cold climate, probably that of Tartary ; and I am inclined to think that it will ripen its fruit very early in the open air in this country.

In the last summer, and in the present year, I have supplied the old plant rather freely with manure in a liquid state ; and it is now growing with very great vigour, and will afford me a large number of buds and cuttings. Being wholly ignorant of the habits of the species, and fearful of destroying the only tree I possessed, I proceeded with much more caution than usual in the use of liquid manure ; for I generally use it very freely, and without apprehension of ill effects, experience having satisfied me that plants of all kinds, even heaths*, very often perish through want of food, and that they very rarely suffer from excess of it, when their roots are confined to the narrow limits of a pot.

* A plant of heath (Erica australis, I believe) was placed under my care in the spring of 1823, with a request that I would treat it in any way I wished. It was then about eight inches high, and growing in a small quantity of peat earth and sand ; and in that it continued to grow with very little increase of size till the following spring. From that period it was regularly supplied with water, which, though clear, was considerably tinged with an infusion of pigeon's dung. I was apprehensive this kind of food would prove fatal to it ; but far from this being the result, the plant grew with excessive health and vigour, emitting very numerous branches, eight of which exceeded eighteen inches each in length. It was then taken away by the owner of it, and I have not since seen or heard of it, but it left me in a state of luxuriant health. How far other species of this genus will bear being thus abundantly fed with liquid manure, is an interesting question to the gardener.

LXIII.—AN ACCOUNT OF SOME IMPROVEMENTS IN THE CONSTRUCTION
OF HOTBEDS.

[*Read before the* Horticultural Society, *July 3rd*, 1827.]

I submit an account of a small addition which I have made in the
machinery of a common hotbed, from the use of which I believe that
every gardener who has occasion to raise cucumbers and other plants in
winter, or very early in the spring, will be able to derive very considerable
advantages. At these periods of the year, it is not easy to give the plants
a sufficiently high temperature, with proper change of air, however well
the bed may have been constructed, and with whatever care the material
which composes it may have been prepared ; and the sudden changes of
temperature which often occur in the climate of England will frequently
subject the roots of the plants to be injured by excess of heat, and the
mould, when lying upon horse-dung, to be what is called by the gardener
barned, that is, I believe, so much impregnated with ammonia, that the
roots of the plants cannot retain life in it. Another defect of the common
hotbed is, that whilst its interior part is excessively hot, so little heat
ascends through the mould, that a covering of glass alone does not afford sufficient protection to any tender plant in very cold weather, during the night.

By means of the machinery which I shall proceed to describe and to
recommend, abundant air may be given at all times, and so high a
temperature preserved, that, with a hotbed of a very moderate degree
of strength, the most tender plant will be perfectly protected without any
other covering than that of an ordinary glass-light during the severest
frost of our climate, provided the spaces where the panes of glass overlap
each other be perfectly closed.

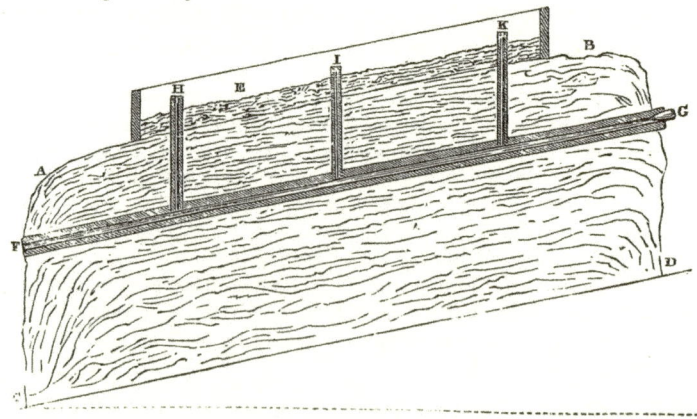

The annexed design will give a sufficiently accurate representation of the apparatus which I have above recommended. A, B, C, D, is a hot-bed, resting upon an inclined plane of earth. E, the frame; F, G, a pipe, made of a slender oak pole; and H, I, K, smaller pipes fixed into the larger one, through which the air which enters the latter at F ascends into the hotbed. The tube of the large pipe is one inch and a half, and that of the smaller three-quarters of an inch diameter. The smaller tubes have near their upper ends two horizontal apertures, through which the heated air passes laterally into the frame. I consider three of the large pipes to be fully sufficient to give heated air to a bed twenty feet long; the heated air entering at all times very rapidly, and consequently always keeping all within the frame in motion. The larger pipes might, I conceive, be with advantage made of cast-iron.

If the heat of the air be at any time excessive, it may be lessened by opening the end of the tube at G, where it is usually kept closed. The hotbed in which I have placed the above-described kind of tubes is composed almost wholly of leaves; but the mass of these is great, and the temperature in consequence high. I immersed a deep pot into the leaves, and caused the heated air of the tube K to ascend into it, having previously shortened the tube, and fitted it accurately to the aperture of the pot, placing a thermometer, with some eggs of the common domestic fowl within it, with the view of ascertaining whether these could be hatched by such means. I have not yet seen the result; but the temperature of the ascending current of air which arises into the pot, and of course into the frame, appears never to have varied during fifteen days more than three degrees, the lowest temperature being 101°, and the highest 104°; and it has, of course, been nicely adapted to both the purposes for which it was intended.

I have formerly ascertained, that the power of a current of heated air when made to enter a pit, or chamber of any kind, was found greatly to exceed the calculation which I had previously made; and in the last winter, very contrary to my expectations, a very feeble current of air, the temperature of which was below 50°, proved sufficient to preserve geraniums which were placed close to the glass in the severest frost from receiving the slightest injury.

The operation of a hotbed into which a pipe is introduced in the manner above mentioned has been observed by me only during the spring and part of the summer of the present year; but the results have been so satisfactory, that I can, with the utmost confidence, recommend the machinery which I have described, particularly when tender plants of any species are to be raised in cold seasons of the year.

LXIV.—ON THE CULTURE OF THE POTATO.

[*Read before the* HORTICULTURAL SOCIETY, *July* 1*st*, 1828.]

WHATEVER may have been the amount of the advantages or injury which the British Empire has sustained by the very widely-extended culture of the potato, it is obvious that under present existing circumstances it must continue to be very extensively cultivated ; for though it is a calamity to have a numerous population who are compelled by poverty to live chiefly upon potatoes, it would certainly be a much greater calamity to have the same population without their having potatoes to eat.

Under this view of the subject, I have been led to endeavour to ascertain, by a course of experiments, the mode of culture by which the largest and most regular produce of potatoes, and of the best quality, may be obtained from the least extent and value of ground ; and having succeeded best by deviating rather widely from the ordinary rules of culture, I send the following account of the results of my experiments. These were made upon different varieties of potatoes ; but as the results were in all cases nearly the same, I think that I shall most readily cause the practice I recommend to be understood by describing minutely the treatment of a single variety only, which I received from the Horticultural Society, under the name of Lankman's potato.

The soil in which I proposed to plant being very shallow, and lying upon a rock, I collected it with a plough into high ridges of four feet wide, to give it an artificial depth. A deep furrow was then made along the centre and highest part of each ridge ; and in the bottom of this, whole potatoes, the lightest of which did not weigh less than four ounces, were deposited, at only six inches' distance from the centre of one to the centre of another. Manure, in the ordinary quantity, was then introduced, and mould was added, sufficient to cover the potatoes rather more deeply than is generally done.

The stems of potatoes, as of other plants, rise perpendicularly under the influence of their unerring guide, gravitation, so long as they continue to be concealed beneath the soil ; but as soon as they rise above it, they are, to a considerable extent, under the control of another agent, light. Each inclines in whatever direction it receives the greatest quantity of that fluid, and consequently each avoids, and appears to shun, the shade of every contiguous plant. The old tubers being large and under the mode of culture recommended rather deeply buried in the ground, the young plants in the early part of the summer never suffer from want of moisture ; and being abundantly nourished, they soon extend themselves in every direction till they meet those of the contiguous rows, which they do not overshadow, on account of the width of the intervals.

The stems being abundantly fed, owing to the size of the old tubers, rise from the ground with great strength and luxurance, support well their foliage, and a larger breadth of this is thus, I think, exposed to the light during the whole season than under any other mode of culture which I have seen ; and as the plants acquire a very large size early in the summer, the tubers, of even very late varieties, arrive at a state of perfect maturity early in the autumn.

Having found my crops of potatoes to be in the last three years, during which alone I have accurately adopted the mode of culture above described, much greater than they had ever previously been, as well as of excellent quality, I was led to ascertain the amount in weight which an acre of ground such as I have described, the soil of which was naturally poor and shallow, would produce. A colony of rabbits had, however, in the last year done a good deal of damage, and pheasants had eaten many of the tubers which the rabbits had exposed to view ; but the remaining produce per acre exceeded five hundred and thirty-nine bushels of eighty-two pounds each,—two pounds being allowed in every bushel on account of a very small quantity of earth which adhered to them.

The preceding experiments were made with a large and productive variety of potato only; but I am much inclined to think that I have raised, and shall raise in the present year, 1828, nearly as large a produce per acre of a very well-known small early variety, the ash-leaved kidney potato. Of this variety I selected in the present spring the largest tubers which I could cause to be produced in the last year ; and I have planted them nearly in contact with each other in the rows, and with intervals, on account of the shortness of their stems, of only two feet between the rows. The plants at present display an unusual degree of strength and vigour of growth, arising from the very large size (for that variety) of the planted tubers ; and as large a breadth of foliage is exposed to the light by the small, as could be exposed by a large variety ; and as I have always found the amount of the produce, under any given external circumstance, to be regulated by the extent of foliage which was exposed to light, I think it probable that I shall obtain as large, or very nearly as large, a crop from the small variety in the present year as I obtained from the large variety in the last. I have uniformly found that, to obtain crops of potatoes of great weight and excellence, the period of planting should never be later than the beginning of March.

March 23, 1829.—Somewhat contrary to my expectations, the produce of the small early potato exceeded very considerably that of the large one above mentioned ; being per acre 665 bushels of 82 pounds. It is usually calculated by farmers that eighty pounds of potatoes, though eaten raw, after they have begun to germinate, will afford two pounds of pork; and I doubt much if the haulm, and the whole of the manure made by the hogs, were restored to the ground, whether it would be in any degree impoverished. I am not satisfied that it would not be enriched,—an important subject for consideration in a country of which the produce is at present unequal to support its inhabitants, and which produce is, I confidently believe and fear, growing gradually less, whilst the number of its inhabitants is rapidly increasing.

LXV.—ON THE CULTIVATION OF THE PINE-APPLE.

[Read before the HORTICULTURAL SOCIETY, *Aug. 19th, 1828.]*

I HAVE now completed a long course of experiments upon the culture of the pine-apple in the dry stove, the object of which has been to ascertain the means by which that species of fruit might be most advantageously grown, and particularly at those periods of the year when the scarcity of other fruits gives it an additional value. In these experiments I have endeavoured to ascertain the effects of excess of drought and of moisture, and of very high and of very low temperature. I have, of course, sacrificed many plants in experiments which I neither found nor expected to find successful; but from these I have derived information which I believe will prove useful to the cultivators and advantageous to the consumers of that species of fruit *.

The effects of a very dry atmosphere necessarily were an inspissated

---

* I have, in a communication last year to the Horticultural Society, shown that the mould in pots circumstanced as those which contain my pine-apple plants are, acquires a temperature very nearly equal to that of the aggregate temperature of the air in the house, but not subject to such extensive variations. Thus, if the highest temperature of the air within the house during the day be 90° and the lowest during the night be 70°, the temperature of the mould in the pots will nearly approximate the arithmetical mean 80° : and surely the intelligent gardeners of the present day must be fully sensible that mould at eighty degrees is warm enough without the aid of the irregular and ungovernable heat of a bark-bed, whatever their ignorant predecessors who first introduced the bark-bed into the pine-stove may have thought.

state of the sap of the plant ; and this, as it does in all other similar cases, led to the formation of blossom-buds and of fruit ; and it thus operated upon some pine-apple plants to such an extent as to cause even the scions from their roots to rise from the soil with an embryo pine-apple upon the head of each, and every plant to show fruit in a very short time, whatever were its state and age.

Very low temperature, under the influence of much light, by retarding and diminishing the expenditure of sap in the growth of the plants, comparatively with its creation, produced nearly similar effects, and caused an injuriously early appearance of fruit.

Very high temperature, if accompanied with a sufficiently humid state of the atmosphere, I found beneficial at all seasons of the year under a curvilinear iron-roofed house ; for this admitted as much light even in the middle of winter as the pine-apple plants appeared to require.

Many months previously to the publication of Mr. Daniel's very excellent communication in the Transactions of this Society (Vol. VI. page 1), and without being in any degree acquaintedwith his opinions, I had placed unglazed shallow earthen pans upon the flues of my curvilinear-roofed stove, such as he has recommended, nearly in contact with each other ; and I had increased the dampness of the air within the house by keeping the ground, which is not paved, constantly very wet. The effects of excess of humidity in the air of the house were, as might have been anticipated, diametrically opposite to those which had resulted from drought ; and the plants grew so rapidly as to become soon too large for the spaces allotted to them, without indicating at any season of the year a disposition to show fruit. By subjecting these plants to the influences of the drier atmosphere, their exuberance of growth was soon checked ; and the production of fruit immediately followed in every season of the year, provided that a sufficiently high temperature was given.

I have never cultivated the white Providence pine-apple, because I never thought it worth culture ; nor any of the large varieties, excepting a very few of the Enville ; and I have scarcely ever had a plant which has not fruited within less than twenty months of the period at which the sucker was taken from the parent plant ; and the suckers were invariably taken off at the same time with the fruit. The utmost horizontal space which I have ever allowed to any plant has not exceeded twenty-three by twenty-four inches during the latter half of its life, and less than half that space during the preceding part of it ; and I in consequence have never had a pine-apple which has weighed quite four pounds *. But I possess at the present moment succession plants of the greatest excellence, and such as I could cause to bear fruit of very great weight, if I chose to give them age and space ; for comparatively with the age and spaces allotted to the plants in my fruiting-house, the fruit of my older plants is of very large size, and in every respect exceedingly perfect. I also obtain a regular succession of produce without having ever many pine-apples ripe at the same period of the year ; and I can venture confidently to assert that I could without difficulty, in properly constructed stoves, cause crops of pine-apples to ripen regularly, and without failure, at any appointed period of the year. Some varieties of the pine-apple appear to me to be capable of acquiring a very high state of perfection under a curvilinear iron roof in the most unfavourable seasons of the year; and the most excellent fruit of the species, in my estimation, which I have ever seen has been that of the St. Vincent's or green olive in the middle of winter : and my guests have, in more than one instance, unanimously coincided with me in opinion.

I have raised as many succession plants as I have wanted (and I have used a very large number comparatively with the extent of my stoves), by placing my suckers and young plants to take root and grow over the flues between the larger plants ; but crowns and suckers never emit roots more freely, nor afford better plants, than they do when placed in a common hotbed.

I often plant suckers without detaching them from the roots and stems of the parent plants; and for the purpose of receiving such roots and long stems, I employ pots which vary in depth from eighteen to twenty-two inches with a cylindrical diameter of eleven inches only. Much time is thus gained ; for plants thus raised, if properly managed, will afford good fruit at a year old ; and they are capable whilst young of being very closely packed together.

Under a curvilinear iron roof, it will be necessary to shade the pine-apple plants during the first bright days of the spring, or the healthful verdant colour of their leaves will be tarnished ; and also to shade the plants during the long and bright days of summer from ten o'clock in the morning to three in the afternoon, or the fruit will ripen with injurious rapidity at that season. For this purpose I employ a net, of the kind I use to cover cherry-trees, doubled.

* Since the above was written, I sent a black Jamaica pine-apple to the Horticultural Society, the produce of a plant which was some months less than two years old, and which was confined to the space above mentioned, which exceeded $4\frac{1}{4}$ lbs. in weight; but I have had no other quite so heavy.

The gardener who has never cultivated pine-apples in a dry stove, should bear in mind that in giving water he should put as much at once into each pot as will moisten the mould to the bottom of it, and avoid watering very frequently.

There are in different parts of England enormous heaps of coal-dust lying at the tops of the pits of no value whatever, and in situations where pine-apples might be conveyed within three days to London by water carriage; and I am perfectly confident that these may be raised by the mode of culture recommended in this, and former communications, at less than half the expense now incurred; and I do not entertain the slightest doubt, that as large, and even larger pine-apples, may be raised without, than with a hot-bed of any kind. Nothing can be more easy than the act of giving a more regular and uniform warmth to the roots than that which can be given by the ever varying heat of a bark bed; and a sufficiently humid state in the atmosphere of the house may be regularly produced by many different means.

Some gardeners however have, as I have been informed, wholly failed in attempts to cultivate pine-apples without the aid of a bark bed; and one case of this kind has come within my own observation. In this (and probably in all others) the failure obviously arose from want of sufficient humidity in the atmosphere of the house; for the plants not only grow best, but the fruit acquires, I think, its highest state of perfection, when ripened in damp air, provided that there be a sufficient change of it, and that too much water be not given to the roots of the plants. A very dry state of the air in the stove is noxious, I believe, to almost every species of plant, and particularly to the pine-apple *.

Whenever it is wished that pine-apples should be produced of very large size, it will obviously be necessary to restrain the plants from bearing fruit till they have acquired a greater age than mine have ever been permitted to acquire; and in such case it will be beneficial to remove the plants annually into larger pots. This, when the pots, as well as the plants, are large, will not very easily be done without danger of injury to the roots. It has been my custom to remove melon plants of large size; and to preserve the roots of these from injury in trans-planting, I have had baskets, of loose texture and coarse workmanship, and consequently of very low price, made to fit the pots from which the

---

* Very dry air appears to me to be particularly injurious, when it is made to come into contact with the roots through the sides of a porous and unglazed earthen pot: I suspect, owing to causes pointed out by M. Dutrochet; see *L'agent immédiat du mouvement vital;* and *Nouvelles Recherches sur l'Endosmose et l'Exosmose.*

melon plants were to be removed ; if such baskets were to be introduced into the pots in which the pine-apple plants were placed in the autumn of one year, they would remain sufficiently sound till the following autumn to enable the gardener to remove plants of the largest size without any danger of injury to their roots. It will also be necessary when fruit of the largest size is required, to place the plants, at all periods of their growth, at considerable distances from each other, because the leaves of the pine-apple plants act less efficiently in the generation of sap, in proportion as they are made to take a perpendicular direction ; and this direction they are compelled to take when they are laterally much shaded ; for the leaves of this plant, like the stems of potatoe plants, as I have remarked in the last communication * which I had the honour to address to this Society, are subject to the conflicting influence of gravitation † and of light, the one labouring to give a perpendicular, the other a horizontal direction to the leaves ; and the comparative power of one agent increasing as that of the other decreases.

I shall conclude the present communication with an account of a very simple and efficient method of destroying the different species of insects that infest the pine-apple plant, which I have practised during the last two years with perfect success. Pine-apple plants are not at all injured by having water at the temperature of 150° of Fahrenheit's scale thrown upon and into them with a syringe. The mealy bug does not appear to be injured by a single washing, or immersion for a short time in water of the above-mentioned temperature ; but if the application be repeated three or four times on as many successive days, it wholly disappears. My gardener has, I have reason to believe, used water of a higher tempera-ture than 150° without any injury to the plants ; but as hot water, when applied in the way above-mentioned, will operate accordingly to the compound ratio of its quantity and temperature, I would recommend the gardener, when he first uses it, to apply it to a worthless plant, and not to use water of quite so high a temperature as 150°.

Having some red spiders upon the leaves of a fig-tree in the stove, I endeavoured to ascertain the effects of hot water upon these. The first application of it appeared only to render them more alert and active ; a

* See page 300.

† The influence of gravitation upon the forms of plants is still greater than I have inferred in my paper in the Philosophical Transactions upon that subject. M. Dutrochet, having used very superior machinery to that employed by me, discovered, that if a seed be made to revolve upon its own axis,.and its axis of rotation made to dip only a degree and a half below the hori-zontal line, the roots will always take the descending, and the germs the ascending line, of that axis.

second appeared to have diminished their numbers very considerably; and after a third application I could not discern any. Whether they died, or marched off only, I am ignorant; and the period at which I remove my fig-trees into the open air having arrived, I had no further opportunity of trying the experiment. I applied the water to the mature and somewhat old leaves only of the fig-trees*.

LXVI.—UPON THE SUPPOSED CHANGES OF THE CLIMATE OF ENGLAND.

[Read before the HORTICULTURAL SOCIETY, *May 5th,* 1829.]

THERE are, I believe, few persons who have noticed, and who can recollect, the state of the climate of England half a century ago, who will not be found to agree in opinion that considerable changes have taken place in it; and that our winters are now generally warmer than they were at that period. The opinions of such persons would be entitled to very little attention if they were adduced to prove that our climate has grown colder, because they themselves being far advanced in life, and therefore less patient of cold, and being also incapable of bearing the same degree of exercise which kept them warm in youth, might be readily drawn to conclude that the severity of our winters has increased. But when their evidence tends to prove that our winters have grown warmer, it cannot, I think, reasonably be rejected. My own habits and pursuits, from a very early period of my life to the present time, have led me to expose myself much to the weather in all seasons of the year, and under all circumstances; and no doubt whatever remains in my mind but that our winters are generally a good deal less severe than formerly, our springs more cold and ungenial, our summers, and particularly the latter parts of them, as warm at least as they formerly were, and our autumns considerably warmer; and I think that I can point out some physical causes, and adduce some rather strong facts, in support of these opinions.

The subject is one of much importance to the horticulturist, as it points out to him in what respects he ought to deviate from the practice of

* During the last season, several specimens of the fruit of the pine-apple, managed as above described, were sent to the Society by Mr. Knight. They were all, without exception, of the very best quality in point of flavour; they were universally destitute of fibre; and in every respect as perfectly grown as any I ever saw of the same size.—March 30, 1829.—*Jos. Sabine, Secretary.*

his predecessors, and the expediency of creating, or selecting, such varieties of different species of fruits as are well adapted to the present state of his climate.

As the chief object of this communication is to direct the attention of the gardener to the subject of fruit trees, I shall begin my observations upon that part of the year in which the blossom-buds of the succeeding year are generally formed and closed up (though much change of struc-ture within them subsequently takes place), that is, in the latter end of May. Within the last fifty years very extensive tracts of ground, which were previously covered with trees, have been cleared, and much waste land has been inclosed and cultivated; and by means of trenches and ditches, and other improvements in agriculture and covered drains, the water which falls from the clouds, and that which arises in excess out of the ground, has been more rapidly and more efficiently carried off than at previous periods. The quantity of water which our rivers contain and carry to the sea in summer and autumn is, in consequence, as I have witnessed in many instances, greatly diminished; and upon the estate where I was born, and which I now possess, my title-deeds, and the form of the ground, prove a mill to have stood, in the reign of Queen Elizabeth, and probably at a good deal later period, in a situation to which sufficient water to turn a mill-wheel one day in a month cannot now be obtained in the latter part of the summer and autumn. Under these circum-stances the ground must necessarily become much more dry in the end of May than it could have been previously to its having been inclosed and drained and cultivated; and it must consequently absorb and retain much more of the warm summer rain (for but little usually flows off) than it did in an uncultivated state; and as water in cooling is known to give out much heat to surrounding bodies, much warmth must be communicated to the ground; and this cannot fail to affect the tempera-ture of the following autumn. The warm autumnal rains, in conjunction with those of the summer, must necessarily operate powerfully upon the temperature of the succeeding winter; and, consistently with this hypo-thesis, I have observed that during the last forty years, when the weather of the summer and autumn has been very wet, the succeeding winter has been in the climate of this vicinity generally mild. And that when north-east winds have prevailed after such wet seasons the weather in the winter has been cold and cloudy, but without severe frost, probably in part owing to the ground upon the opposite shores of the Continent being in a state similar to that on this side the Channel.

I was first led to notice the preceding effects by having observed, many years ago, that some trees of the common laurel, which grew in a very high and cold situation, and which usually lost a very large portion of the annual wood, in more than one winter totally escaped all injury after such wet seasons, though their annual wood did not appear more mature in the end of November, than it would have been, in a warm and favourable situation and season, in the end of July ; and I thought the whole of it must have inevitably perished.

Supposing the ground to contain less water in the commencement of winter, on account of the operation of the drains above-mentioned, as it almost always will, and generally must do, more of the water afforded by dissolving snows, and the cold rains of winter, will be necessarily absorbed by it ; and in the end of February, however dry the ground may have been at the winter solstice, it will almost always be found saturated with water derived from those unfavourable sources ; and as the influence of the sun is as powerful on the last day of February, as on the 15th day of October, and as it is almost wholly the high temperature of the ground in the latter period which occasions the different temperature of the air in those opposite seasons, I think it can scarcely be doubted that if the soil have been rendered more cold by having absorbed a larger portion of water at very near the freezing temperature, the weather of the spring must be, to some extent, injuriously affected. But whether it be owing to the preceding or other causes, I feel most perfectly confident that the weather in the spring has been considerably less favourable to the blossoms of fruit trees, and to vegetation generally, during the last thirty years, than it was in the preceding period of the same duration ; and I shall in conclusion adduce one fact, the evidence of which I think cannot easily be controverted. The Herefordshire farmers formerly calculated upon having a full crop of acorns upon the oaks, which grew dispersed over their farms, once in three years ; but a good crop of acorns is now a thing of rare occurrence, upon the value of which the farmer has almost wholly ceased to calculate, even upon those farms which contain extensive groves of oaks. The trees nevertheless blossom annually very freely, but no fruit is produced. Many causes may be assigned for the diminished produce of orchards, and of fruit trees generally ; but the blossoms of the oak must be now as capable of bearing cold as they were half-a-century ago, and their failing to produce acorns can only be attributed to the agency of some external cause ; and I am wholly unable to conjecture any such cause except the above-mentioned.

LXVII.—AN ACCOUNT OF AN ECONOMICAL METHOD OF OBTAINING
VERY EARLY CROPS OF NEW POTATOES.

[*Read before the* HORTICULTURAL SOCIETY, *May 4th*, 1830.]

I COMMUNICATE the following account of a method of raising very early
crops of potatoes, which I have practised during the last two years, and
which will, I believe, be found to point out the means of obtaining that
vegetable at much less expense than by any other now practised, and in
a state of great perfection.

It is well known to every gardener, that potatoes which have been
buried sufficiently deep in the soil to render them secure from injury by
frost, usually vegetate very strongly in the succeeding spring; and I was
thence led to hope that by planting in September large tubers which had
ripened early in the preceding summer, and had by a period of rest
become excitable, I should be able to cause roots and stems to be emitted
to some extent in the autumn; and that these, by being well defended
from frost through winter, might operate so as to afford me a very early
produce. The experiment was not successful. The tubers vegetated
almost immediately, and the stems just reached the surface of the ground,
when they were destroyed by frost; and although the ground was immediately so well covered as securely to exclude frost from it, not a single
plant appeared in the following spring. I therefore concluded that the
experiment had totally failed, and that the tubers planted, after once
vegetating, had perished.

Late in the following summer, however, I observed that a very large
number of rather strong potatoe plants rose through the soil, precisely
where I had deposited the large tubers in the preceding autumn: and
the appearance of these perfectly satisfied me that I had erred in supposing those to have perished. The experiment was therefore repeated
in the autumn of 1828; and the result in the succeeding spring was the
same, not a single plant appearing above the soil; but upon examination
I found beneath it, in June, a very abundant crop of excellent young
potatoes, which attained maturity at least a month earlier than those
raised at the same time, in the same soil and situation, in the usual way.
It now became obvious, that a similar crop of young potatoes had been
produced in the preceding year; and that these, having remained at rest
till late in the summer, had become excitable, and had produced the
numerous plants above-mentioned. The tubers planted were of the

largest size which I could obtain of the variety, the ash-leaved kidney potatoe.

Similar experiments were made in the last autumn; but the temperature of the ground was so low, owing to the excessive coldness of the preceding summer, that not a single tuber vegetated. A part were therefore taken up, and made to vegetate by means of artificial heat, till they had emitted stems about three inches long, when they were taken from the soil, and the further progress of vegetation arrested. In the middle of January these were put into a pot with some barren sandy soil, and placed in the pine-stove, and supplied moderately with water till the middle of March. At that period I discovered that small new potatoes had been abundantly generated, and water was not subsequently given till the middle of April; when I found the pot to contain very well-grown young potatoes, which were without any other defect than that of not being, to my taste, sufficiently mature. The requisite degree of artificial heat to insure success in experiments similar to the preceding may, of course, be obtained from a variety of different sources, which I need not point out; and not improbably, I think, by means of a temperate hot-bed, the surface of the mould of which might be applied to other purposes; but I should prefer clean and barren sand for the tubers to be placed in, as those could not receive early benefit from a rich soil, and their produce might be injured in quality.

The largest crops of early potatoes will usually be obtained from tubers which have ripened late, and somewhat imperfectly, in the preceding year; but it is quite essential to the success of the preceding experiment, that the tubers which are planted in autumn should have ripened early in the foregoing summer; for otherwise they will not be found sufficiently excitable in autumn. It is also necessary that they should be of large size, otherwise the young potatoes which they afford will be small; and it will be advantageous, if the tubers to be planted have been detached from their parent plants upon their having just attained their full growth.

I believe, but I am not prepared to speak upon the evidence of experiment, that the best and the most economical mode of treating the old tubers, after their progress of vegetation has been arrested by cold, will be to put them into such heaps as are usually seen in the gardens of cottagers, and to cover them with mould; as a very large quantity would occupy only a small space, and their produce would there probably acquire a more early maturity, and might be collected at any time with little trouble.

A writer in Mr. Loudon's Gardener's Magazine has recommended the exposure of such potatoes as are intended for planting to the sun, as soon as they acquire their full growth, till they attain a green colour; and I am inclined to think the process may prove in some degree advantageous, for the action of the sun and air certainly causes chemical changes to take place in their component parts ; and chemical changes are the precursors and concomitants of excitability, if not the cause and source of it.   I am also inclined to think that similar treatment would be beneficial in the culture of all those varieties of the potatoe which do not naturally vegetate till late in the spring.

I am not prepared to say what weight of new potatoes may be obtained from any given weight of old ; but I have reason to think that the young will be equal to the weight of one-third at least of the old; and as I have shown, in a communication two years ago,* that more than thirty-five thousand pounds of our best and earliest variety of potatoe now cultivated may be obtained from an acre of ground, the mode of culture recommended will not be found expensive, (where artificial heat is not employed,) comparatively with the usual price of new potatoes early in the season.   Hogs, if hungry, will eat the old tubers, when the young have been taken away ; but those probably contain little nutriment, and their value therefore may not be worth calculating.

Two early varieties only of potatoe have been the subjects of the above-stated experiments : but there does not appear any reason to doubt that similar success may be obtained with all other early kinds.

LXVIII.—AN ACCOUNT OF A METHOD OF OBTAINING VERY EARLY CROPS OF GREEN PEAS.

[*Read before the* HORTICULTURAL SOCIETY, *May* 18*th*, 1830.]

THERE is scarcely any vegetable which is so much sought after as the pea in its green state early in the season, nor probably any one, in the culture of which so much labour is usually expended in vain.   For a very small portion only of the plants obtained from seeds sown early in the autumn survive the winter and early spring, and many of those which survive exist in a feeble and unhealthy state, and consequently afford but a very small produce.   Much more certain and abundant, and generally as early, crops of green peas, may be obtained by raising the plants under glass early in the spring, and transferring them to the open border when

* See p. 301.

they are about four or five inches high. I have also raised my plants in semi-cylindrical tiles, such as are usually employed in draining ground, and by previously depositing a little straw or litter longitudinally upon the bottoms of these, I have been enabled to slide out the plants into the appointed rows, without at all injuring or disturbing their roots. But 1 have ascertained, in the present spring, that I can obtain, by the following means, an abundant crop of peas at a much earlier period than I formerly thought possible, and at little expense or trouble.

Having found it impracticable to raise melons worth bringing to table before the days become long, and light abundant, I never plant my melon-seeds till the end of February, nor put the plants into the beds or pots in which they are to remain to bear fruit, till the end of March or beginning of April. The frames and lights were consequently out of employment in January and February in the present spring ; and I had also a heap of oak-leaves unemployed, which had been collected for the purpose of making hot-beds, and to which use they have subsequently been applied in March. With those a hot-bed was made in the middle of January, into which pots of about nine inches diameter were placed, at the distance of one foot from centre to centre. In each of these pots a couple of dozen peas were put in a circular row ; and around them was planted a row of numerous slender twigs, one foot above the surface of the mould. Thus circumstanced, the peas grew very freely, and soon attached themselves by means of their tendrils firmly to their supports ; and in the middle of March they had become fourteen inches high, and nearly in contact with the glass roof, which had been previously raised a little. They were then transferred to the open border, and some manure was given, and very numerous sticks were employed to afford them some degree of protection. This transplantation and removal from the pots did not appear to injure them in any degree ; and in the end of March many of their blossoms were so far advanced that they had shed their pollen. On the second day of April a frost of almost unprecedented severity occurred, having been preceded by an incessant fall of snow of forty hours' duration; and I anticipated the total destruction of my crop of peas. I was, however, very agreeably disappointed in finding that little or no greater injury had been sustained by plants of sixteen than by those of four inches high : and on the 26th of April, when I last saw them, they were at least three weeks earlier than any I had ever previously been able to raise ; and that, in a high and cold situation, some of the pods were above an inch and a half long.

An interval of nine inches was left between each pot of plants, which intervals soon ceased to be visible ; and a prospect of an abundant crop was afforded. I therefore conceive myself to have raised an exceedingly early and valuable crop of peas, without any loss of time to my melons ; plants of which, of proper size and age, and growing in pots, had been made ready to occupy the frames whence the peas were taken.

LXIX.—UPON THE CULTIVATION OF THE PERSIAN VARIETIES OF THE MELON.

[*Read before the* HORTICULTURAL SOCIETY, *May 1st*, 1831.]

I sent to the Horticultural Society, in the last season, a couple of Ispahan Melons ; one in August, which, I had the pleasure to hear was thought very excellent : and the other (which did not ripen till the latter end of October) not more inferior than might have been anticipated, on account of the diminished powers of the sun in the latter period. Both were the produce of very ill-treated plants : but both had the advantages of very excellent machinery ; and the effects of the management were so singular, that a statement of them may prove alike interesting to the mere practical, and to the physiological horticulturist

Having, during several years, observed, that fine Persian melons were preferred at my table to almost every other species of fruit, I was led to erect, early in the last spring, a small forcing-house for the almost exclusive culture of them, and by means of heat obtained from fire only, under an impression that in some seasons and states of the weather, the power of commanding a dry atmosphere, and high temperature, would prove highly beneficial to the quality of the fruit. This forcing-house consists of a back wall nearly nine feet high, and of a front wall nearly six feet high, inclosing a horizontal space of nine feet wide ; and the house is thirty feet long. It might as well have been forty feet long ; but the smaller size was sufficient for my purpose. The fire-place is at the east end, very near the front wall, and the flue passes to the other end of the house within four inches of the front wall, and returns back again, leaving a space of eight inches only between the advancing and returning course of it; and the smoke escapes at the north-east corner of the building. The front flue is composed of bricks laid flat, as I wished to have a temperate permanent heat, and the returning flue of bricks standing on their edges, as is usual ; the space between the flues is filled with fragments of burned bricks, which absorb much water, and gradually give out moisture to the air of the house.   Air is admitted through apertures in the front wall, which are four inches wide, and nearly three in height ; and which are situated level with the top of the flues, and are eighteen inches distant from each other.   The air escapes through similar apertures near the top of the back wall.   These apertures are left open, or partially or wholly closed, as circumstances require.   Thirty-two pots are placed upon the flues described above, each being sixteen inches wide at least, and four-teen inches deep ; but they are raised by an intervening piece of stone and brick out of actual contact with the flues.   Into each of these pots one melon plant is put, which in its subsequent growth is trained upon a trellis placed about fourteen inches distant from the glass, and each plant is permitted to bear one melon only.   Each might be made to bear more, but if they should be as large as Ispahan melons are when perfect, they would certainly be of inferior quality.   The height from the ground at which the trellis is placed is such that I can with convenience walk under it, and of course discover without difficulty the first appearance of red spiders, or other noxious insects.

When I left the country to come to London in the last spring, my plants were growing most luxuriantly ; and their appearance was every-thing that I wished.   But during my absence a few red spiders appeared upon one of the plants, as I had anticipated, and my gardener, in consequence, and in obedience to my instructions, sprinkled the under surfaces of the leaves frequently, and rather freely, with water.   By these measures the increase and spreading of the red spider was effectually prevented ; but on my return from London, I found that my plants had wholly ceased to grow, though their appearance was healthy ; and subse-quently all the fruit dropped off either before or soon after their blossom had expanded.   I in consequence immediately ordered other plants to be raised, still, however, entertaining hopes of preserving those I had.   But those hopes were not realised ; and I was obliged to throw away the whole of them, with the exception of one, which was more healthy than the others, and which lived to produce the first fruit sent to the Society. That appeared to be, as it proved, of good quality ; but it was defective in size ; its weight seemed little, if anything, more than five pounds.

My second family of plants were treated nearly as the first had been, and with the same approaching results ; but I was led by the discoveries of M. Dutrochet to change my mode of management, and, I believe, to discover the cause of the preceding failure.   This eminent physiologist had discovered that if a lighter fluid be in contact with one side of an animal or vegetable membrane, and a denser fluid with the opposite side, the lighter fluid will rush into the denser through the membrane, though that be under other circumstances impervious to it. The force with which the lighter fluid, in some of the experiments of M. Dutrochet, rushed through animal membranes into the denser, appears to be exceedingly wonderful. He found, that under such circumstances water would pass upwards through three folds of the substance of a recently-extracted animal bladder, and in opposition to the perpendicular pressure of forty-five inches of quicksilver; which is nearly equivalent to a pressure of twenty-two and a half pounds upon a square inch of surface. This power in vegetable membranes to transmit the lighter into the denser fluid is, I think, probably in active operation during the ascent of the sap of trees in the spring; for it is through the cellular substance, and not through the tubes of the alburnum, that the sap ascends, or its ascent would be prevented, which it is not, by intersection of those tubes; and those tubes are also dry at midsummer, when the sap is rising to supply moisture to the leaves in great abundance. Previously to the discoveries of M. Dutrochet, I had shown that the sap of trees is lightest, or least dense, near the ground; and that in any particular tree, the weight of the sap increases as its distance from the ground through the course of the alburnum increases: and I had also proved that saccharine matter exists in considerable quantity in the sap in the spring, in cases where no vestige of it can be discovered in winter: and sugar was the material employed by M. Dutrochet to form his denser fluid. These facts were not in any degree known to M. Dutrochet when he made his discoveries, and he therefore was certainly not led in any degree by me in making them.

The sap in the leaves of my melon plants was certainly a denser fluid than the water with which they were sprinkled; and therefore I imagine that the latter fluid passed in injurious excess into the cells and vessels, and that the ingress and circulation of the proper fluid, which ought to have continued to ascend from the roots, was to a great extent prevented, and that the creation of the true or living sap of the plant almost wholly ceased. The plant consequently, I conclude, ceased to grow, and the fruit fell off, owing to want of proper nutriment. Soon after I had ceased to sprinkle the under surfaces of the leaves, the young fruit began to set well, and the plants to grow, but never with very great vigour; and the fruit, though its quality was exceedingly good, was smaller a good deal than I conceived it would have been if the under sides of the leaves had not been so frequently wetted. The weather was, however, very unfavourable, and the fruit, I entertain no doubt, would have been larger, if the foliage of these plants had received the benefit of more light. I have mentioned, in a former communication upon the culture of the melon, that a single melon or gourd will put in requisition, during the period of its rapid growth, the services of the most distant leaf, and cause the most distant blossom to fall off abortively. But I was, at that period, wholly unprepared to offer any conjecture whatever respecting the power by which the sap generated in very distant leaves could be conveyed to the extent indicated to the fruit.

The above mentioned discoveries of M. Dutrochet appear to me to have thrown some light upon this mysterious point; for if the fluid within the fruit be denser than that in the leaves and stems, (and in certain states at least of the growth of the fruit it certainly is so,) the lighter fluid must rush into the denser; and that the sap flows in very large quantity into the growing melon, can I think scarcely be doubted. I am well satisfied that a very large quantity of the sap of the plant, or more properly of the aqueous part of that fluid, passes through the fruit into the vessels of the plant again; but by what means it can be propelled, I am wholly at a loss to conjecture. Much must, I conceive, be done by some operation of the fruit itself; for it is totally absurd to suppose that a distant leaf can, by any mode of action properly its own, cause the true sap which it generates, to flow to and into the fruit. Previously to the maturity of my late crop of melons, I had prepared some strong cucumber plants, which I had protected from the frost; and these being brought into the place whence the melon plants had been taken, afforded me a crop of fine cucumbers in November and December. I have now cucumber plants growing in great health and vigour, from which I do not entertain any doubt of obtaining an abundant crop of cucumbers in March and the beginning of April, when it is my intention to introduce strong Ispahan melon plants; and I feel confident that, by having a proper plant ready to supply the place of every one which affords a ripe fruit, I shall be able to obtain two abundant crops of excellent melons within the same season*. If these expectations should prove to be well founded, I conceive that forcing-houses, such as I have described, for the culture of very early cucumbers and Persian melons, might be erected with advantage in those districts in which coals are raised; for the dust of coals is all that is wanted, and in fact is preferable; and cucumbers can be sent to a very considerable distance without suffering much, and melons without suffering any deterioration.

The best varieties of Persian melons are, I believe, very subject to burst when raised in this country; and I imagine that they very frequently

* I shall obtain three successions in the present season.

do so in their native country ; for Sir Harford Jones Brydges informed me, that he had heard the Persian gardeners express fear when a horse was ridden at a rapid pace near the melon beds, that the vibration of the soil would cause the melons to burst. It occurred to me in the last summer, that melons might possibly be made more safe from accidents of this kind, if I raised their points higher than their stems, and thus caused gravitation, which operates very powerfully upon the form and growth of plants, to assist in carrying away any excess of fluid, which the fruit, from any cause, might happen at any period to contain. I consequently gave to every melon an elevation of thirty degrees, and not one of those failed to ripen in a whole and perfect state ; but whether owing to any action of gravitation or not, I am, of course, unprepared to decide : the experiment, however, appears worth repeating. I suspect melons frequently burst owing to the injurious effects of the pressure of their weight upon their lower sides; for when I have suffered them to hang down perpendicularly, they have always ripened well ; but the Ispahan melons, under such circumstances, assumed forms nearly similar to those of cucumbers swollen at their points, and such forms are to my eyes very unpleasing.

LXX.—ON THE POTATOE.

[*Read before the* HORTICULTURAL SOCIETY, *February* 1st, 1831.]

IF the potatoe could only be employed, as it has chiefly been, to afford vegetable food to mankind, its improvement would be an exceedingly important object; for, circumstanced as this country is, it must necessarily constitute a large part of the food of the poorer classes ; and it is consumed in large quantities at the tables of the affluent and luxurious. But I am convinced, by the evidence of experiments which I have been some years in making, that the potatoe plant, under proper management, is capable of causing to be brought to market a much greater weight of vegetable food, from any given extent of ground, than any other plant which we possess, with equal profit to the farmer. The Swedish turnip may, in certain seasons and when the soil is favourable, rival, and perhaps excel it ; but a total failure of crops of that plant is an event of no unfrequent occurrence, and partial failures occur in almost every season; whilst by proper culture, and selection of varieties which vegetate and acquire maturity in successive parts of summer and autumn, there is not any crop which I conceive to be so certain as that of potatoes ; and it has the advantage of being generally most abundant, when the crops of wheat are defective : that is, in wet seasons *. And, I think, I shall be able to adduce some strong facts in support of my opinion, that by a greatly extended culture of the potatoe, for the purpose of supplying the markets with vegetable food, a more abundant and more wholesome supply of food for the use of the labouring classes of society may be obtained than wheat can ever afford, and, I believe, of a more palatable kind to the greater number of persons. I can just recollect the time when the potatoe was unknown to the peasantry of Herefordshire, whose gardens were then almost exclusively occupied by different varieties of the cabbage. Their food at that period chiefly consisted of bread and cheese, with the produce of their gardens, and tea was unknown to them. About sixty-six years ago, before the potatoe was introduced into their gardens, agues had been so extremely prevalent, that the periods in which they, or their families, had been afflicted with that disorder, were the eras to which I usually heard them refer in speaking of past events ; and I recollect being cautioned by them frequently not to stand exposed to the sun in May, lest I should get an ague. The potatoe was then cultivated in small quantities in the gardens of gentlemen ; but it was not thought to afford wholesome nutriment, and was supposed by many to possess deleterious qualities. The prejudices of all parties, however, disappeared so rapidly, that within ten years the potatoe had almost wholly driven the cabbage from the garden of the cottagers. Within the same period, ague, the previously prevalent disease of the country, disappeared ; and no other species of disease became prevalent. I adduce this fact, as evidence only, that the introduction of the potatoe was not injurious to the health of the peasantry at that period ; but whether its production was, or was not, instrumental in causing the disappearance of ague, I will not venture to give an opinion. I am, however, confident, that neither draining the soil (for that was not done), nor any change in the general habits of the peasantry, had taken place, to which their improved health could be attributed.

Bread is well known to constitute the chief food of the French peasantry. They are a very temperate race of men ; and they possess

---

* Failures of crops of potatoes occur in Ireland, because the excessive poverty of the peasantry compels them to plant their ground generally with less than one-fifth of the proper quantity of potatoes ; and all the Irish varieties which I have seen have been unproductive, though generally of exceedingly good quality ; the Irish mode of culture is also, I have reason to believe, excessively bad.

the advantages of a very fine and dry climate.   Yet the duration of life amongst them is very short, scarcely exceeding two-thirds of the average duration of life in England, and in some districts much less. Dr. Hawkins, in his Medical Statistics, states upon the authority of M. Villerme, that in the department of Indre, " one-fourth of the children born die within the first year, and half between fifteen and twenty, and that three-fourths are dead within the space of fifty years." Having inquired of a very eminent French physiologist, M. Dutrochet, who is resident in the department of Indre, the cause of this extraordinary mortality, he stated it to be their food, which consisted chiefly of bread ; and of which he calculated every adult peasant to eat two pounds a day.   And he added, without having received any leading question from me, or in any degree knowing my opinion upon the subject, that if the peasantry of his country would substitute (which they could do) a small quantity of animal food with potatoes instead of so much bread, they would live much longer, and with much better health.   I am inclined to pay much deference to M. Dutrochet's opinion ; for he combines the advantages of a regular medical education with great acuteness of mind, and I believe him to be as well acquainted with the general laws of organic life as any person living: and I think his opinion deserves some support from the well known fact, that the duration of human life has been much greater in England during the last sixty years than in the preceding period of the same duration.   Bread made of wheat, when taken in large quantities, has probably, more than any other article of food in use in this country, the effect of overloading the alimentary canal ; and the general practice of the French physicians points out the prevalence of diseases thence arising amongst their patients.

I do not, however, think or mean to say, that potatoes alone are proper food for any human being : but I feel confident, that four ounces of meat, with as large a quantity of good potatoes as would wholly take away the sensation of hunger, would afford, during twenty-four hours, more efficient nutriment than could be derived from bread in any quantity, and might be obtained at much less expense.

I now proceed to give an account of the result of the experiment abovementioned, which, I hope, will be found sufficiently interesting to attract the attention of the Members of this Society.   It has been proved by many other persons, as well as by myself, that if all the blossoms of a potatoe plant be picked off, as soon as they become visible, the quantity of tubers will be considerably increased, particularly if the variety be one which produces seeds ; and I have shown that the cause why early varieties of the potatoe do not afford blossoms is the preternaturally early disposition of the plant to generate its tuberous roots. The early varieties are of dwarfish growth, and therefore improper for extensive field culture; but I have found that by cross-breeding between those, and varieties of tall and luxuriant growth, I can communicate to the latter the habit of producing tubers only, without blossom ; with, I have reason to hope, considerable advantages. I now possess a good many of such varieties, selected from a very great number, which prove totally worthless ; but many of those varieties which do not produce blossom, have other defects, which render them of little value. The stems of some of these are not strong and rigid enough to support themselves and their foliage ; and they are consequently beaten down by rain and winds. The foliage of one stem consequently often becomes so placed as to shade the foliage of another; and as the whole material of the tubers is formed of living matter, which is generated in the leaves only, and as all leaves which are shaded become inefficient and useless, a sufficient degree of strength and rigidity in the stems to enable them to retain their foliage in its first position is very important ; though I believe that this circum= stance has not hitherto attracted the attention of any cultivator of the potatoe.

The tubers of other varieties, which were in all other respects appa= rently good, were defective in specific gravity, and consequently aqueous and worthless; and in others, veins of a red colour extended in to the body of the tubers, and gave an unpleasant colour to their meal, which was in some other respects of very good quality. But I have obtained several varieties which do not blossom, and which are, as far as I am at present capable of judging, without any particular defect ; though I am far from thinking I possess any variety which has even approximated to the greatest state of perfection which the species is capable of attaining.

I have succeeded in obtaining, as I wished, some varieties which vege= tate early, and others late, in the spring. Those of the first mentioned habit will generally be found to afford the largest produce by having the advantages of a longer summer ; but it is desirable to possess varieties of less excitable habits, because such usually remain good till a later period in the spring, when good vegetables are not always readily obtainable. I have also succeeded in obtaining varieties which do not vegetate till late in the spring, and which, nevertheless, acquire perfect or rather early maturity in autumn, and there are probably climates in which such varieties would be peculiarly valuable ; and the ductility and obedience of this species of plant to human will is so great, that I doubt whether, by the creation and selection of proper varieties, as abundant a produce might not be obtained within the limits of the frigid zone as in the torrid zone, of which the potatoe is a native. The weather in some parts of the coast of Norway, within the limits of the frigid zone, is very warm and bright during a period, which I believe to be quite long enough to ripen any early variety of the potatoe perfectly.

It is my wish to send in the spring one or two potatoes of each of the varieties which I think likely to prove valuable; and I shall be happy subsequently to send a quantity of any which may be approved.

In raising varieties of the potatoe from seeds it is always expedient to use artificial heat. I have trained up a young seedling plant in a somewhat shaded situation in the stove till it has been between four and five feet high, and then removed it to the open ground in the beginning of May, covering its stem during almost its whole length lightly with mould, and by such means I have obtained within the first year nearly a peck of potatoes from a single plant. But I usually sow the seeds in a hotbed early in March, and, after having given them one transplantation in the hotbed, I have gradually exposed them to the open air, and planted them out in the middle of May: and, by immersing their stems rather deeply into the ground, I have within the same season usually seen each variety in such a state of maturity as has enabled me to judge, with a good deal of accuracy, respecting its future merits.

I stated, in a former communication two years ago, that I had obtained from a small plantation of the early ash-leaved kidney potatoe a produce equivalent to that of 665 bushels, of 80 pounds each, per acre; and my crop of that variety in the present year was to a small extent greater. By a mistake of my workmen I was prevented ascertaining with accuracy the produce per acre of a plantation of Lankman's potatoe; but one of my friends having made a plantation of that variety precisely in conformity with the instructions given in my former communication to this society, I requested that he would send me an accurate account of the produce; which I have reason to believe he did, for its amount very nearly agreed with my calculation upon viewing the growing crop about six weeks before it was collected. The situation in which this crop grew was high and cold, and the ground was not rich, but the part where the potatoes to be weighed were selected was perfectly dry, and afforded a much better crop than the remainder of the field; which was planted with several different varieties. I calculated the produce of the selected part to be 600 bushels per acre, and the report I received, and which I believe to have been perfectly accurate, stated it to be 628. If this produce be eaten by hogs, or cows, or sheep (for all are equally fond of potatoes), I entertain no doubt whatever that it will afford twenty times as much animal food as the same extent of the same ground would have yielded in permanent pasture; and I am perfectly satisfied upon the evidence of facts which I have recently ascertained that, if the whole of the manure afforded by the crops of potatoes above-mentioned be returned to the field, it will be capable of affording as good, and even a better, crop in the present year than it did in the last; and that as long a succession of at least equally good crops might be obtained as the cultivator might choose, and with benefit to the soil of the field. Should this conclusion prove correct, a very interesting question arises, viz.—whether the spade husbandry might not be introduced upon a few acres of ground surrounding, on all sides, the cottages of day-labourers, to and from every part of which the manure and the produce might be conveyed without the necessity of a horse being ever employed. A single man might easily manage four statute acres thus situated, with the assistance of his family; and if nothing were taken away from the ground except animal food, I feel confident that the ground might be made to become gradually more and more productive, with great benefit to the possessor of the soil, and to the labouring classes, wherever the supply is found to exceed the demand for labour.

LXXI.—ON THE MEANS OF PROLONGING THE DURATION OF VALUABLE VARIETIES OF FRUITS.

[*Read before the* HORTICULTURAL SOCIETY, *May 3rd*, 1831.]

THE fact that all trees of the same variety of fruit, where each tree partakes necessarily of one common life, are in their habits strongly connected with those of the first original tree of the variety, is, I think, placed beyond the reach of controversy. None can be made to produce blossoms or fruit till the original tree has attained its age of puberty; and, under our ordinary modes of propagation by grafts and buds, all become subject within no very distant period to the debilities and diseases of old age. It is therefore desirable that the planter should know at what periods of their existence varieties of fruits are most productive and eligible; and by what means (if any exist) the deterioration of valuable varieties may be prevented or retarded. I was formerly inclined to believe that grafts taken from very young seedling trees, as soon as the qualities of their produce could be known, would show more disposition to grow than to produce fruit, and I had previously satisfied myself that the blossoms of old and debilitated varieties of fruits were extremely impatient of cold and unfavourable weather; and I was thence led to infer that each variety possessed its greatest value in its middle age. But subsequent experiment and observation have compelled me to draw a different conclusion; and I believe that in vegetable, as in animal life, the most prolific period is that which immediately succeeds the age of puberty.

I have made a good many experiments with a view of ascertaining this point, of which the following are amongst the most satisfactory. I took in the summer of 1828 some buds from the extremities of the leading branches of seedling pear-trees, which, being nearly twenty years old, had in the preceding autumn produced their first fruit. The buds were in July inserted in stocks, which had sprung from seeds in the preceding spring, and were then only four months old. The trees are consequently three years old now, dating from the period when they sprang from the ground; and many of them, though they have not been transplanted or subjected to any peculiar mode of treatment, have produced blossoms, some of them very abundantly and vigorously, in the present spring. I never previously saw, and I do not think that any other person has seen, in this climate fruit produced by pear-trees at so early an age. I had previously made the same experiment with apple-trees with the same results.

Some branches of a plum-tree which had not attained the age of puberty were employed as layers, and these, as I expected they would, very freely emitted roots; but, very contrary to my expectations, I found that the young shoots which these layers had produced afforded in the following spring much blossom. The variety of plum which was the subject of this experiment is, I have reason to believe, exceedingly productive of blossom; but I doubt much whether such blossoms would have appeared if the variety had been a century old. The only inference, however, which I wish to draw from the foregoing premises is, that grafts or buds taken from the bearing branches of very young seedling trees afford trees capable of bearing freely at a very early age; as it would be waste of time to offer facts or arguments in proof that such trees would continue to grow with health and vigour.

Any information which the gardener might derive from knowledge of the preceding facts would be of very little value if every part of seedling trees were in the same degree affected by age; but it is not so; and the decay of the powers of life in the roots of seedling trees is exceeding slow comparatively with the bearing branches. Scions, obtained from the roots of pear-trees of two hundred years old, afford grafts which grow with great vigour; and which in many cases are covered with thorns like young seedling stocks, whilst other grafts taken at the same time from the extremities of the branches of such trees present a totally different character, and a very slow and unhealthy growth. I do not, however, conceive that any scion which thus springs from the root of an old tree possesses all the powers of a young seedling tree, but it certainly possesses no inconsiderable portion of such powers; and I have proved such scions to be capable of affording healthy trees of a considerable size.

If grafts or buds were taken from such scions, on their first emission, much time would elapse before any blossom would be produced: but if buds were not taken from such scions till the branches attained the age of puberty, no loss of time whatever would subsequently occur.

The branches of the plum-tree, in the experiment above-mentioned, emitted roots just at the period when they had attained the age of puberty; and I do not doubt, but that scions from the roots of these will spring from the soil in full possession of all the powers attached to the branches from which they derived their existence. My own experience leads me to think that trees of the pear, the apple, and the plum, might be better raised by layers and cuttings of the roots, than by the methods usually practised, and at less expense.

The garden of the Society contains many varieties of fruits, which I believe to be extremely valuable as well as new; and the preservation of these permanently in their pristine and present state of health and vigour, appears to be an object of great importance. And the decay of many varieties (such as the Cornish gilliflower-apple, which in my estimation is and always was without a rival in the climate of England) might be greatly retarded by propagating it from scions which have recently sprung from the trunks of old trees, in obedience to the instructions of Virgil (whose authority is however generally of little value), and probably of Hyginus, "summa ne pete flagella."

LXXII.—UPON GRAFTING THE WALNUT-TREE.

[*Read before the* HORTICULTURAL SOCIETY, *April 17th*, 1832.]

THE walnut-tree appears hitherto to have effectually baffled, under all ordinary circumstances, the art of the grafter. The inserted scions wither and die, without apparently making any effort to unite themselves to the stock, or to draw nutriment from it; and consequently the value of every superior variety has been limited by its use to the possessor of the original seedling-tree. It is true that a part of the seedling offspring of every fine variety generally inherits a portion of its good qualities; but I have found it extremely difficult to obtain from seed good varieties of sufficiently early habits to ripen well in this vicinity, except in very warm seasons; and I doubt much whether the value of the crop of walnuts, throughout the British Islands, be one-third as great as it would be if proper varieties were everywhere planted.

It must, however, be admitted, that, amongst fruit-trees in general, ungrafted seedling plants usually afford the finest trees: but if the grafts be taken from young seedlings, or from scions which have sprung out of the trunks, or large branches, of trees of greater age, and those be varieties of luxuriant and healthy growth, the vigour and durability of the future tree will not be much diminished. The more early production of fruit, by grafted trees, will necessarily, to some extent, impede their growth; because a portion of their sap must be expended in giving nourishment to such fruit: but the largest pear-trees which I have ever seen must have sprung from grafts taken from trees of considerable age. One of these, which grows upon an estate that belongs to me, a Barland pear-tree, (an old variety now nearly expended,) has been known to afford, in the same season, two hundred and seventy-five gallons of perry.

The walnut-tree may be propagated with more success by budding. I have succeeded tolerably well in some seasons, and in one season perfectly well; but in several others not a single inserted bud has been found alive in the following year, though all had been inserted with the greatest care.

I therefore communicate the following mode of grafting the walnut-tree, which I found in the last season most perfectly successful under many unfavourable circumstances; and which mode, for reasons which I shall proceed to state, will, I believe, point out the means of propagating some other species of trees with facility, which have not hitherto been so propagated without difficulty and uncertainty.

The fluid which the seeds of the walnut-tree contain, when that is fully prepared to germinate in the spring, and which was deposited within it for the purpose of affording nutriment to the seminal buds, or plumule, in the preceding autumn, is sweet, as in a great many other kinds of seeds : but during germination this becomes, in the seed of the walnut-tree, bitter and acrid. Similar changes take place in the sap which is deposited, for analogous purposes, in the bark and wood of the walnut-tree, during the germination of its buds; and I was led by the discoveries of M. Dutrochet to infer the probability, that the sap during, and subsequent to, its chemical changes, might acquire new and more extensive vital powers. I therefore resolved to suffer the buds of my grafts, and those of the stocks, to which I proposed to apply them, to unfold, and to grow during a week or ten days ; then to destroy all the young shoots and foliage, and to graft at a subsequent period. A very severe frost in the morning of the 7th of May saved me the trouble of destroying the young shoots ; but it deranged my experiment by killing much of the slender annual wood, which I proposed to use for grafts ; so that I found some difficulty in choosing proper grafts. The swelling of the small, and previously almost invisible, buds, within a few days enabled me to distinguish the living wood from that which had been killed by the frost, and the stocks were grafted upon the 18th day of May. My grafter had more than once been previously employed by me to graft walnut-trees in various ways, and never having in any degree succeeded, he did not seem at all pleased with the task assigned him, and very confidently foretold that every graft would die : and I subsequently found that he had insured, to some extent, the truth of his prophesy, by having applied grafts which were actually dead. The whole number employed was twenty-eight, and out of these twenty-two grew well ; generally very vigorously, many producing shoots of nearly a yard long, and of very great strength ; and the length of the longest shoot exceeding a yard and five inches. The grafts were attached to the young (annual) wood of stocks, which were between five and eight feet high ; and in all cases they were placed to stand astride the stocks, one division being in some instances introduced between the bark and the wood ; and both divisions being, in others, fitted to the wood or bark in the ordinary way. Both modes of operating were equally successful. In each of these methods of grafting it is advantageous to pare away almost all the wood of both the divisions of the grafts ; and therefore the wide dimensions of the medulla in the young shoots of the walnut-tree do not present any inconvenience to the grafter.

No difficulties will henceforth, I conclude, occur in propagating varieties of walnuts by grafting; and I am much inclined to believe, that different species and varieties of oaks may be successfully grafted by the same mode of management.

The art of grafting our common fruit trees has been so long, and so extensively practised, that it may reasonably be supposed to be, at this late period, incapable of much improvement. But, nevertheless, I am much inclined to believe that a good deal is still to be learned; and it would not afford matter of much astonishment to me, if it should be proved that branches provided with blossom-buds might be transferred with success from one side of the Atlantic to the other, to afford fruit in the following season. The results of some experiments, which I made in the last winter, and present spring, induce me to think this practicable, though I am not yet prepared to decide that it is so.

LXXIII.—ON THE BENEFICIAL EFFECTS OF THE ACCUMULATION OF SAP IN ANNUAL PLANTS.

[*Read before the* HORTICULTURAL SOCIETY, *December 20th*, 1830.]

BIENNIAL plants very obviously form in one season the sap, which they expend in the following season in the production of blossoms and seeds ; and the capacity of the reservoirs they form is greater or less, in proportion as external circumstances are more or less favourable. Trees also (as I conceive myself to have satisfactorily proved in the Philosophical Transactions) generate in a preceding season, or seasons, the sap which feeds, in the spring, their unfolding blossoms and young leaves. Annual plants, on the contrary, possess no such reservoirs ; and they must generate, in each season, all the sap which they can expend, exclusively of the very small portion derived from the seeds from which they spring. But by appropriate management, and creation of varieties, annual plants may be made to accumulate, in one period of their lives, the sap which they expend in another, with very great advantages to the cultivator.

The first produced female blossoms of the melon-plant, particularly of the larger and superior varieties, do not often set ; and if they set, the fruit they afford never attains as large a size, or as much excellence, as the same plants, at a more mature age, would have given to it under the same external circumstances. This, I imagine, arises not only from the different quantity, but from the different qualities of the sap in the young and in the more mature plant; for I have found the sap of very young birch and sycamore trees to be specifically much lighter, and to contain much less saccharine matter, than the sap of trees of greater age of the same species, and growing in the same soil, and in the same seasons. Under the influence of abundant light, in those climates in which the melon was placed by nature, the first formed fruit probably acquires a high state of perfection, possibly greater than it can ever be made to acquire in less favourable climates. But this I am much disposed to question, and to believe that, by proper management, the melon may be made to acquire in the climate of England a degree of excellence which it is very rarely found to possess in any climate, and that the degeneracy of the finest varieties may be totally prevented.

Very young plants of the sweet melon of Ispahan (the variety which till within the present year I have chiefly cultivated) very rarely show fruit; and in my melon-house I never suffer a lateral shoot or blossom of this variety to be produced at a less distance from the root than that of the fourteenth or fifteenth joint above the seed-leaves: and when I am anxious to obtain the fruit and seeds in the highest state of perfection, I do not suffer a blossom to be produced nearer the root than its eighteenth or twentieth joint. Under this mode of management, the expenditure of sap, being confined to the extremity of a single stem, is very small comparatively with the creation of it; and it consequently accumulates, and the fruit is therefore most abundantly nourished,—I conceive more abundantly than it usually is in any natural climate: and its growth is always enormously rapid.

The striped and green Hoosainee melon-plants, of which I received seeds from the Horticultural Society in the last spring, being much disposed to bear fruit, produced blossoms at their third joints; but being desirous of obtaining the fruit and seeds of those varieties in the highest possible state of perfection, I subjected those varieties to the same mode of management, and I believe with the best success, though I am ignorant of the merits of those varieties under other circumstances.

The fruit of the striped Hoosainee melon-plant requires a very long period to attain maturity after it has attained its full growth, and after it has apparently ceased to draw much nourishment from the plant. During this period, I conceived that the plants, having all their foliage in a perfectly healthy state, must be in the act of generating much more sap than they were expending, and I therefore suffered two plants, from which I took off the fruit in the end of August, to remain wholly unpruned.   Much fruit was in consequence soon offered, and I obtained very good melons for any season, and perfectly well grown, in the latter end of the last month (November), which fruit, I do not entertain any doubt, was chiefly nourished by sap generated in the month of August.

The quality of some Ispahan melons, which I have sent to the Society, has afforded, I believe, satisfactory evidence that that variety has not become deteriorated by having been raised through many successive generations in the unfavourable climate of this place : but the following statement, I think, affords strong evidence that, like other highly improved varieties, it does degenerate under our ordinary modes of culture. Sir Harford Jones Brydges, from whom I, many years ago, first received seeds of this variety, informed me, in the beginning of the last year, that it had so much degenerated and diminished in size, that he had ceased to cultivate it.   He then received a few seeds from me, from which he assured me, in the last month, that he had obtained melons in the present year, scarcely inferior to any he had eaten in Persia;—conclusive evidence, I think, that the finest Persian varieties of the melon do not necessarily degenerate in the climate of England.

Every gardener who has been in the habit of raising cucumbers in winter perfectly well knows the advantages of raising his plants in July or August, and preventing their expending themselves in the production of blossoms or fruit till they have been introduced into the stove.   The general opinion of gardeners is, that such plants succeed best only because their stems are more firm and ligneous than those of young plants ; but I feel confident that the real cause of their succeeding best is the existence of accumulated sap within them.   I have a melon-plant now growing in the stove, which sprang from a seed sown in the end of July, but upon which no fruit was made to set till the 1st day of November. The plant possesses abundant foliage, and the fruit has grown tolerably well, and it will, I conclude, be ripe about Christmas.   Upon the 23d of October I placed a blossom, which had been produced by a Dampsha melon-plant, from which I had a few days before taken the fruit, within the distance of an inch of a very warm flue, where the temperature of the air was never below 86°.   In this situation the fruit set well, and grew with most extraordinary rapidity, though it was so near the front wall, and so far (nearly three feet) from the glass, that no direct ray of the sun could fall upon it.   At the end of seven days precisely from the period when the pollen was put into the flower, I measured the fruit, when it was seven inches long, and seven inches and a half in circumference.   On the 10th day the fruit suddenly ceased to grow, having apparently exhausted the reservoir whence it drew nutriment, and the plant withered; on the fourteenth day the fruit was gathered, when it weighed very nearly a pound and a half. If the days had been long, and the weather bright, the creation of sap would, I conclude, have nearly kept pace with the very rapid expenditure of it ; and the plant would not have died, as it apparently did, of exhaustion.

By delaying the period of sowing the seeds of many species of plants (the turnip and some varieties of the cabbage afford examples), those which would have afforded flowers and seeds within the same season form reservoirs of accumulated sap in autumn, which becomes, during winter, the food of man and other animals.

Proportionably late varieties of different species of annual plants generate, in one part of their lives, the sap which they expend in another. I, every season, plant in the beginning of June, and a little earlier, a large quantity of the very late variety of pea which bears my name ; and by supplying the plants abundantly with water I prevent (as I have stated in a communication to the Society many years ago), to a very great extent, the injurious effects of mildew : and by these means I regularly obtain a most abundant supply of peas in September and October, and of better quality than I can obtain in the month of June. In this case the sap which is prepared in the summer is obviously expended in the autumn.

The good effects which I have proved to arise from planting large tubers of the potatoe-plant obviously spring from the large accumulation of sap in them. Fed by means of this, not only a large breadth of foliage is produced and exposed to sight more early in the year ; but that foliage contains much disposable organisable matter, which once formed a part of the parent tuber. Any person who will pay close attention to the growth of produce of early crops of potatoes, which have sprung from large tubers, will readily obtain ample evidence of the truth of this position. The variation in the comparative growth of fruits of different species in similar seasons frequently arises, I have good reason to believe, from the more or less perfect state of the reservoir formed in the preceding year ; and every experienced gardener knows that under any given external circumstances, the blossom of his fruit trees sets best when the preceding season has been warm and bright, and when his trees, in such season, have not expended their sap in supporting heavy crops of fruit.

Note by the Secretary.—The quality of the Ispahan melons referred to in the preceding paper was found, when the fruit was tasted at the house of the Society, to be of the highest excellence which it is supposed that the melon is capable of attaining in this country.

LXXIV.—ON THE ADVANTAGES OF IRRIGATING GARDEN GROUNDS
BY MEANS OF TANKS OR PONDS.

[Read before the HORTICULTURAL SOCIETY, *August 7th,* 1832.]

THE quantity of water which may be given with advantage to plants of almost every kind, during warm and bright weather, is, I believe, very much greater than any gardener, who has not seen the result, will be inclined to suppose possible; and it is greater than I myself could have believed upon any other evidence than that of actual experience.

My garden, in common with many others, is supplied with water by springs, which rise in a more elevated situation; and this circumstance afforded me the means of making a small pond, from which I can cause the water to flow out over every part of my garden whenever I wish. I am thus enabled to irrigate my strawberry beds whilst in flower, and my alpine strawberry beds, and plants of every other kind, through every part of the summer; and I cause a stream to flow down the rows of celery and along the rows of brocoli, and other plants which are planted out in summer, with very great advantage. But the most extensive and beneficial use which I make of the power to irrigate my garden by the means above mentioned is in supplying my late crops of peas abundantly with water; by which the ill effects of mildew are almost wholly prevented, and my table is most abundantly supplied with very excellent peas through the month of October, as I have stated in a former communication. Several of my friends, who have caused large quantities of water to be carried, have obtained abundant crops late in the autumn of the variety of pea which bears my name; but they have complained that the birds have eaten the whole crop. This will almost always occur where means are not taken to prevent it: but there are only two species of bird which ever break open the pods of green peas, the large black-headed and the blue titmouse (the Parus major and Parus cæruleus of Linnæus), and both these are very easily caught. The coal titmouse, the nuthatch, the chaffinch, and the robin, will eat the peas when the pods are opened; but neither of these ever break them. For the purpose of taking such birds, I employ a little trap, which I invented when a school-boy, and which secures without injuring them, and enables me to release the unoffending; and I do not find the smallest difficulty in preserving my crops of peas in any season.

When water is delivered in the usual quantity from the watering-pan, its effects, for a short time, are almost always beneficial, by wetting the surface of the ground. But if water thus given be not continued regularly, injurious effects frequently follow ; for the roots of plants (as I have shown in the Philosophical Transactions, in a paper upon the causes which direct the roots *) extend themselves most rapidly wherever they find proper moisture and food ; and if the surface alone be wetted, the roots extend themselves superficially only, and the plants consequently become more subject to injury from drought than they would have been if no water had been given to them ; a circumstance which can scarcely have escaped the notice of any observant gardener. When, on the contrary, the soil is irrigated in the manner above recommended, it is wetted to a great depth ; and a single watering once in eight or ten days is, in almost all cases, fully sufficient.

I have found the advantage of being able to command, by the means above-mentioned, abundant water at all seasons, and at very small expense, so great, that I feel confident that a market gardener could, in many cases, afford to give as much rent for one acre as he could under ordinary circumstances give for two acres ; for he would not only be able generally to command more abundant crops, but, by possessing exclusive advantages, he would often, in unfavourable seasons, be enabled to raise abundant crops of articles which, in such seasons, usually take a very high price. In selecting the site of a garden the advantage of irrigating it, by the means above-mentioned, may very frequently be obtained ; and the number of gardens above which a small tank or pond might be easily made is probably much greater than at a first view will be supposed.

It may be objected that excess of rain is more often injurious in the climate of England than drought ; but in wet seasons plants suffer owing to want of light, and generally of warmth ; and I feel confident that if the same quantity of rain, which the soil receives in our wettest summer, were to fall only between the hours of nine in the evening and three in the following morning, and the sun were to shine brightly and warmly through the whole of the days, no injurious effects would follow ; and every experienced gardener knows with what luxuriance and rapidity plants of every species grow in hot and bright weather, after the ground has been drenched with water by thunder-storms.

* See above, p. 157.

LXXV.—ON THE CULTURE OF THE POTATOE.

[*Read before the* HORTICULTURAL SOCIETY, *March 19th,* 1833.]

I HAVE so often addressed communications to this Society upon the culture of the potatoe, that many of its members may not improbably think that more than a sufficient extent of the pages of our Transactions have been already devoted to that subject. It would certainly not be difficult to find one more entertaining; but if the farmer can be made to derive such information from our Transactions as will enable him to cause the same space of ground which now affords one bushel of potatoes to afford two, and the peasant to cause the half acre which now supplies his table with potatoes to afford him in addition a considerable weight of animal food, few subjects can be more important; and therefore, conceiving myself to be prepared to communicate some further useful information, I venture to address another communication upon the same subject.

The fact that every variety of potatoe when it has been long propagated from parts of its tuberous roots becomes less productive, is, I believe, unquestionable. I have often witnessed the progressive decay of vigour, and the different effects of the influence of age, upon many different varieties. The quality of some has remained perfectly good, after the produce in quantity has become highly defective; whilst in others that has disappeared with the vigour of the plant. I brought to this place a single tuber of Lankman's potatoe soon after that was imported: the produce of that variety was then, and continued during some successive years, very great; but its vigour was gradually diminished; and in the last year its produce was at least one third (more than seven tons per acre) less than I obtained from the same soil, and under in every respect the same management, from other varieties of nearly similar habits, but which had recently sprung from seed. The propagation of expended varieties, therefore, appears to me to be one of the causes why the crops of potatoes generally have been found so much less than those which I have stated to have been produced here. I have received letters within a few months from persons in different parts of the kingdom, informing me that they have been unable to obtain by any mode of culture above two hundred and fifty or three hundred bushels of potatoes from an acre of good and well-manured ground. I have in answer desired to know the age of the varieties cultivated; but upon that point I have uniformly found my correspondents totally uninformed; communicating to me, however, the important intelligence that the same varieties bore more abundantly at a former period, and often that the quality of the former produce was superior. When I first stated, in a former communication, that I had obtained a produce equivalent to six hundred and seventy bushels of eighty pounds per acre, I found some difficulty in obtaining credit for the accuracy of my statement, though I then felt perfectly confident that by first obtaining varieties better adapted to my purpose, I should be able to raise much heavier crops ; and the following statement, in support of which I am prepared to adduce the most unquestionable evidence, will prove that my confidence was perfectly well founded.

I planted in my garden, in the last season, some tubers of a variety of potatoe of very early habits, but possessing more vigour of growth than is usually seen in such varieties. The soil in which they were planted was in good condition, but not richer than the soils of gardens usually are, and the manure which it had received consisted chiefly of decayed oak leaves, which I prefer to other manures, because it never communicates a strong taste or flavour to any vegetable. No previous preparation was given to the soil, and the spot where the plantation was made was not fixed upon till the day of planting ; and no manure of any kind was then given. Owing to the variety being of a very excitable habit, I planted the tubers at least nine inches deep in the soil, and I subsequently raised the mould in ridges three inches high to prevent the young plants sustaining injury from frost ; but no subsequent moulding was given. I anticipated from the previous produce of the variety, which I had raised by cross-breeding from two early varieties in 1830, a very extraordinary crop ; and I therefore invited several gardeners and farmers to witness the amount of it ; and I procured the attendance of the two most eminent agriculturists of the vicinity, who were tenants to other gentlemen. The external rows (two deep), and the external plants at the ends of all the remaining rows, were taken away, and the produce of the interior part of the plantation was alone selected ; and that was pronounced to be fully equivalent to nine hundred and sixty-four bushels and forty-three pounds, or 34 tons 8 cwt. 107 lbs. per statute acre. Still larger crops may, I feel satisfied, be obtained, and my opinion is, that more than a thousand bushels of potatoes may, and will be, obtained from an acre of ground.

An opinion is I believe generally prevalent, that varieties of potatoes of very high and luxuriant growth are capable of affording per acre the greatest weight of produce : but this is certainly erroneous. Such will grow in poorer soil, and, requiring wider intervals between the rows, are better calculated for culture with the plough; and therefore, perhaps, their produce may be raised at as little or less cost per bushel, though that is, I think, very questionable. Much time and much labour of the plant must be expended in raising the nutriment absorbed from the soil into the leaves upon the top of a very tall stem, and down again to the roots and tubers.

The potatoes, in the extraordinary crop of which I have above spoken, were not washed, and therefore a deduction must be made for a portion of soil which adhered to them: but that was small, owing to the dryness and nature of the soil. Supposing a deduction of one hundred and sixty-four bushels be made in the above-mentioned account, and to afford potatoes sufficient to plant the acre of ground again, eight hundred bushels would still remain; and these, if judiciously given to proper animals, would certainly give twelve hundred pounds of animal food.

For this purpose early varieties of potatoes possess great advantages; because all our domesticated animals thrive most on potatoes after these have begun to germinate: and if those of early, and of course of very excitable habits, be taken up and collected into heaps, as soon as they have acquired maturity, they will germinate in autumn, and be fit for use, without being boiled, through the winter. Potatoes of such varieties are, however, wholly unfit for human food late in the spring; and for such purpose those of later and less excitable habits must be cultivated. Of such kinds in the last season, which was not favourable, owing to the plants having suffered injury from drought, I obtained a produce varying from twenty to twenty-four tons per acre, the soil being naturally light and poor, and not more highly manured than would have been necessary for a crop of Swedish turnips.

LXXVI.—UPON THE CAUSES OF THE PREMATURE DEATH OF PARTS OF THE BRANCHES OF THE MOOR-PARK APRICOT, AND SOME OTHER WALL FRUIT-TREES.

[*Read before the* HORTICULTURAL SOCIETY, *June 2nd*, 1835.]

THE branches of all trees, during much the larger portion of the periods in which they continue to live, are in their natural situations kept in continual motion, by the action of wind upon them; and of this motion their stems and superficial roots partake, whenever the gales of wind are even moderately strong: and I have shown, in the

Philosophical Transactions, that the forms of all large and old trees must have been much modified by this agent. The motions of the circulating fluids, and sap of the tree, are also greatly influenced and governed by it; and whenever any part of the root, the stem, or the branches of a tree are bent by winds or other agents, an additional quantity of alburnum is there deposited; and the form of the tree becomes necessarily well adapted to its situation, whether that be exposed or sheltered. If exposed to frequent and strong agitation, its stem and branches will be short and rigid, and its superficial roots will be large and strong; and if sheltered, its growth will be in every part more feeble and slender. I have much reason to believe, upon the evidence of subsequent experiments, that the widely-extended branches of large timber-trees would be wholly incapable of supporting their foliage when wetted with rain, if the proportions of their parts were not to be extensively changed and their strength greatly augmented by the operation of winds upon them during their previous growth. Exercise, therefore, appears to be productive of somewhat analogous effects upon vegetable and upon animal life, and to be nearly as essential to the growth of large trees as to that of animals.

Whenever the branches of a tree are bound to a wall, they wholly lose the kind of exercise above described, which nature obviously intended them to receive; and many ill consequences generally follow—not, however, to the same extent, nor precisely of the same kind, to trees of different species and habits. When a standard plum or peach tree is permitted to take its natural form of growth, its sap flows freely and most abundantly to the extremities of its branches, and it continues to flow freely through the same branches during the whole life of the tree: but when the branches are bound to a wall, and are no longer agitated by winds, each branch becomes in a few years what Duhamel calls " usée," that is, debilitated and sapless, owing apparently to its being no longer properly pervious to the ascending sap. This obstruction to its ascent causes luxuriant shoots to spring from the lower parts of the tree; and these are in succession made to occupy the places of the debilitated older branches by the process which the gardener calls " cutting in."

The branches of the apricot, and particularly of the Moor-park varieties, often die suddenly, owing to the same cause, with much more inconvenience and loss very frequently to the gardener; for trees of this species do not usually afford him the means of filling up vacancies upon his wall, as those of the peach and plum do.

The pear-tree better retains its health and vigour, when trained to a wall, than those of either of the preceding species, or than the cherry-tree ; but the proper course of its sap is nevertheless greatly deranged ; and it is difficult, and in some varieties almost impossible, to cause it to flow properly to the extremities or nearly to the extremities of its branches.   Much the larger part of it is generally expended in the production of what are called "foreright" useless shoots ; and the quantity of fruit which is afforded by the central parts of an old pear-tree, when trained to a wall, is usually very small.

The vine alone amongst fruit-trees appears capable of being bound and trained to a great distance upon a wall without sustaining any injury, its sap continuing to flow freely and abundantly to its very distant branches.   Owing to a peculiarity of structure and habit which is con-fined to those species of trees from which nature has withheld the power of supporting their own branches, the alburnum of all plants of this habit is (as far as I have had opportunities of observing) excessively light or porous ; and not being intended by nature to support its own weight, or that of any part of the foliage of the tree, does not acquire with age any increased solidity, like that of trees of a different habit ; and on this account probably it never, how long soever deprived of exercise, loses in any degree its power of transmitting the ascending sap.   The alburnum of those trees which nature has caused to support themselves without external aid, becomes annually more firm and solid, and con-sequently less well adapted to afford a passage to the ascending sap, and as heart-wood it is totally impervious to that fluid.   Whenever the branches of such trees are wholly deprived of exercise, too rapid an increase of the solidity of the alburnum probably takes place ; and it in consequence ceases to be capable of properly executing its office.   I have, of course, never had an opportunity of examining the character of the alburnum of the Glycine sinensis, of which the garden of this Society contains so splendid a tree ; but I do not entertain a shadow of a doubt of its being extremely light and porous, like that of other trailing and creeping plants which depend for support upon other bodies.

LXXVII.—ON THE MEANS EMPLOYED IN RAISING A TREE OF THE
IMPERATRICE NECTARINE.

[*Read before the* HORTICULTURAL SOCIETY, *February 3rd,* 1835.]

I WAS informed in the last spring that the Society's garden did not contain a tree of the Impératrice nectarine, and that it was wished to obtain one. I in consequence promised that I would raise and send one as soon as I could; and I believe that the means which I employed in raising a tree of that variety will prove that I have not lost time in proceeding to perform my promise.

The tree which I send is composed of an almond-stock which sprang from seed early in the last spring, into which two buds were inserted on opposite sides in the end of April; and as soon as those had properly united themselves to the stock, that was removed from the forcing-house, and placed under a north wall. After a few days it was headed down, and brought again into the forcing-house, when the two inserted buds vegetated, and each produced a lateral branch, which has acquired the length of about two feet six inches, and has formed a few blossom-buds. I had previously, early in the spring, grafted an almond-stock which was a year old with the Impératrice nectarine, with the intention of obtaining a tree to send to you; but it acquired, early in the summer, too large a size; and it was consequently planted out to fill up a vacancy upon my south wall, where it has produced two branches, each of which is more than six feet long; and it has covered fifty square feet of the wall with much excellent bearing-wood. I have never witnessed such rapidity and excellence of growth in a peach or nectarine tree, planted at the usual periods.

The almond as a stock for the peach and nectarine possesses, I think, every good quality, except that of bearing transplantation very well, and in that respect alone it is inferior to the plum-stock. I have, on this account, sent the little plant above mentioned in the pot in which the almond was first planted.

In the soil and climate of this place the Impératrice nectarine is, in my estimation and in that of a great many other persons who have tasted it, the best fruit of its family. It presents, I think, a greater concentration of taste and flavour than is found in any other variety which I have cultivated. It is inferior in size to the Downton nectarine: but that, in favourable seasons, is here very large; one measured in circumference nine inches, and several of them exceeded eight inches and seven lines.   I named it the Impératrice nectarine, because the first fruits which I saw shrivelled much upon the tree; but those have not subsequently done so more than some other varieties of nectarines.

I will request that the little tree sent may be planted in fresh unmanured soil without having the branches shortened, and so superficially that a part of its roots may remain permanently visible above the soil.   The fruit which it will produce will not be nearly as good as that of an older tree; and it is therefore my wish that some buds should be taken from it in the next season, and inserted into the branches of more mature trees.

LXXVIII.—ON THE PROPAGATION OF TREES BY CUTTINGS IN SUMMER.

[*Read before the* HORTICULTURAL SOCIETY, *April 3rd,* 1838.]

WHEN a cutting of any deciduous tree is planted in autumn, or winter, or spring, it contains within it a portion of the true, as it has been called, or vital sap, of the tree of which it once formed a part.   This fluid, relatively to plants, is very closely analogous to the arterial blood of animals ; and I shall therefore, to distinguish it from the watery fluid, which rises abundantly through the alburnum, call it the *arterial sap* of the tree. Cuttings of some species of trees very freely emit roots and leaves ; whilst others usually produce a few leaves only and then die ; and others scarcely exhibit any signs of life : but no cutting ever possesses the power of regenerating, and adding to itself vitally, a single particle of matter, till it has acquired mature and efficient foliage.   A part of the arterial sap previously in the cutting assumes an organic solid form ; and the cutting in consequence necessarily becomes, to some extent, exhausted.

Summer cuttings possess the advantage of having mature and efficient foliage ; but such foliage is easily injured or destroyed, and if it be not carefully and skilfully managed, it dies.   These cuttings (such as I have usually seen employed) have some mature and efficient foliage, and other foliage, which is young and growing ; and consequently two distinct processes are going on at the same time within them, which operate in opposition to each other.   By the mature leaves, carbon, under the influence of light, is taken up from the surrounding atmosphere, and arterial sap is generated.   The young and immature leaves, on the contrary, vitiate the air in which they grow by throwing off carbon ; and they expend, in adding to their own bulk, that which ought to be expended in the creation of shoots. This circumstance respecting the different operations of immature and mature leaves upon the surrounding air presented itself to the early labourers in pneumatic chemistry. Dr. Priestley noticed the discharge of oxygen gas, or dephlogisticated air (as it was then called), from mature leaves; Scheele making, as he supposed, a similar experiment upon the young leaves of germinating beans, found these to vitiate air in which they grew. These results were then supposed to be widely at variance with each other; but subsequent experience has proved both philosophers to have been equally correct.

I possess many young seedling trees of the Ulmus campestris, or suberosa, or glabra, for the widely-varying characters of my seedling trees satisfy me that these three supposed species are varieties only of a single species. One of these seedling plants presented a form of growth which induced me to wish to propagate from it. It shows a strong disposition to aspire to a very great height with a single straight stem, and with only very small lateral branches, and to be therefore calculated to afford sound timber of great length and bulk, which is peculiarly valuable, and difficult to be obtained, for the keels of large ships; and the original tree is growing with very great rapidity in a poor soil and cold climate.

The stem of this tree near the ground presented, in July, many very slender shoots about three inches long. These were then pulled off and reduced to about an inch in length, with a single mature leaf upon the upper end of each; and the cuttings were then planted so deeply in the soil, that the buds at the bases of the leaves were but just visible above the surface of the soil. The cuttings were then covered with bell-glasses in pots, and put upon the flue of a hothouse, and subjected to a temperature of about 80°. Water was very abundantly given; but the under surfaces of the leaves were not wetted. These were in the slightest degree faded, though they were wholly exposed to the sun; and roots were emitted in about fifteen days. I subjected a few cuttings, taken from the bearing-branches of a mulberry-tree, to the same mode of management, and with the same result; and I think it extremely probable that the different varieties of camellia, and trees of almost every species, exclusive of the fir tribe, might be propagated with perfect success and facility by the same means.

Evergreen trees of some species possess the power of ripening their fruit during winter. The common ivy and the loquat are well known examples of this; and this circumstance, combined with many others, led me to infer that the leaves of such trees possess in a second year the same, or nearly the same power, as in the first. I therefore planted, about a month ago, some cuttings of the old double-blossomed white and Warratah camellia, having reduced the wood to little more than half an inch in length, and cut it off obliquely, so as to present a long surface of it; and I reduced it further by paring it very thin, at and near to its lower extremities. The leaves continue to look perfectly fresh; and the buds in more than one instance have produced shoots of more than an inch in length, and apparently possessing perfect health and much vigour. Water has been very abundantly given; because I conceived that the flow of arterial sap from the leaf would be so great, comparatively with the quantity of the bark and alburnum of the cuttings, as to preclude the possibility of the rooting of these.

The cuttings above described present, in the organisation, a considerable resemblance to seedling trees at different periods of the growth of the latter. The bud very closely resembles the plumule; and the leaf, the cotyledon, extended into a seed-leaf; and the organ which has been and is called a radicle, is certainly a caudex, and not a root. It is capable of being made to extend, in some cases, to more than two hundred times its first length, between two articulations; a power which is not possessed in any degree by the roots of trees. Whether the caudex of the cuttings of camellia, above mentioned, have emitted, or will or will not emit roots, I am not yet prepared to decide; but I entertain very confident hopes of success.

APPENDIX:

CONTAINING

PAPERS ON ANIMAL ECONOMY.

======

I.—ON THE COMPARATIVE INFLUENCE OF MALE AND FEMALE PARENTS ON THEIR OFFSPRING.

[*Read before the* ROYAL SOCIETY, *June* 22, 1809.]

I HAVE been engaged, during many years, in experiments on fruit-trees, of which the object has been to discover the best means of forming new varieties, that may be found better calculated for the climate of Britain than those at present cultivated. In this inquiry my efforts have been always most successful, when I propagated from the males of one variety and the females of another; and I was enabled, by the same means, to ascertain more accurately than had previously been done the comparative influence of the male and female parent on the character of the offspring. The analogy that subsists between plants and animals, in almost everything which respects generation, induced me also to attend very minutely to similar experiments in which I engaged on some species of animals; and as the repetition of such experiments would necessarily require a very considerable space of time, and as the results seem to lead to conclusions that may be of public utility, I have thought the following account sufficiently interesting to induce me to address it to you.

Linnæus conceived that the character of the male parent predominated in the exterior parts both of plants and animals; and the same opinions have been generally entertained by more modern naturalists. But the Swedish philosopher appears to have been misled by the striking predominance of the character of the male parent in male animals, and to have drawn his conclusions somewhat too generally: for I have observed that seedling plants, when propagated from male and female parents of distinct characters and permanent habits, generally, though with some few exceptions, inherit much more of the character of the female than of the male parent; and the same remark is applicable in some respects to the animal world, as I shall point out in the succeeding narrative.

My experiments were made on many different species of fruit-trees; but most extensively, and under the most advantageous circumstances, on the apple-tree; and as the results were all in unison with each other, it will be necessary to trouble you only with an account of some of the experiments which were made on that species of fruit-tree.

The apple, or crab of England, and of Siberia, however dissimilar in habit and character, appear to constitute a single species only; in which much variation has been effected by the influence of climate on successive generations: for the two varieties rarely breed together, and the off-spring, whether raised from the seeds of the Siberian or British variety, were prolific to a most exuberant extent. But there was a very consider-able degree of dissimilarity in the appearance of the offspring; and the leaves and general habits of each presented an obvious prevalence of the character of the female parent. The buds of those plants which had sprung from the seeds of the cultivated apple did not unfold quite so early in the spring; and their fruits generally exceeded very consider-ably in size those which were produced by the trees which derived their existence from the seeds of the Siberian crab. There was also a preva-lence of the character of the female parent in the form of the fruit; but the same degree of prevalence did not extend to the quality and flavour of the fruit; for the richest apple that I have ever seen, and which afforded expressed juice of much higher specific gravity than any other, sprang from a seed of yellow Siberian crab.

The prevalence of the character of the female parent in the preceding cases may possibly be suspected to have arisen from some error or neg-lect of accuracy in making the experiments; but I do not conceive that any such errors could have existed; for the trees of each variety were trained to walls, where they blossomed much before any others of the same species, and the stamina were always carefully extracted, whilst immature, from every blossom, which I intended to afford seeds. The remaining blossoms of the trees were also totally destroyed, and no other blossoms, except those from which the pollen was taken, were ever un-folded in the neighbourhood, in the season when the experiments were made; and I have also invariably declined to draw any conclusion from the appearance of a plant in which I could not certainly distinguish some portion of the features and character of the supposed male parent.

It is perhaps also proper to state, that the predominance of the character of the female parent could scarcely have arisen from any defective action of the pollen ; for, except in cases where superfœtation took place, I have invariably found the effect of a very large or a very small quantity of pollen to be invariably the same in its influence on the offspring ; and in the greater part of the experiments from which I have drawn the preceding conclusions, more than ten times as much pollen was deposited on the stigmata as could have been deposited in unmutilated blossoms by the ordinary means employed by nature.

In all attempts to discriminate the different influence of the male and female parent on the offspring of animals many difficulties present themselves, owing to the intermixtures which have been made of the different breeds of domesticated animals of every species, and the consequent absence of all hereditary permanency in the character of each variety. For under these circumstances, the offspring will be very frequently found to show little resemblance either to its male or female parent, either in form, or stature, or colour. It will therefore be necessary, before I enter on the subject of viviparous animals, to observe that when I apply the terms large and small to the male or female parent, I extend the meaning of those terms to the parentage from which the male and female descend, and not to the size of the individual only which becomes the immediate parent of the offspring.

Mr. Cline has observed, in a communication to the Board of Agriculture, that if the male and female parent differ considerably in size, the dimensions of the fœtus at the birth will be regulated much more by the size of the female than of the male parent ; and if the meaning of the terms large and small be extended to the varieties as well as to the individuals, his remark is perfectly just. But experience compels me wholly to reject the inference that he has drawn respecting the advantages of propagating from large, in preference to small females.

Nature has given to the offspring of many animals (those of the sheep, the cow, and the mare afford familiar examples) the power at an early age to accompany their parents in flight; and the legs of such animals are very nearly of the same length at the birth, as when they have attained their perfect growth. When the female parent is large, and the fœtus consequently so, the offspring will be large at its birth in proportion to the bulk it will ultimately attain, and its legs will thence be long comparatively with the depth of the chest and shoulders. When, on the contrary, the female is small, and the fœtus so, at the birth, the length of the legs of the young animal will be short com-

paratively with the depth of its chest and shoulders; and an animal
in the latter form will be greatly preferable, either for the purposes of
labour, or of food to mankind.   I have seen this difference in the influence of the male and female parent on the offspring very strikingly
exemplified in the result of an attempt to obtain very large mules from
the male ass and the mare.   The largest females that could be procured
were selected, and the forms of the offspring, at the birth, were perfectly
consistent with the theory of Mr. Cline ; they were remarkably large :
and I observed that the length of their legs, when they were only a few
days old, very nearly equalled that of the legs of their female parents.
I examined the same animals when five years old, and in the depth of
their chests and shoulders they very little exceeded their male parent ;
and they were consequently of little or no value ; whilst other mules
which were obtained from the same male parent (a Spanish ass), but
from mares of small stature, were perfectly well proportioned.   I have
never seen the little mule which is propagated from the female ass and
the horse, nor even a delineation or description of its form ; but I do
not entertain any doubt that its chest and shoulders are excessively
deep and strong, comparatively with the length of its legs, and that, on
account of this peculiarity in its form, it has been so frequently shown
on the Continent, under the name of a jumart, as the pretended offspring
of the mare and the bull.

In opposing the theory advanced by Mr. Cline, it is not by any means
my intention to enter the lists with him as a physiologist ; but as a
farmer and breeder of animals of different species, I have probably had
many advantages which he has not possessed ; and my conclusions have
been drawn from very extensive, and, I believe, accurate observation.

There is another respect in which the powers of the female appear to
be prevalent in their influence on the offspring, and that is relative to
its sex.   In several species of domesticated, or cultivated animal (I believe
in all), particular females are found to produce a very large majority,
and sometimes all their offspring, of the same sex; and I have proved
repeatedly, that, by dividing a herd of thirty cows into three equal parts,
I could calculate, with confidence, upon a large majority of females from
one part, of males from another, and upon nearly an equal number of
males and females from the remainder.   I frequently endeavoured to
change these habits by changing the male, but always without success ;
and I have in some instances observed the offspring of one sex, though
obtained from different males, to exceed those of the other in the proportion of five or six, and even seven to one.   When, on the contrary,

I have attended to the numerous offspring of a single bull, or ram, or horse, I have never seen any considerable difference in the number of offspring of either sex. I am therefore disposed to believe that the sex of the offspring is given by the female parent ; and the probability of this seems obvious in fishes, and several other species of animals which breed in water ; and though the evidence afforded by the facts adduced is not by any means of sufficient weight to decide the question, it probably much exceeds all that can be placed in the opposite scale.

In oviparous animals, I have had reason to think the influence of the female parent quite as great as amongst the viviparous tribes, though my observations have been more limited and less conclusive. In viviparous animals, the size of the fœtus is affected by the influence of the male parent, and, in some instances, not inconsiderably ; but the size and form of the eggs of birds do not appear to be in any degree changed or modified by the influence of the male, and therefore the size of the offspring at the birth must be regulated wholly by the female parent ; and this circumstance permanently affects the form and character of the offspring. The eggs of birds, and those of fishes and insects (if such can properly be called eggs), appear to resemble the seeds of plants, in having their forms and bulk wholly regulated by the female parent ; but nevertheless their formation appears to depend on very different laws. For the eggs both of birds and of fishes and insects attain their perfect size in total independence of the male, and the cicatricula, the vitellus, and the chalazæ have appeared (I believe) to the most accurate observers, to be as well organised in the unimpregnated, as in the impregnated egg : in the seed, on the contrary, everything relative to its internal organisation appears dependent on the male parent. Spallanzani has, however, stated, that many plants produced well-organised seeds, and even seeds which vegetated perfectly, under circumstances in which it is not easy to conceive how the pollen of the male plant or flower could have been present. But the Italian naturalist appears to have blundered most egregiously in his experiment; or (which I conceive to be more probable) he became the dupe of the refined malice of his countrymen ; for I repeated his experiments under very favourable circumstances, and with the closest attention, but I failed to obtain a single seed. The gourd alone produced apparently perfect fruit, and the seed-coats acquired their natural size and form; and in this respect the growth of its seeds appeared to be, like that of eggs, wholly independent of the influence of the male. But the *seed-coats* of the gourd were perfectly *empty*, and I could not discover, at any period of their growth, the slightest vestige either of cotyledons, or plumule, nor of anything that appeared to correspond with internal organisation of a seed of the same plant under different circumstances. Spallanzani, has not I believe, mentioned the species of gourd upon which he made his experiments: the common, or orange gourd of our gardens, was the subject of mine.

In comparing the mode of the formation and growth of eggs with the observations I had previously made on the growth of seeds, I have been favoured with the very able assistance of Mr. Carlisle, for which I have on this, as on many other occasions, to acknowledge much obligation.

II.—ON THE ECONOMY OF BEES.

[*Read before the* ROYAL SOCIETY, *May* 14*th*, 1807.]

IN the prosecution of those experiments on trees, accounts of which you have so often done me the honour to present to the Royal Society, my residence has necessarily been almost wholly confined to the same spot; and I have thence been induced to pay considerable attention to the economy of bees amongst other objects; and as some interesting circumstances in the habit of these singular insects appear to have come under my observation, and to have escaped the notice of former writers, I take the liberty to communicate my observations to you.

It is, I believe, generally supposed that each hive or swarm of these insects remains at all times wholly unconnected with other colonies in the vicinity, and that the bee never distinguishes a stranger from an enemy. The circumstances which I shall proceed to state will, however, tend to prove that these opinions are not well founded, and that a friendly intercourse not unfrequently takes place between different colonies, and is productive of very important consequences in their political economy.

Passing through one of my orchards rather late in the evening in the month of August in the year 1801, I observed that several bees passed me in a direct line from the hives in my own garden to those in the garden of a cottager, which was about a hundred yards distant from it. As it was considerably later in the evening than the time when bees usually cease to labour, I concluded that something more than ordinary was going forward. Going first to my own garden, and then to that of the cottager, I found a very considerable degree of bustle and agitation to prevail in one hive in each: every bee as it arrived seemed to be stopped and questioned at the mouth of each hive, but I could not discover anything like actual resistance or hostility to take place; though I was much inclined to believe the intercourse between the hives to be hostile and predatory. The same kind of intercourse continued, in a greater or less degree, during eight succeeding days; and though I watched them very closely, nothing occurred to induce me to suppose that their intercourse was not of an amicable kind. On the tenth morning, however, their friendship ended, as sudden and violent friendships often do, in a quarrel; and they fought most furiously, and after this there was no more visiting.

Two years subsequent to this period I observed the same kind of intercourse to take place between two hives of my own bees, which were about two hundred yards distant from each other; they passed from each hive to the other just as they did in the preceding instance, and a similar degree of agitation was observable. In this instance, however, their friendship appeared to be of much shorter duration, for they fought most desperately on the fifth day; and then, as in the last-mentioned case, all further visiting ceased.

I have some reason to believe that the kind of intercourse I have described, which I have often seen and which is by no means uncommon, not unfrequently ends in a junction of the two swarms; for one instance came under my observation many years ago in which the labouring bees, under circumstances perfectly similar to those I have described, wholly disappeared, leaving the drones in peaceable possession of the hive, but without anything to live upon. I have also reasons for believing that whenever a junction of two swarms, with their property, is agreed upon, that which proposes to remove, immediately or soon afterwards unites with the other swarm, and returns to the deserted hive during the day only to carry off the honey; for having examined at night a hive from which I suspected the bees to be migrating, I found it without a single inhabitant. I was led to make the examination by information I had received from a very accurate observer, that all the bees would then be absent. A very considerable quantity of honey was in this instance left in the hive without any guards to defend it; but I conclude that the bees would have returned for it, had it remained till the next day. Whenever the bees quit their habitation in this way, I have always observed some fighting to take place; but I conceived it to be between the bees of the adjoining hives and those which were removing; the former being attracted by the scent of the honey which the latter were carrying off.

On the farm which I occupy there were formerly many old decayed trees, the cavities of which were frequently occupied by swarms of bees; and when these were destroyed, a board was generally fitted to the aperture which had been made to extract the honey; and the cavity was thus prepared for the reception of another swarm in the succeeding season. Whenever a swarm came, I constantly observed that, about fourteen days previous to their arrival, a small number of bees, varying from twenty to fifty, were every day employed in examining and apparently in keeping possession of the cavity; for if molested, they showed evident signs of displeasure, though they never employed their stings in defending their proposed habitation. Their examination was not confined to the cavity, but extended to the external parts of the tree above; and every dead knot particularly arrested their attention, as if they had been apprehensive of being injured by moisture which this might admit into the cavity below; and they apparently did not leave any part of the bark near the cavity unexamined. A part of the colony which purposed to emigrate appeared in this case to have been delegated to search for a proper habitation; and the individual who succeeded must have apparently had some means of conveying information of his success to others; for it cannot be supposed that fifty bees should each accidentally meet at and fix upon the same cavity, at a mile distant from their hive; which I have frequently observed them to do in a wood where several trees were adapted for their reception; and indeed I observed that they almost uniformly selected that cavity which I thought best adapted to their use.

It not unfrequently happened that swarms of my own bees took possession of these cavities, and such swarms were in several instances followed from my garden to the trees; and they were observed to deviate very little from the direct line between the one point and the other; which seems to indicate that those bees which had formerly acted as purveyors now became guides.

Two instances came under my own observation in which a swarm was received into a cavity of which another swarm had previous possession. In the first instance I arrived with the swarm, and I could not discover that the least opposition was made to their entrance: in the second instance, observing the direction that the swarm took, I used all the expedition I could to arrive first at the tree to which I supposed they were going, whilst a servant followed them; and a descent of ground being in my favour, and the wind against them, I succeeded in arriving at the tree some seconds before them; and I am perfectly confident that not the least resistance was opposed to their entrance.

Now it does not appear probable that animals so much attached to their property as bees are, so jealous of all approach towards it, and so ready to sacrifice their lives in defence of it, should suffer a colony of strangers, with whose intentions they were unacquainted, to take possession without making some effort to defend it : nor does it seem much more probable that the same animals which spent so much time in examining their future habitation in the cases I have mentioned, should have attempted in this case to enter without knowing whether there was space sufficient to contain them, and without any examination at all. I must therefore infer that some previous intercourse had taken place between the two swarms, and that those in the possession of the cavities were not unacquainted with the intentions of their guests; though the formation of anything like an agreement between the different parties be scarcely consistent with the limitations generally supposed to be fixed by nature to the instinctive powers of the brute creation.

Brutes have evidently language ; but it is a language of passion only, and not of ideas. They express to each other sentiments of love, of fear, and of anger ; but they appear to be wholly incapable of transmitting to each other any ideas they have received from the impression of external objects. They convey to other animals of their species, on the approach of an enemy, a sentiment of danger ; but they appear wholly incapable of communicating what the enemy is, or the kind of danger apprehended. A language of more extensive use seems, from the preceding circumstances, to have been given to bees ; and if it be not in some degree a language of ideas, it appears to be something very similar.

When a swarm of bees issue from the parent hive, they generally soon settle on some neighbouring bush or tree ; and as in this situation they are generally not at all defended from rain or cold, it is often inferred that they are less amply gifted with those instinctive powers that direct to self-preservation than many other animals. But their object in settling soon after they leave the hive is apparently nothing more than to collect their numbers ; and they have generally, I believe always, another place to which they intend subsequently to go : and if the situation they select be not perfectly adapted to secure them from injuries, it is probably, in almost all instances, the best they can discover. For I have very often observed that when one of my hives was nearly ready to swarm, one of the hollow trees I have mentioned (and generally that best adapted for the accommodation of a swarm) was every day occupied by a small number of bees ; but that after the swarm had issued from that hive, and had taken possession of another, the tree was wholly deserted ; whence I inferred that the swarm which would have taken possession of the cavity of that tree had relinquished their intended migration when a hive was offered them at home. And I am much disposed to doubt, whether it be not rather habit, produced by domestication, during many successive generations, than anything inherent in the nature of bees, which induces them to accept a hive, when offered them, in preference to the situation they have previously chosen : for I have noticed the disposition to migrate to exist in a much greater degree in some families of bees than in others; and the offspring of domesticated animals inherit, in a very remarkable manner, the acquired habits of their parents. In all animals this is observable; but in the dog it exists to a wonderful extent ; and the offspring appears to inherit not only the passions and propensities, but even the resentments, of the family from which it springs. I ascertained by repeated experiment that a terrier whose parents had been in the habit of fighting with polecats will instantly show every mark of anger when he first perceives the scent of that animal, though the animal itself be wholly concealed from his sight. A young spaniel brought up with the terriers showed no marks whatever of emotion at the scent of the polecat ; but it pursued a woodcock, the first time it saw one, with clamour and exultation : and a young pointer, which I am certain had never seen a partridge, stood trembling with anxiety, its eyes fixed and its muscles rigid, when conducted into the midst of a covey of those birds. Yet each of these dogs are mere varieties of the same species ; and to that species none of these habits are given by nature. The peculiarities of character can therefore be traced to no other source than the acquired habits of the parents, which are inherited by the offspring, and become what I shall call instinctive hereditary propensities. These propensities or modifications of the natural instinctive powers of animals are capable of endless variation and change ; and hence their habits soon become adapted to different countries and different states of domestication, the acquired habits of the parents being transferred hereditarily to the offspring. Bees, like other animals, are probably susceptible of these changes of habit, and thence, when accustomed through many generations to the hive, in a country which does not afford hollow trees or other habitations adapted to their purpose, they may become more dependent on man, and rely on his care wholly for an habitation ; but in situations where the cavities of trees present to them the means of providing for themselves, I have found that they will discover such trees in the closest recesses of the woods, and at an extraordinary distance from their hives ; and that they will keep possession of such cavities in the manner I have stated :

and I am confident that, under such circumstances, a swarm never issues from the parent hive without having previously selected some such place to retire to.

It has been remarked by Mr. John Hunter, that the matter which bees carry on their thighs is the farina of plants with which they feed their young, and not the substance with which they make their combs; and his statement is, I believe, perfectly correct: but I have observed that they will also carry other things on their thighs. I frequently covered the decorticated parts of trees, on which I was making experiments, with a cement composed of bees-wax and turpentine; and in the autumn I have frequently observed a great number of bees employed in carrying off this substance. They detached it from the tree with their forceps, and the little portion thus obtained was then transferred by the first to the second leg, by which it was deposited on the thigh of the third: the farina of plants is collected and transferred in the same manner. This mixture of wax and turpentine did not, however, appear to have been employed in the formation of combs, but only to attach the hive to the board on which it was placed, and probably to exclude other insects, and air during winter. Whilst the bees were employed in the collection of this substance, I had many opportunities of observing the peaceful and patient disposition of them as individuals, which Mr. Hunter has also, in some measure, noticed. When one bee had collected its load, and was just prepared to take flight, another often came behind it, and despoiled it of all it had collected. A second, and even a third, load was collected and lost in the same manner; and still the patient insect pursued its labour, without betraying any symptoms of impatience or resentment. When, however, the hive is approached, the bee appears often to be the most irritable of all animals; but a circumstance I have observed amongst many other species of insects, whose habits are in many respects similar to those of bees, induces me to believe that the readiness of the bees to attack those who approach their hives does not in any degree spring either from the sense of injury or apprehensions of the individual who makes the attack. If a nest of wasps be approached without alarming its inhabitants, and all communication be suddenly cut off between those out of the nest and those within it, no provocation will induce the former to defend their nest or themselves. But if one escape from within, it comes with a very different temper, and appears commissioned to avenge public wrongs, and prepared to sacrifice its life in the execution of its orders. I discovered the circumstance, that wasps thus excluded from their nest would neither defend it nor themselves, at a very early period of my life; and I profited so often by the discovery as a schoolboy, that I am quite certain of the fact I state; and I do not entertain any doubt, though I speak from experiments less accurately made, that the actions of bees under similar circumstances would be the same*.

III.—ON SOME CIRCUMSTANCES RELATING TO THE ECONOMY OF BEES.

[Read before the ROYAL SOCIETY, *May 22nd, 1828.]*

IN a paper which I had the honour to address to the Royal Society about twenty years ago (in the year 1807) upon the Economy of Bees, I stated, that having adapted cavities in hollow trees for the reception of swarms of those insects, I had observed that several days previous to the arrival of a swarm, a considerable number of bees were constantly employed in examining the state of the tree, and particularly of every dead knot above the cavity which appeared likely to admit water into it. At that period it appeared to me rather extraordinary, that animals so industrious as bees, and so much disposed to make the best use of their time, should, at that important season of the year, waste so much of it in apparently useless repetitions of the same act: for I, at that time, supposed that on different days, and at different periods of the same day, I saw only the same individuals. But in a case which at a subsequent period came under my observation, where the cavity into which the bees

---

* A curious circumstance relative to wasps attracted the notice of some of my friends last year, and has not, I believe, been satisfactorily accounted for. A greater number of female wasps were observed in different parts of the kingdom, in the spring and early part of the summer of that year, than at almost any former period; yet scarcely any nests, or labouring wasps, were seen in the following autumn; the cause of which I believe I can explain. Attending to some peach-trees in my garden, late in the autumn of the year 1805, on which I had been making experiments, I noticed, during many successive days, a vast number of female wasps, which appeared to have been attracted there by the shelter and warmth of a south wall; but I did not observe any males. At length, during a warm gleam in the middle of one of the days, a single male appeared, and selected a female close to me; and this was the only male I saw in that season. The male wasp, which is readily distinguishable from the female and labourer, by his long antennæ and shining wings, and by a blacker and more slender body, is rarely seen out of the nest, except in very warm days, like the drone bee; and the nests of wasps, though very abundant in the year 1805, were not formed till remarkably late in the season; and thence I conclude that the males had not acquired maturity till the weather had ceased to be warm, and that the females, in consequence, retired to their long winter sleep without having had any intercourse with them.

apparently proposed to enter was not more than a quarter of a mile distant from the hive whence a swarm were prepared to emigrate, I witnessed a very rapid change of the individuals who visited their future contemplated habitation ; and the number which in the course of three days entered it, appeared to me to be fully equal to constitute a very large swarm : and upon the evidence of these and other facts, which I shall proceed to state, I am much disposed to infer, that not a single labouring bee ever emigrates in a swarm without having seen the future proposed habitation of that swarm. That the queen-bee has also always seen her future habitation, I am also much inclined to believe, as she is well known to absent herself from the hive some time previously to the emigration of a swarm : though her object may be to meet a male of another hive ; for I much doubt whether she ever receives the embraces of a brother. The results of some of Huber's experiments are very favourable to this conclusion, as is the otherwise excessive number of male bees ; and in both the animal and vegetable world, nature has taken very ample means of facilitating what the breeders of improved varieties of domesticated animals call cross-breeding.

I have also been led by the following facts to believe, that not only the future permanent habitation of each swarm, but the place where they temporarily settle, apparently to collect their numbers, soon after they quit their hive, is known also to each individual. Different families of domesticated animals of every species present some peculiarities of disposition and habit ; and the swarms of the family of bees which were the subject of my experiments showed, I think, more than an ordinary disposition to unite, by two apparently joining the same queen. My attention was consequently attracted to the circumstances which preceded such unions.

The simultaneous movements and agitation of two hives had during several days led me to expect that a junction of their swarms was contemplated ; and the two ultimately issued out almost at the same moment, and instantly united, as I had concluded they would. The weather was excessively hot ; and I put them into a hive which was scarcely large enough to hold them, affording them no further shelter from the sun than I thought just sufficient to prevent the melting of their combs. This occurred upon the first day of June, and in the morning of the twenty-third a very large swarm emigrated. There was in this, I believe, nothing very extraordinary or peculiar, except the excessive expedition apparently employed in raising a second queen.

In the following year two other hives presented similar indications that their swarms would unite; and being anxious to ascertain whether such unions were accidental, or the consequence of previous arrangements, I paid very close attention to their proceedings, and the following singular circumstances came under my observation :—After both hives had given frequent indications that a swarm was ready to issue from each of them, one swarm only rose, and that, after hovering in the air during a much longer time than ordinary, settled upon, and around, a bush about twenty-five yards distant from the hive whence they had issued ; but instead of collecting together into a compact mass, as they usually do, they remained thinly dispersed, scarcely two being anywhere in contact with each other. In this state they continued nearly half an hour motionless, and apparently discontented and sulky ; and they then gradually began to rise and return home, not apparently in obedience to any command or signal ; for they did not rise more abundantly at any one point of time than at another, but each individual seemed to go when tired of waiting.

The next morning a swarm issued from the other hive, and proceeded to the bush upon and around which the other swarm had settled on the preceding day, collecting themselves into a mass as they usually do when their queen is present. This was precisely what I had anticipated, but I was much disappointed that no movement or agitation took place in the other hive. Within a very few minutes, however, and very soon after the swarm above mentioned had fully settled, a very large number of bees suddenly rushed from the hive to which the swarm had returned on the preceding day, and proceeded so directly to the swarm which had just settled, that their course was marked through its whole extent by a perfectly visible dark and narrow line, and they united themselves, without hovering a single instant, to the other swarm. These circumstances, conjointly with others which I have stated in my former communication upon this subject, satisfied me that these unions are generally, if not always, the result of previous and perfectly well understood arrangements, though it is not easy to conjecture how such arrangements can be made.

I shall proceed to state a few circumstances which appear to throw light upon some of the phænomena observable in the mode of breeding of bees. It has long been known that these animals possess the power of raising a queen-bee from any recently-deposited egg which under ordinary circumstances would have produced a labouring bee ; but whether this power extends to those eggs which, when deposited in larger cells, afford male or drone bees, has not, I believe, been accurately ascertained. The following circumstances lead me to believe that sex is not given to the eggs of birds, or to the spawn of fishes or insects, at any very early period of their growth.

I selected early in winter four female birds of the common duck, which I kept apart from any male bird of that or any kindred species, till the period of their laying eggs approached. One was then killed, and the largest of its eggs was found to be three lines in diameter. A musk drake (Anas moschata) was then put into company with the three remaining ducks; and from these I obtained a numerous offspring, six out of seven of which proved to be males, as the result of similar previous experiments (but in which the male of another species had been introduced at a period when the growth of the eggs was less advanced,) had led me to expect. I repeated the experiment often, and always with nearly the same result, a large majority of male birds being uniformly produced; and hence I conclude that the eggs of birds in early periods of their growth are without sex.

I have never possessed means of obtaining mule fishes; but one kind of fish, which I think is obviously a mule, is found in many rivers where the common river-trout abounds, and where a solitary salmon is sometimes seen. These formerly existed, in some seasons, in considerable numbers, in the river which passes near my residence; but since salmon have become scarce, they have wholly disappeared. I had formerly opportunities of examining a large number of them, without having ever found a single female. I have subsequently found them in large numbers in small mountain rivulets in Wales, below, but never above, the lowest cataract. They are readily distinguished from the young salmon, by their form being intermediate between that of a trout and of a salmon; by their being all, or nearly all, males; and by their remaining through the summer and autumn in the rivers, long after the young salmon have descended to the sea: they leave the fresh water with the first winter floods, and I believe are not known ever to return. In the north of England they are distinguished by the name of wrackriders, and by that of samlets in some other parts. If these be mules, as I do not entertain any doubt that they are, the spawn of fishes must be without sex when it is deposited by the female; and I am much disposed to entertain the same opinion respecting the spawn (for it is more properly spawn than eggs) of bees.

I have frequently witnessed some somewhat analogous circumstances in the vegetable world, respecting the sexes of the blossoms of plants; and I can at any time succeed in causing several kinds of monœcious plants to produce solely male or solely female blossoms. If heat be, comparatively with the quantity of light which the plant receives, excessive, male flowers only appear; but if light be in excess, female flowers alone will be produced :—the experiments necessary must of course be made with skill and accuracy.

In a former communication to the Royal Society, "Upon the comparative influence of the male and female parent upon the character of the offspring," I have inferred, from facts there stated, that the sex of the offspring of some species of animals is given by the female parent. Subsequent experience and observation have strengthened my belief in the truth of this inference : but I believe the power of the female parent to be rather strongly influential than positive, and that external causes operate which (I have some reason to suspect) are not in all cases wholly beyond the reach of human control.

IV.—ON THE HEREDITARY INSTINCTIVE PROPENSITIES OF ANIMALS.

[*Read before the* ROYAL SOCIETY, *May* 25*th*, 1837.]

IN a communication which I had the honour many years ago to address to this Society upon the Economy of Bees, I gave an opinion that families of those insects, in common with those of every species of domesticated animal, are to a greater or less extent governed by a power which I have there called " an instinctive hereditary propensity ;" that is, by an irresistible propensity to do that which their predecessors of the same family have been taught or constrained to do, through many successive generations. In that communication I stated that a young terrier whose parents had been much employed in destroying polecats, and a young springing spaniel whose ancestry through many generations had been employed in finding woodcocks, were reared together as companions, the terrier not having been permitted to see a polecat, or any other animal of similar character, and the spaniel having been prevented seeing a woodcock, or other kind of game ; and that the terrier evinced, as soon as it perceived the *scent* of the polecat, very violent anger ; and as soon as it *saw* the polecat, attacked it with the same degree of fury as its parents would have done. The young spaniel, on the contrary, looked on with indifference ; but it pursued the first woodcock which it ever saw with joy and exultation, of which its companion, the terrier, did not in any degree partake.

I had at that period made a great many analogous experiments, and I have subsequently made a considerable number, chiefly upon one variety of dog, namely, that which is generally used in search of woodcocks, and is usually called the springing spaniel. These experiments were commenced nearly sixty years ago, and occupied a good deal of my attention during more than twenty years, and to a less extent nearly to the present time ; and as it does not appear to me probable that any person is now likely to investigate this subject as laboriously, or through so long a period, I have been induced to believe that the facts which I am prepared to communicate may be thought to deserve to be recorded in the Transactions of this Society.

At the period in which my experiments commenced, well-bred and well-taught springing spaniels were abundant, and I readily obtained possession of as many as I wanted.   I had at first no other object in view than that of obtaining dogs of great excellence; but within a very short time some facts came under my observation which very strongly arrested my attention.   In several instances young and wholly inexperienced dogs appeared very nearly as expert in finding woodcocks as their experienced parents.   The woods in which I was accustomed to shoot did not contain pheasants, nor much game of any other kind, and I therefore resolved never to shoot at anything except woodcocks, conceiving that by so doing the hereditary propensities above mentioned would become more obvious and decided in the young and untaught animals; and I had the satisfaction, in more than one instance, to see some of those find as many woodcocks, and give tongue as correctly, as the best of my older dogs.

Woodcocks are driven in frosty weather, as is well known, to seek their food in springs and rills of unfrozen water, and I found that my old dogs knew about as well as I did the degree of frost which would drive the woodcocks to such places ; and this knowledge proved very troublesome to me, for I could not sufficiently restrain them.   I therefore left the old experienced dogs at home, and took only the wholly inexperienced young dogs ; but, to my astonishment, some of these, in several instances, confined themselves as closely to the unfrozen grounds as their parents would have done.   When I first observed this, I suspected that woodcocks might have been upon the unfrozen ground during the preceding night, but I could not discover (as I think I should have done had this been the case) any traces of their having been there ; and as I could not do so, I was led to conclude that the young dogs were guided by feelings and propensities similar to those of their parents.

The subjects of my observation in these cases were all the offspring of well-instructed parents, of five or six years old or more; and I thought it not improbable that instinctive hereditary propensities might be stronger in these than in the offspring of very young and inexperienced parents. Experience proved this opinion to be well founded, and led me to believe that these propensities might be made to cease to exist, and others be given; and that the same breed of dogs which displayed so strongly an hereditary disposition to hunt after woodcocks might be made ultimately to display a similar propensity to hunt after truffles; and it may, I think, be reasonably doubted whether any dog, having the habits and propensities of the springing spaniel, would ever have been known if the art of shooting birds on wing had not been acquired.

I possess one young spaniel of which the male parent, apparently a well-bred springing spaniel, had been taught to do a great number of very extraordinary tricks (some of which I previously thought it impossible that a dog could be made to learn), and of which the female parent was a well-taught springing spaniel; and the puppy had been taught before it came into my possession a part of the accomplishments of its male parent. This animal possessed a very singular degree of acuteness and cunning, and in some cases appeared to be guided by something more nearly allied to reason than I have ever witnessed in any of the inferior animals. In one instance I had walked out with my gun and a servant, without any dog, and having seen a woodcock, I sent for the dog above mentioned, which the servant brought to me. A month afterwards I sent my servant for it again, under similar circumstances, when it acted as if it had inferred that the track by which the servant had come from me would lead it to me. It left my servant within twenty yards of my house, and was with me in a very few minutes, though the distance which it had to run exceeded a mile. I repeated this experiment at different times, and after considerable intervals, and uniformly with the same results—the dog always coming to me without the servant. I could mention several other instances nearly as singular of the sagacity of this animal, which I imagined to have derived its extraordinary powers, in some degree, from the highly-cultivated intellect of its male parent.

I have witnessed within the period above mentioned, of nearly sixty years, a very great change in the habits of the woodcock. In the first part of that time, when it had recently arrived in the autumn, it was very tame; it usually chuckled when disturbed, and took only a very short flight. It is now, and has been during many years, comparatively a very wild bird, which generally rises in silence and takes a comparatively long flight, excited, I conceive, by increased hereditary fear of man.

I procured a puppy of a breed of setters, which had through many generations been employed in setting partridges for the flight-net only, and of whose exploits I had heard many very extraordinary accounts. I employed it as a pointer in shooting partridges; and for finding coveys of those birds in the open field I never saw its equal, or in its manner of setting them; but it would never set its game amongst brakes or hedge-rows. Whenever it found a bird in such a situation, it invariably sat down in the same attitude, and alternately looked into the bush and at me, seeming to think that setting partridges in such situations was not a part of its duty.

It is well known that very young pointers, of slow and indolent breeds, will point partridges without any previous instruction or practice. I took one of those to a spot where I had just seen a covey of small partridges alight, in August; and amongst them I threw a piece of bread to induce the dog to move from my heels, which it had very little disposition to do at any time, except in search of something to eat. On getting amongst the partridges and perceiving the scent of them, its eyes became suddenly fixed and its muscles rigid, and it stood trembling with anxiety during some minutes. I then caused the birds to take wing, at sight of which it exhibited strong symptoms of fear, and none of pleasure. A young springing spaniel, under the same circumstances, would have displayed much joy and exultation; and I do not doubt but that the young pointer would have done so too, if none of its ancestry had ever been beaten for springing partridges improperly.

The most extraordinary instance of the power of instinctive hereditary propensity which I have ever witnessed, came under my observation in the case of a young dog of a variety usually called retrievers. The proper office of these dogs is that of finding and recovering wounded game; but they are often employed for more extensive purposes, and are found to possess very great sagacity. I obtained a very young puppy* of this family, which was said to be exceedingly well bred, and had been brought to me from a distant county. I had walked up the side of the river which passes by my house, in search of wild ducks, when the dog above mentioned followed me unobserved, and contrary to my wishes; for it was too young for service, not being then quite ten months old. It had not received any other instruction than that of being taught to bring

---

* It was only one month old when it came into the author's possession.

any floating body off a pond, and I do not think that it had ever done this more than three or four times. It walked very quietly behind my gamekeeper upon the opposite side of the river, and it looked on with apparent indifference whilst I killed a couple of mallards and a widgeon; but it leaped into the river instantly upon the gamekeeper pointing out the birds to it; and it brought them on shore, and to the feet of the gamekeeper, just as well as the best-instructed old dog could have done. I subsequently shot a snipe, which fell into the middle of a large nearly stagnant pool of water, which was partially frozen over. I called the dog from the other side of the river, and caused it to see the snipe, which could not be done without difficulty; but as soon as it saw it, it swam to it, brought it to me, laid it down at my feet, and again swam through the river to my gamekeeper. I never saw a dog of any age acquit itself so well, yet it was most certainly wholly untaught. I state the circumstances with reluctance, and not without hesitation; because I doubt whether I could myself believe them to be well founded, upon any other evidence than that of my own senses: the statement is nevertheless most perfectly correct.

I could add an account of a great many more experiments and observations which were made with other varieties of dogs, and upon other species of animals; but as all the facts which I have noticed are confirmations of the truth of the conclusions which I have drawn from those above stated, I shall state the result of one other experiment only, and that solely because it tends to establish a fact which appears to me to be of a good deal of importance.

I stated in a communication to this Society many years ago, " Upon the comparative influence of the male and of the female parent upon the offspring of some species of animals," that in cases where nature intended the offspring to accompany its parent in flight at an early age, the influence of the parent of one sex upon the form of the offspring differed very widely from that of the other parent; and that when the female parents were of small size and of a small breed, and of permanent habits, and the male of a large size and large breed, and of permanent habits, the length of the legs of the fœtus was given by those of the family of the female parent. I imported some Norwegian pony mares with the intention of obtaining cross-bred animals between them and the London dray-horse; having satisfied myself that the experiment might be made without danger or injury to the smaller animal. The bodies and shoulders of the cross-bred animals which I have obtained are excessively deep, comparatively with the length of their legs, which remains unchanged, except that the joints being greatly larger, on account of the greatly increased strength of the legs, and being of the same form, necessarily occupy a little more space. The strength of these animals appears to be very great; I believe that they will prove capable of drawing, particularly up-hill, as heavy weights as the London dray-horses, provided that they be made to draw from a proper level; and I am quite confident that they will prove capable of bearing much more long-continued labour and living upon much less food.

The hereditary propensities of the offspring of the Norwegian ponies, whether full or half-bred, are very singular. Their ancestry have been in the habit of obeying the *voice of their riders*, and not the bridle ; and the horse-breakers complain, and certainly with very good reason, that it is impossible to give them what is called a mouth : they are nevertheless exceedingly docile, and more than ordinarily obedient where they understand the commands of their master. They appear also to be as incapable of understanding the use of hedges as they are of bridles, for they will walk deliberately, and much at their ease, through a strong hedge ; and I therefore conclude that the Norwegian horses are not in the habit of being restrained by hedges similar to those of England.

The male and female parent appear to possess similar powers of transferring to their offspring their hereditary feelings and propensities, except in cases where mule offsprings are produced. In such cases, I think that I have witnessed a decided prevalence of the power of the male parent. The organisation of the mule which is obtained by cross-breeding between the horse and the ass is well known to be regulated to a much greater extent by the male than by the female parent ; and its disposition is, I have some reason to believe, to a very great extent given by its male parent. I have noticed this in the mule which is the offspring of a female ass. I have seen a few only of these animals ; but those which I have seen presented the expression of countenance of the horse, and were perfect horses in temper, and perfectly without the sullenness and obstinacy of the more common mule. The results of such violations of the ordinary laws of nature appear to be very various in different species of animals ; and I should not here have introduced the subject, but that the characters of mules have in many instances misled the judgment of physiologists in their estimates of the comparative influence in ordinary cases of the male and the female upon the offspring.

Whenever I have obtained cross-bred animals by propagating from families of dogs of different permanent habits, the hereditary propensities of the offspring have been very irregular, sometimes those of the male and at other times those of the female parent being prevalent; and in one instance I saw a very young dog, a mixture of the springing spaniel and setter, which dropped upon crossing the track of a partridge, as its male parent would have done, and sprang the bird in silence; but the same dog having within a couple of hours afterwards found a woodcock, gave tongue very freely, and just as its female parent would have done. Such cross-bred animals are, however, usually worthless; and the experiments and observations which I have made upon them have not been very numerous or interesting.

INDEX.

———◆———

THE END.

LONDON :
BRADBURY AND EVANS, PRINTERS, WHITEFRIARS.

For EU product safety concerns, contact us at Calle de José Abascal, 56–1°,
28003 Madrid, Spain or eugpsr@cambridge.org.

www.ingramcontent.com/pod-product-compliance
Ingram Content Group UK Ltd.
Pitfield, Milton Keynes, MK11 3LW, UK
UKHW010853090126
466816UK00011B/208